Biopolymer-Based Nanosystem for Drug Delivery

Biopolymer-Based Nanosystem for Drug Delivery

Guest Editors

János Borbély
Ildikó Bácskay

MDPI Basel • Beijing • Wuhan • Barcelona • Belgrade • Novi Sad • Cluj • Manchester

Guest Editors
János Borbély
Doctoral School of Clinical Medicine
University of Debrecen
Debrecen
Hungary

Ildikó Bácskay
Department of Pharmaceutical Technology
University of Debrecen
Debrecen
Hungary

Editorial Office
MDPI AG
Grosspeteranlage 5
4052 Basel, Switzerland

This is a reprint of the Special Issue, published open access by the journal *Pharmaceutics* (ISSN 1999-4923), freely accessible at: https://www.mdpi.com/journal/pharmaceutics/special_issues/biopolymer_delivery.

For citation purposes, cite each article independently as indicated on the article page online and as indicated below:

Lastname, A.A.; Lastname, B.B. Article Title. *Journal Name* **Year**, *Volume Number*, Page Range.

ISBN 978-3-7258-5139-3 (Hbk)
ISBN 978-3-7258-5140-9 (PDF)
https://doi.org/10.3390/books978-3-7258-5140-9

Contents

About the Editors

János Borbély

János Borbély is a professor at the Doctoral School of Clinical Medicine, University of Debrecen, Hungary. He is also affiliated with the Department of Physical Chemistry and Materials Science. His general interests lie in colloid and interface science, nanostructured materials, and their applications in biomedical and pharmaceutical fields. His current research focuses on the development of functionalized polymer-based nanocarriers, hydrogels, and hybrid systems for drug delivery and controlled release. He is also engaged in projects related to environmental and industrial applications of polymer-based materials. Prof. Borbély's scientific interests include polymer chemistry, nanoparticle synthesis, rheology, and the physicochemical characterization of nanostructured systems. He has authored numerous peer-reviewed publications and book chapters and has been recognized with several awards for his contributions to science and education. He actively collaborates with interdisciplinary and international research teams.

Ildikó Bácskay

Ildikó Bácskay is a full professor at the Department of Pharmaceutical Technology, Faculty of Pharmacy,University of Debrecen, Hungary. Her general interests include innovative dosage forms and pharmaceutical nanotechnology. Her current research projects focus on the development of biopolymer-based nanosystems for targeted drug delivery, permeability enhancement, and improving the bioavailability of poorly soluble drugs. She is also involved in the formulation of nasal, buccal, and transdermal delivery systems. Her scientific interests span drug transport across biological barriers, nanoparticle characterization, and cell-based in vitro models. Prof. Bácskay has published extensively in peer-reviewed journals and actively collaborates with national and international research groups. She has received multiple awards for her scientific and teaching excellence, including recognition from the Hungarian Society for Pharmaceutical Sciences.

Preface

This Reprint brings together recent scientific contributions on biopolymer-based nanosystems for drug delivery, highlighting innovative approaches in formulation, characterization, and therapeutic application. The aim is to provide a comprehensive overview of how natural and synthetic biopolymers can be utilized to develop efficient, safe, and targeted nanocarriers. It is intended for researchers, clinicians, and professionals in pharmaceutics, nanotechnology, and biomedical sciences. We thank all contributing authors for their valuable work and express our sincere gratitude to the reviewers for their expert insights and the editorial team for their support in making this collection possible.

János Borbély and Ildikó Bácskay

Guest Editors

pharmaceutics

MDPI

Article

Effect of Copolymer Properties on the Phase Behavior of Ibuprofen–PLA/PLGA Mixtures

Anton Iemtsev [1], Martin Klajmon [1], Fatima Hassouna [2] and Michal Fulem [1,*]

[1] Department of Physical Chemistry, University of Chemistry and Technology, Prague, Technická 5, 166 28 Prague, Czech Republic
[2] Faculty of Chemical Engineering, University of Chemistry and Technology, Prague, Technická 5, 166 28 Prague, Czech Republic
* Correspondence: fulemm@vscht.cz

Abstract: Prediction of compatibility of the active pharmaceutical ingredient (API) with the polymeric carrier plays an essential role in designing drug delivery systems and estimating their long-term physical stability. A key element in deducing API–polymer compatibility is knowledge of a complete phase diagram, i.e., the solubility of crystalline API in polymer and mutual miscibility of API and polymer. In this work, the phase behavior of ibuprofen (IBU) with different grades of poly(D,L-lactide-co-glycolide) (PLGA) and polylactide (PLA), varying in composition of PLGA and molecular weight of PLGA and PLA, was investigated experimentally using calorimetry and computationally by the perturbed-chain statistical associating fluid theory (PC-SAFT) equation of state (EOS). The phase diagrams constructed based on a PC-SAFT EOS modeling optimized using the solubility data demonstrated low solubility at typical storage temperature (25 °C) and limited miscibility (i.e., presence of the amorphous–amorphous phase separation region) of IBU with all polymers studied. The ability of PC-SAFT EOS to capture the experimentally observed trends in the phase behavior of IBU–PLA/PLGA systems with respect to copolymer composition and molecular weight was thoroughly investigated and evaluated.

Keywords: biodegradable polymers; PLGA; PLA; amorphous solid dispersion; API–polymer compatibility; PC-SAFT; phase diagrams

1. Introduction

The poor bioavailability of many active pharmaceutical ingredients (APIs) represents one of the major challenges to successful pharmaceutical development [1]. Several strategies to tackle this problem have been developed, including the formation of polymeric amorphous solid dispersions (ASDs) [2], which consist of amorphous API molecularly dispersed in the polymer matrix, and the preparation of polymeric micro- and nanoparticles [3]. The design of these drug formulations is still governed to a large extent by trial-and-error approaches. An influential parameter determining the successful polymeric drug formulations, their physical stability, and performance-related characteristics is the compatibility of API with the polymer. The compatibility between API and polymer, which at the molecular level reflects the strength of their intermolecular interactions compared to API–API and polymer–polymer cohesive interactions, affects the phase behavior of API–polymer mixtures, such as the size of the immiscibility region (i.e., liquid–liquid equilibria (LLE)) or the solubility of crystalline API in the polymeric matrix (solid–liquid equilibrium (SLE)). SLE determines the maximum concentration of amorphous API that can be thermodynamically stabilized against the recrystallization in the polymer matrix, while LLE identifies the conditions (temperature and concentration) at which liquid–liquid or amorphous phase separation (APS) can occur. Both the recrystallization of API and APS can lead to undesirable heterogeneities of ASD formulations. An initial step in the polymeric drug formulation development is the selection of a suitable polymeric excipient, which must be biocompatible and offer the desired formulation properties such as optimal API loading and physical stability over time. The latter properties are directly related to the compatibility of drug formulation components. As the number of initially identified biocompatible polymers for a given API can be high, it is highly desirable to develop efficient computational tools that would be capable of estimating or ranking the polymer compatibility with a given API.

Among various polymeric carriers, aliphatic polyesters have received a growing interest in the development of efficient drug delivery systems due to their unique properties including biocompatibility, biodegradability, and low toxicity. Within the biodegradable polyesters, poly(D,L-lactide-co-glycolide) (PLGA) and polylactide (PLA) represent particular relevance in drug delivery [4]. They have been FDA-approved [5]. Their degradation rate by hydrolysis is controlled by regulating the polymer molecular weight, the composition of monomer units in the copolymer (e.g., lactide (LA) to glycolide (GA) ratio in PLGA), and the type of stereo-forms of lactide units (i.e., D- and L-lactide) [6]. PLGA and PLA have proven themselves as effective polymeric carriers in the development of amorphous solid dispersions (ASDs) [7,8] and micro- or nanoparticle-based drug products formulation [9–13].

In our previous work [14], in which the phase behavior of various model APIs with two PLGA copolymers possessing a similar molecular weight (about 10,000 g/mol) but different LA to GA ratios (50:50 and 75:25) was studied, it was shown that the perturbed-chain statistical associating fluid theory (PC-SAFT) equation of state (EOS) in connection with calorimetric measurements of SLE curve is capable of predicting a complete phase diagram including amorphous phase separation (APS), which was found to be in agreement with a long-term behavior of the prepared formulations. The systems of ibuprofen (IBU) with both PLGA polymers showed very limited miscibility, which was demonstrated experimentally by APS during calorimetric measurements and long-term physical stability studies. This system is therefore ideal for exploring the capability of modeling tools, such as the perturbed-chain statistical associating fluid theory (PC-SAFT) equation of state (EOS), to capture the effect of copolymer molecular weight and composition on the size of immiscibility region (i.e., LLE), which is in API–polymer systems demonstrated by amorphous phase separation (APS). We note that, in general, LLE predictions using thermodynamic models are fairly challenging. The situation becomes even more complicated in the case of asymmetric mixtures such as API–polymer mixtures. In this work, the effect of composition and molecular weight of a range of PLA/PLGA polymers on the phase behavior (both SLE and LLE) of IBU was thoroughly investigated experimentally by calorimetry and computationally using PC-SAFT EOS.

2. Materials and Methods

2.1. Materials

Racemic IBU was kindly donated by Zentiva Group (Prague, Czech Republic). Two PLA commercial grades (Purasorb PDL 02A and PDL 04A) and two PLGA commercial grades (PDLG 5004A and PDLG 7504A) consisting of 50:50 and 75:25 of LA:GA units were graciously provided by Corbion (Gorinchem, The Netherlands). All chemicals were used without further purification. The molecular structures of the used chemicals are depicted in Figure 1, and their selected physicochemical properties are summarized in Section 3.1.

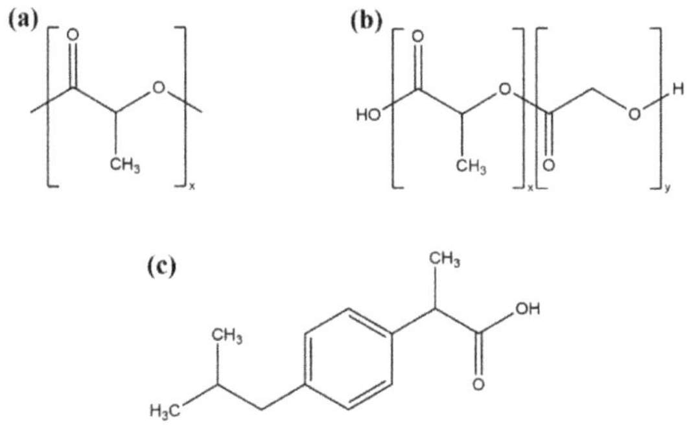

Figure 1. Chemical structures of used compounds: (**a**) PLA, (**b**) PLGA, and (**c**) IBU.

2.2. Experimental Methods

2.2.1. Differential Scanning Calorimetry

The differential scanning calorimetry (DSC) analysis was performed to determine the thermal properties of pure chemical compounds, measure SLE and glass transition temperatures (T_g), and map APS regions. TA Q1000 DSC from TA Instruments, Inc. (New Castle, DE, USA) calorimeter was used. All scans were carried out in the preselected temperature range (−90 to 150 °C) and nitrogen gas purge (50 mL min^{-1}).

The samples (5–10 mg) were placed in aluminum pans, which were tightly sealed and pinned in the lid to permit the evaporation of residual moisture that may be present in the polymer. A thorough calibration of temperature and enthalpy was performed using five reference materials: tin, indium, naphthalene, gallium, and water.

2.2.2. Calorimetric Determination of SLE Curve and Mapping of LLE Region

The solubility temperature (T_s) of IBU in PLA/PLGA polymers was experimentally determined using DSC. Several DSC protocols are described in the literature and generally applied for the measurement of the T_s: recrystallization (demixing) method [15], melting point depression (MPD) method [16], annealing method [17], melting enthalpy method [18], and step-wise dissolution method (S-WD) [19] (a DSC protocol recently developed in our research group). As the IBU–PLA/PLGA mixtures exhibited APS over a wide range of compositions, the S-WD protocol could not be applied in this work as it derives the solubility from the measured T_g of a homogenous amorphous system. The MPD method, the most frequently applied DSC protocol, was therefore selected and used in this study. The MPD procedure consists of measuring the melting temperature of API in the polymer matrix, which decreases with the increasing polymer concentration due to the reduction of API chemical potential in the API–polymer mixture.

The solubility temperatures T_s of IBU in PLA and PLGA polymers were determined from the melting endotherms of corresponding physical mixtures using the melting peak top (following Höhne et al. [20] recommendation), while the melting temperature (T_m) of the pure IBU was taken as the onset temperature of the melting peak. The physical mixtures were obtained by mild mixing of IBU with PLA/PLGA polymers in predetermined ratios for about 10 min using a pestle and mortar. T_s of IBU in the PLGA and PLA polymers were measured at the heating rate (β) of 1 °C min^{-1} (according to Höhne et al. [20] advising the application of β lower than 2 °C min^{-1}) during the first heating ramp, which was followed by the cooling (β = 20 °C min^{-1}) of the melt and second heating ramp (β = 10 °C min^{-1}) for the T_g determination. To map the immiscibility regions, i.e., the temperature-composition regions where APS occurs, IBU–PLGA and IBU–PLA mixtures with compositions ranging from 10 to 70 wt.% IBU were subject to successive heating–cooling–heating cycles during which T_g values of the system were measured. The upper temperature of the initial heating cycle was set slightly above T_m of IBU to 80 °C. Upper heating temperatures were subsequently increased by 10 °C in each successive cycle up to reaching the final temperature of 150 °C. By this procedure, the temperature-concentration region from 10 to 70 wt.% IBU and from 80 °C to 150 °C was screened for each binary system for APS. If only a single T_g value was obtained during T_g determination, positioned between the value for pure amorphous IBU and polymer, the system was considered to consist of a single homogenous amorphous phase. If two T_g values were detected, the system was considered separated into two amorphous phases—IBU-rich and polymer-rich amorphous phases.

2.3. *Modeling Approaches*

2.3.1. Modeling of Glass-Transition Temperature

The Kwei equation [21] was used to model T_g lines for IBU–PLA/PLGA mixtures:

$$T_g = \frac{w_1 T_{g1} + k w_2 T_{g2}}{w_1 + k w_2} + q w_1 w_2 \tag{1}$$

where T_{g1} and T_{g2} are the glass-transition temperatures of the pure components, w_1 and w_2 are the mass fractions of each component, and k and q are parameters that can be obtained by fitting the experimental T_g values for a given binary mixture.

The Kwei equation (Equation (1)) can also be used with the parameter k estimated using the Simha–Boyer rule [22], which requires knowledge of T_g and density values for both pure components, followed by determination of the parameter q by fitting it to experimental binary T_g data. Since this approach leads to higher deviations from experimental T_g data [23], a simultaneous fitting of both parameters k and q to experimental T_g data was used to model T_g lines in this work.

2.3.2. SLE and LLE Modeling Using PC-SAFT EOS

Assuming that the crystalline phase consists of only pure IBU, the solubility of IBU in PLA/PLGA polymers x^{L}_{IBU} (IBU mole fraction in the liquid/amorphous phase) can be calculated using the following rigorous thermodynamic relationship:

$$x^{L}_{IBU} = \frac{1}{\gamma^{L}_{IBU}} \exp\left[-\frac{\Delta_{fus}H}{RT}\left(1 - \frac{T}{T_m}\right) - \frac{1}{RT}\int_{T_m}^{T} \Delta_{fus}C_p \mathrm{d}T + \frac{1}{R}\int_{T_m}^{T} \frac{\Delta_{fus}C_p}{T}\mathrm{d}T \right] \tag{2}$$

where T is the absolute temperature; T_m and $\Delta_{fus}H$ are the melting temperature and fusion enthalpy, respectively, of pure crystalline IBU; $\Delta_{fus}C_p$ is the difference between the isobaric molar heat capacity of the liquid and crystalline phases of IBU; R is the universal gas constant; and γ^{L}_{IBU} is the activity coefficient of IBU in the liquid IBU–polymer phase. As can be seen from Equation (2), SLE calculations require two types of thermodynamic data as the input: (i) thermodynamic fusion data for pure IBU (T_m, $\Delta_{fus}H$, and $\Delta_{fus}C_p$) and (ii) the activity coefficient γ^{L}_{IBU} reflecting the intermolecular interaction between the solute (IBU) and solvent (PLA/PLGA polymers). The fusion thermodynamic properties of pure solute are typically obtained experimentally using calorimetry, while the activity coefficient is calculated using an appropriate thermodynamic model (e.g., EOS).

For LLE calculations, which determine the miscibility regions in IBU–PLA/PLGA systems, the method of alternating tangents [24] was used in this study. When applying this method, only the composition of one of the coexisting liquid phases is sought, making it an efficient approach for LLE calculations in asymmetric mixtures such as API–polymer mixtures. The obtained binodal points have to satisfy the equilibrium condition for two coexisting phases (L1 and L2) given by the following equations:

$$\begin{aligned} x^{L1}_{IBU}\gamma^{L1}_{IBU} &= x^{L2}_{IBU}\gamma^{L2}_{IBU} \\ x^{L1}_{polymer}\gamma^{L1}_{polymer} &= x^{L2}_{polymer}\gamma^{L2}_{polymer} \end{aligned} \tag{3}$$

where γ^{L1}_{i} and γ^{L2}_{i} are IBU and polymer activity coefficients in phases L1 and L2 and x^{L1}_{i} and x^{L2}_{i} are their mole fractions, respectively.

The activity coefficients needed for both SLE and LLE calculations were obtained using the PC-SAFT EOS [25,26] in its copolymer version [27]. A brief description of the PC-SAFT EOS is provided below.

Within the PC-SAFT framework, the reduced residual Helmholtz energy (a^{res}) of a system is described as a sum of independent contributions:

$$a^{res} = a^{hc} + a^{disp} + a^{assoc} \tag{4}$$

where a^{hc} is the hard-chain contribution accounting for repulsive forces, a^{disp} is the contribution for dispersion attractive forces, and a^{assoc} is the contribution accounting for association (hydrogen bonding) interactions. In the PC-SAFT concept, each molecule i is represented as a chain of m spherical segments of the same diameter σ_i. These two parameters are used to calculate the contribution a^{hc}. The dispersion contribution a^{disp} is evaluated using the energy parameter ε_i/k (k is the Boltzmann constant) and the association contribution a^{assoc} using the energy parameter $\varepsilon_i{}^{assoc}/k$ and the association volume $\kappa_i{}^{assoc}$. Besides the five pure-component parameters mentioned above, the number of association sites $N_i{}^{assoc}$ (electron donors and acceptors) must be specified based on the molecular structure of a given compound.

The PC-SAFT parameters for IBU and PLA/PLGA polymers utilized in this study are listed in Table 1. The IBU parametrization was performed in our previous study [28] using the liquid density and vapor pressure data for pure IBU in conjunction with the data on IBU solubility in selected organic solvents. IBU was modeled as a self-associating molecule with four association sites (i.e., 2 hydrogen bond (HB) donors and 2 HB acceptors). The PLGA and PLA were modeled via the copolymer version of PC-SAFT [27] in which polymers are considered as a chain of homopolymer units with their individual sets of PC-SAFT parameters. In this modification, the PC-SAFT EOS is able to account for different arrangements of monomers, composition, and molecular weight of copolymers [27,29]. The parameters for homopolymers PLLA and PDLA were obtained from Cocchi et al. [30] and those for PGA from Prudic et al. [29] (see Table 1). The monomer compositions in the copolymers studied are provided in Table 3. Both PLGA and PLA polymers are assumed to be non-associating in these parametrizations. The bond fractions between the monomer unit segments capturing the copolymer arrangements were taken from Prudic et al. [29].

Table 1. PC-SAFT EOS pure-component parameters.

Compound	$(m/M_w)_i$	σ_i (Å)	ε_i/k (K)	ε_i^{assoc}/k (K)	κ_i^{assoc}	N_i^{assoc} [a]
			API			
IBU [b]	0.02636	4.0179	309.40	516.469	0.08946	4 (2, 2)
			PLGA [c]			
PLLA [d]	0.04545	2.920	230.0	0	0	0 (0, 0)
PDLA [d]	0.03699	3.120	240.0	0	0	0 (0, 0)
PGA [e]	0.03130	2.860	233.9	0	0	0 (0, 0)

[a] N_i^{assoc}(D,A) is the total number of hydrogen bonding (HB) sites, D is the number of HB donor sites, and A denotes the number of HB acceptor sites in a molecule. [b] Adopted from [28]. [c] PLGA and PLA properties are summarized in Tables 2 and 3. [d] Adopted from [30]. [e] Adopted from [29].

Table 2. Monomer composition (as mass fractions) in the copolymers studied.

Polymer	$w_{LLA} = w_{DLA}$	w_{GA}
PDLG 5002A/PDLG 5004A	0.277	0.446
PDLG 7502A/PDLG 7504A	0.394	0.212
PDL 02A/PDL 04A	0.500	0

Table 3. Pure-component physiochemical properties.

Compound	M_w [a] (g·mol^{-1})	T_g [b] (°C)	T_m [b] (°C)	ΔH_{fus} [b] (kJ·mol^{-1})	$\Delta C_{p,fus}$ [c] (J·K^{-1}·mol^{-1})
IBU	206.3	-45.0 ± 0.3	75.6 ± 0.3	26.4 ± 0.8	176.16440 − 0.3449480·(T/K)
PDLG 5002A	9877	39.6 ± 0.3	-	-	-
PDLG 5004A	35,900	46.7 ± 0.3	-	-	-
PDLG 7502A	12,900	38.7 ± 0.3	-	-	-
PDLG 7504A	38,800	48.8 ± 0.3	-	-	-
PDL 02A	16,400	47.8 ± 0.3	-	-	-
PDL 04A	40,200	50.4 ± 0.3	-	-	-

[a] The average molecular weight (M_w) of polymers was determined in this work using gel permeation chromatography (Waters, equipped with a differential refractive index detector) based on narrow polystyrene standards. [b] This study except for T_g values for PDLG 5002A and PDLG 7502A, which were determined in the previous study [14]. Reported uncertainties correspond to the combined expanded uncertainties (k = 2, 0.95 level of confidence) in determining temperature and enthalpy values. [c] The difference between the isobaric molar heat capacity of a liquid and a crystal. The reported equation was derived based on experimental heat capacities published in [31].

Activity coefficients γ_i^L can be obtained from the PC-SAFT EOS using the following thermodynamic relationships [25]:

$$\ln \varphi_i^L = a^{res} + \left(\frac{\partial a^{res}}{\partial x_i}\right)_{T,\rho,x_{k\neq i}} - \sum_j \left[x_j\left(\frac{\partial a^{res}}{\partial x_j}\right)_{T,\rho,x_{k\neq j}}\right] + Z - 1 - \ln Z \quad (5)$$

$$\ln \gamma_i^L = \ln \varphi_i^L - \ln \varphi_{0,i}^L \quad (6)$$

where a^{res} is the reduced residual Helmholtz energy (Equation (4)), φ_i^L is the fugacity coefficient of IBU or polymer in the liquid mixture, $\varphi_{0,i}^L$ denotes the fugacity coefficient of the pure liquid, ρ is the molar density of the system, and Z is the compressibility factor.

Besides pure-component parameters, binary interaction parameter k_{ij} correcting the cross-dispersion energy parameter ε_{ij}/k is introduced within the PC-SAFT framework to achieve closer agreement between experimental results for mixtures and phase behavior predictions [25,26]:

$$\frac{\varepsilon_{ij}}{k} = \sqrt{\frac{\varepsilon_i}{k}\frac{\varepsilon_j}{k}}(1 - k_{ij}) \quad (7)$$

In this work, two variants of PC-SAFT calculations are presented: (i) pure predictions with k_{ij} = 0, i.e., the calculation based solely on pure-component parameters; and (ii) calculations with k_{ij} values optimized using experimental SLE data for the one binary system IBU–PDLG 7502A [14]. Based on the performance of pure predictions in terms of identifying phenomena, such as APS, or providing qualitative ordering of SLE data, the suitability of PC-SAFT EOS as a computational screening tool for selecting suitable polymeric carriers for further experimental investigation can be evaluated. In scenario (ii), the ability of the PC-SAFT EOS to predict complete phase diagrams based on optimized k_{ij} values using SLE data for IBU with only one polymer from the PLA/PLGA family was studied. In other words, the transferability of modeling phase behavior using PC-SAFT EOS and its ability to describe the trends in phase behavior within the PLA/PLGA family were investigated.

3. Results

3.1. Physicochemical Properties of Pure Components

The physicochemical properties of the pure materials utilized in this study are presented in Table 2. Both melting temperature and fusion enthalpy of IBU agree well with the previously published values summarized by Štejfa et al. [31]. The amorphous IBU was obtained by quenching the melt using β = 20 °C min^{-1}, and its T_g was measured during subsequent heating by β = 10 °C min^{-1}. The obtained T_g value is in alignment with the results found in the literature, which cover the range from −48 to −45 °C [31]. The recorded T_g of the PLA/PLGA polymers are in good agreement with the general trend typical for this group of polymers described in the literature: T_g increases with increasing LA content (e.g., from 35.7 °C for PLGA 50:50 (LA:GA) to 45.3 °C for the PLA 100:0 (LA:GA) according to Pyo Park et al. [32]) and molecular weight of the polymer (e.g., from 42.2 °C for 8000 g/mol PLGA to 52.6 °C 110,000 g/mol PLGA, in accordance with Lee et al. [33]).

3.2. IBU–PLA/PLGA Phase Diagrams

Experimental T_s and T_g values for binary systems IBU–PDL 5004A, IBU–PDL 7504A, and IBU–PDL 02A/04A are summarized in Tables S1 and S2 in the Supplementary Material (SM). T_s and T_g data for IBU–PDL 5002A and IBU–PDL 7502A were reported in our previous study [14]. The resulting phase diagrams are shown in Figure 2. The T_g line defining the kinetic stability of ASD formulations was modeled using the Kwei equation (Equation (1)) with parameters listed in Table S3 in SM. The PC-SAFT EOS, which was used to model thermodynamic phase behavior (i.e., SLE and LLE) was applied in two variants: (i) with all binary interaction parameter k_{ij}s set to zero, which represents pure prediction based solely on pure-component parameters (represented by dashed lines in Figure 2); and (ii) with k_{ij} values adjusted based on experimental SLE data for one binary system IBU–PDLG 7502A (represented by solid lines in Figure 2).

The solubility of IBU in the studied PLA and PLGA polymers obtained by pure predictions at 25 °C (typical storage temperature) is listed in Table 4. As can be seen, with k_{ij}s = 0, the solubility of IBU was predicted to be almost identical for all polymers examined (regardless of differences in polymer composition and molecular weight) and fell in the range of 21.6 to 22.7 wt.%. The PDL 02A (i.e., lower molecular weight PLA) polymer demonstrated the highest solubility potential for IBU based on pure predictions. Compared to experimental results, the predicted SLE curves considerably overestimated the IBU solubility in all studied polymers. The average absolute relative deviation (AARD) listed in Table S4 ranges from 93 to 157%. Furthermore, and most importantly, PC-SAFT EOS with k_{ij}s = 0 did not (at least qualitatively) predict the presence of the immiscibility region for any system, which was observed experimentally.

Table 4. IBU solubility w_{IBU} (wt.%) in PLA/PLGA polymers at 25 °C calculated using PC-SAFT EOS with k_{ij}s = 0 and optimized k_{ij}s.

Polymer	M_w (g·mol^{-1})	LA Group (mol%)	Solubility w_{IBU} (wt.%, k_{ij}s = 0))	Solubility w_{IBU} (wt.%, Optimized k_{ij}s)
		Lower M_w polymers		
PDLG 5002A	9877	50	21.9	1.4
PDLG 7502A	12,900	75	22.4	1.7
PDL 02A	16,400	100	22.9	2.1
		Higher M_w polymers		
PDLG 5004A	35,900	50	21.6	1.2
PDLG 7504A	38,800	75	22.2	1.6
PDL 04A	40,200	100	22.7	2.0

To achieve an improved description of experimental SLE data using PC-SAFT EOS, the temperature-independent k_{ij}s between IBU–DLA, IBU–LLA, and IBU–GA (PLGA polymers) and between IBU–LLA and IBU–DLA (PLA polymers) groups, determined based on SLE data for IBU–PDLG 7502A systems in our recent study [14], were applied. In brief, to shorten the number of fitted parameters (following the recommendation of Prudic et al. [29]), k_{ij}s between the monomeric units of the homopolymer DLA, LLA, and GA were considered to be zero, and k_{ij}s between IBU–LLA and IBU–DLA were assumed to be identical. The applied k_{ij}s are listed in Table S5.

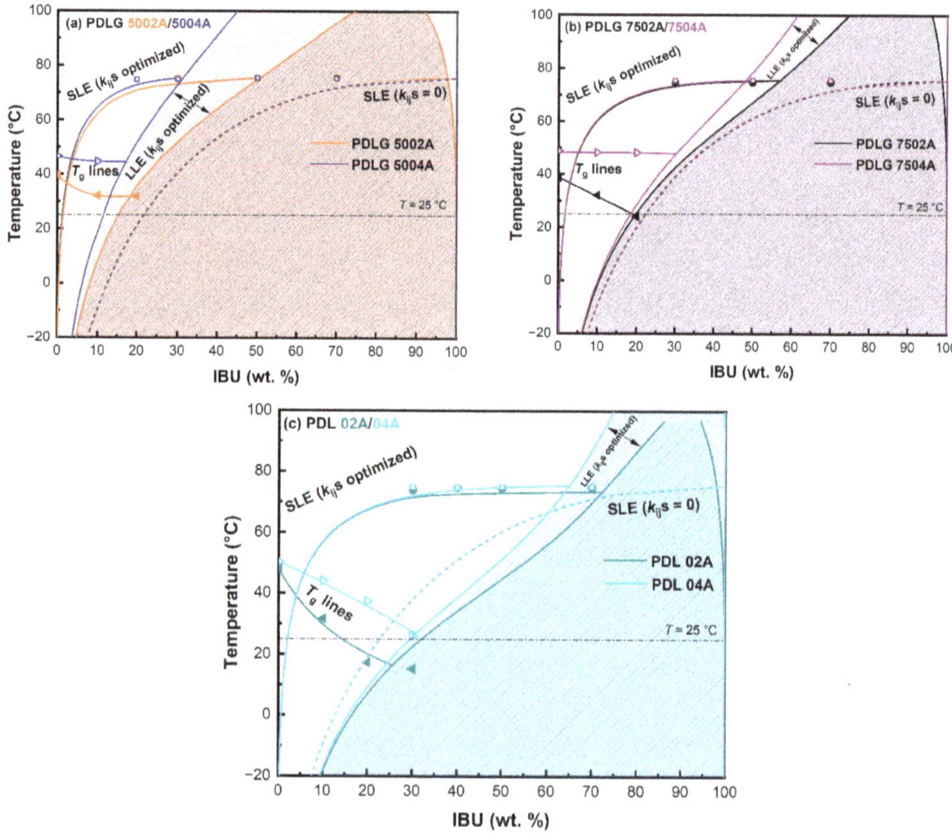

Figure 2. IBU–PLA/PLGA phase diagrams: (**a**) IBU–PDLG 5002A/PDLG 5004A, (**b**) IBU–PDLG 7502A/PDLG 7504A, and (**c**) IBU–PDL 02A/PDL 04A. Symbols represent experimental values and lines represent the calculated properties. T_g lines were modeled by the Kwei equation, Equation (1), and SLE and LLE using the PC-SAFT EOS. Experimental T_g and SLE values are listed in Tables S1 and S2, respectively; parameters of the Kwei equation, Equation (1), are listed in Table S3; and applied binary interaction parameters k_{ij} are listed in Table S5. Data for IBU–PDLG 5002A and IBU–PDLG 7502A were taken from our previous study [14].

Similar to the pure predictions, the impact of polymer molecular weight and composition on the evaluated solubility was relatively modest (see Table 4). The solubility of IBU at 25 °C predicted using the optimized PC-SAFT EOS was found to range from 1.2 to 2.1 wt.%. As for pure predictions, the solubility of IBU in PDL 02A was predicted to be the highest among the studied polymers. The corresponding AARD values lying in the range from 23.3% (PDL 02A) to 33.4% (PDLG 5004A) are shown in Table S4. Such high AARDs can be explained by the incapacity of PC-SAFT together with k_{ij}s (temperature-independent) to accommodate a weak dependence of experimental IBU solubility on the temperature in PLA and PLGA polymers (squares and circles in Figure 2). Nonetheless, the PC-SAFT model with the adjusted k_{ij}s was able to qualitatively predict the presence of liquid–liquid phase splitting events in the IBU–PLA and IBU–PLGA systems, which is in agreement with experimental observations (for details, see Section 3.3).

In accordance with Equation (7), the positive k_{ij} denotes the reduction of API–polymer interactions, which, in turn, leads to a decrease in solubility and miscibility. It can be clearly visible for IBU–PLGA and IBU–PLA systems, where relatively high positive values of

k_{ij} (Table S5) resulted in very limited solubility values (at 25 °C) of the studied systems (see Table 4). Moreover, it can be observed from Figure 2 that, although the polymer composition (i.e., the content of LLA/DLA or GA units) or molecular weight demonstrated only a slight impact on calculated IBU solubility (SLE), the effect on LLE binodal curve was more pronounced, which indicates a higher sensitivity of APS calculations on the model parameters (for details, see Section 3.3). These observations are consistent with the results obtained in the previous study [14] and those by Sadowski and co-authors [29,34].

The corresponding T_g lines were constructed to determine the regions where the IBU–PLGA and IBU–PLA ASD formulations can be stabilized kinetically owing to considerably lower molecular motion at conditions under the T_g line. As can be seen from the phase diagrams depicted in Figure 2, at a typical storage temperature of 25 °C, ASD formulations with drug loadings higher than a couple of weight percentages are estimated to be thermodynamically unstable as their temperature-composition coordinates are located on the right from the SLE curve. Therefore, the physical stability of ASD formulations (with w_{IBU} > 2 wt.% at 25 °C) will be governed by kinetic aspects, i.e., by the difference between the storage temperature and T_g line. Based on T_g lines depicted in the phase diagrams (Figure 2), a higher kinetic stabilization of ASD formulations consisting of higher molecular weight polymers can be deduced, which represents an advantage of using higher molecular weight PLA/PLGA polymers in ASD-based drug delivery systems.

3.3. *Miscibility of IBU with PLA/PLGA Polymers*

For all binary systems studied, APS (LLE) was observed during the solubility measurements using the MPD method. To outline approximate phase boundaries for APS, i.e., to identify temperature-composition regions at which APS occurs, a series of complementary melt quenching calorimetric experiments was conducted. The calorimetric mapping consisted of the determination of T_g for binary IBU–PLA/PLGA mixtures, which were annealed at temperatures above IBU melting temperature for fixed composition and subsequently quenched to −90 °C. The maximum annealing temperature was set to 150 °C to avoid the thermal decomposition of IBU. A single T_g value located between T_g of pure IBU and pure PLA/PLGA polymer was considered to indicate a single homogenous amorphous mixture, while two T_g values observed close to T_g of pure IBU and that of pure polymer were considered to indicate of creation of a two-phase system comprising polymer-rich and IBU-rich phases. The mapping outcomes together with the LLE region predicted using PC-SAFT EOS (optimized k_{ij}s) are depicted schematically in Figure 3. The T_g values determined during the mapping experiments are summarized in Table S2.

Based on the approximate areas of immiscibility regions obtained during the mapping experiments (Figure 3), the following ranking of experimental miscibility of the polymers studied with IBU can be deduced: PDL 04A > PDL 02A > PDLG 7504A > PDLG 7502A > PDLG 5002A > PDLG 5004A. Considering these results, the impact of the polymer composition is clearly visible: the higher LA content, the smaller LLE region (i.e., the higher mutual miscibility of IBU and polymer). At the same time, the impact of the molecular weight of the polymer is not as obvious. In the case of IBU–PDL 04 and IBU–PDLG 7504A systems, the determined LLE region was noticeably smaller compared to the systems of IBU with lower molecular weight analogs (i.e., PDL 02A and PDLG 7502). Hoverer, the situation with PDLG 5002A and PDLG 5004A (i.e., LA to GA ratio 50:50) was the opposite.

The results of LLE modeling using PC-SAFT with optimized k_{ij}s are presented and compared with experimental results in Figure 3 (modeling with k_{ij}s = 0 did not indicate LLE). The equilibrium concentrations predicted using the PC-SAFT for the polymer-rich phase at 25 °C are summarized in Table 5. Based on these results, the following miscibility trend was derived: w_{IBU} (PDL 02A) > w_{IBU} (PDL 04A) > w_{IBU} (PDLG 7502A) > w_{IBU}(PDLG 7504A) > w_{IBU} (PDLG 5002A) > w_{IBU} (PDLG 5004A).

Table 5. Miscibility limit w_{IBU} (wt.%, polymer-rich phase) at 25 °C calculated using the PC-SAFT with k_{ij}s = 0 and optimized k_{ij}s.

Polymer	M_w ($g \cdot mol^{-1}$)	LA Group (mol%)	APS w_{IBU} (wt.%, k_{ij}s = 0)	APS w_{IBU} (wt.%, Optimized k_{ij}s)
		Lower M_w polymers		
PDLG 5002A	9877	50	-	15.4
PDLG 7502A	12,900	75	-	19.8
PDL 02A	16,400	100	-	31.9
		Higher M_w polymers		
PDLG 5004A	35,900	50	-	11.5
PDLG 7504A	38,800	75	-	18.4
PDL 04A	40,200	100	-	29.6

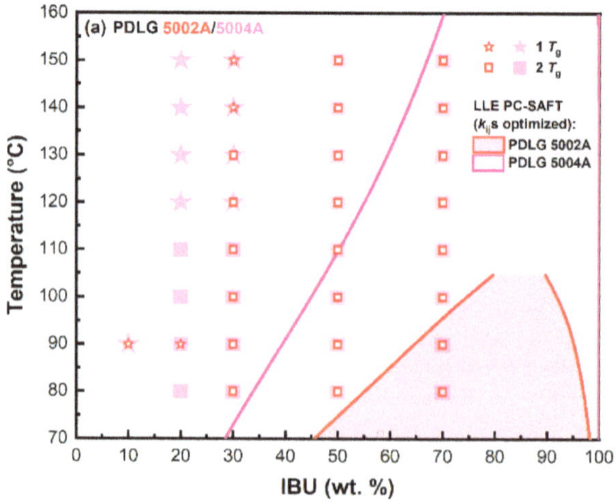

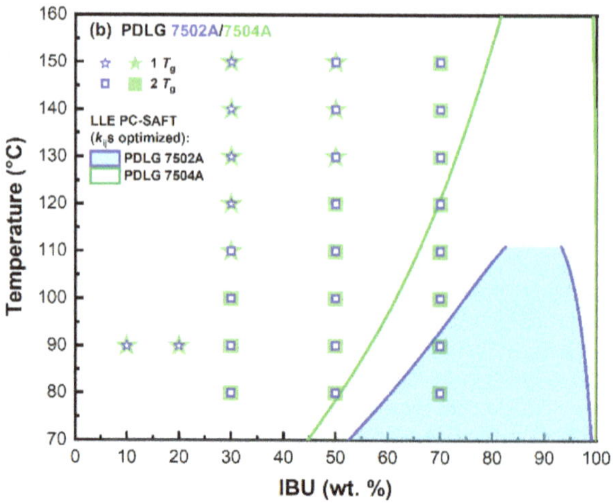

Figure 3. *Cont.*

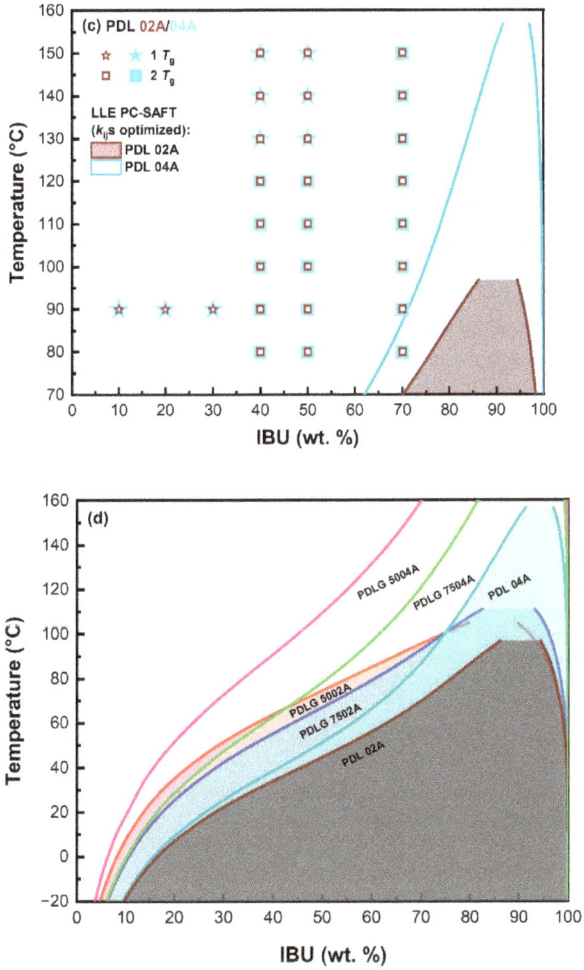

Figure 3. Comparison of LLE areas determined using DSC measurements with those predicted using the PC-SAFT EOS: (**a**) PDLG 5002A/5004A, (**b**) PDLG 7502A/7504A, and (**c**) PDL 02A/04A. (**d**) Comparison of LEE regions modeled by PC-SAFT.

For both lower and higher molecular weight polymers, the miscibility between the IBU and polymer increases with increasing LA content, reflecting a higher number of favorable hydrophobic interactions between IBU and LA. These findings qualitatively agree with the experimental results obtained in this work as well as with the results obtained by Luebbert et al. [34] for low molecular weight PLGA and PLA polymers. Regarding the impact of molecular weight at fixed copolymer composition, the experimental mapping of APS suggests that the mutual miscibility of IBU with the polymer increases with increasing polymer molecular weight for polymers with LA content of 75 and 100 mol.%. LLE calculations by PC-SAFT with optimized k_{ij}s do not capture this trend.

The prediction of the conditions under which the components are miscible is a key parameter in designing homogeneous drug formulations and predicting their long-term physical stability. At the same time, LLE calculations are extremely challenging, especially when considering asymmetrical mixtures such as API–polymer mixtures. As mentioned above, PC-SAFT EOS without optimization using SLE experimental data was unable to signal a possible immiscibility region for the binary systems studied, which is contrary to experimental evidence. This can be considered as a significant limitation if PC-SAFT EOS is to be considered as an in silico screening tool for the selection of suitable polymeric excipients for a given API. In order to detect a possible liquid–liquid split or APS using PC-SAFT EOS, the experimental SLE data should be provided and used for the optimization of binary interaction parameters k_{ij} (we note that, in this work, k_{ij} values were derived based on SLE data for only one binary system, IBU–PDLG 7502A). With the optimized k_{ij} values, PC-SAFT EOS correctly predicted the presence of immiscibility region for all the binary systems studied, although compared to experimental results, the size of heterogonous regions was systematically underestimated.

4. Conclusions

Herein, the thermodynamic phase behavior of IBU with PLA/PLGA polymers of various molecular weights and copolymer compositions was investigated experimentally using calorimetry and computationally using PC-SAFT EOS. Kinetic stability was estimated on the basis of the T_g line modeled using the Kwei equation. The constructed phase diagrams indicate a very low solubility of IBU in PLA/PLGA polymers at a typical storage temperature of 25 °C, limited miscibility of IBU with PLA/PLGA polymers, and increasing kinetic stability with increasing molecular weight of PLA/PLGA polymers. While the effect of copolymer composition or molecular weight on the solubility of IBU was minimal, it had a more pronounced impact on the size of the miscibility region. The PC-SAFT predictions using solely pure-component parameters for IBU and PLA/PLGA polymers (i.e., pure predictions without any experimental input for a given binary system) significantly overestimated IBU solubility and provided no evidence of possible liquid–liquid split or amorphous phase separation. After optimizing binary interaction parameters k_{ij} using SLE experimental data (based on SLE data for one of the six binary systems studied), the PC-SAFT EOS correctly detected a possible formation of a heterogeneous immiscibility region for all the binary systems studied. Using optimized k_{ij} values, the PC-SAFT EOS system was able to capture miscibility trends as a function of copolymer composition, while the effect of molecular weight was only partially captured, and the miscibility region was systematically underestimated compared to experimental results.

Supplementary Materials: The following supporting information can be downloaded at: https://www.mdpi.com/article/10.3390/pharmaceutics15020645/s1, Table S1: T_s values obtained for IBU–PLGA and IBU–PLA systems; Table S2: (a) T_g values obtained for IBU–PDLG 5004A system annealed at 80–150 °C. (b) T_g values obtained for IBU–PDLG 7504A system annealed at 80–150 °C. (c) T_g values obtained for IBU–PDL 02A system annealed at 80–150 °C. (d) T_g values obtained for IBU–PDL 04A system annealed at 80–150 °C; Table S3: Parameters k and q of the Kwei equation ((Equation (1) in the main article) for IBU– PLGA and IBU–PLA systems; Table S4: AARD (w_{API}) values between experimental SLE data and data calculated using PC-SAFT EOS for all IBU–PLGA and IBU–PLA systems; Table S5: PC-SAFT EOS binary interaction parameters, k_{ij}, between the IBU and the different monomer units of PLA/PLGA polymers.

Author Contributions: A.I.: Investigation, Formal analysis, Visualization, Writing—original draft. M.K.: Software, Writing—review and editing. F.H.: Conceptualization, Methodology, Formal analysis, Supervision, Writing—review and editing. M.F.: Conceptualization, Methodology, Supervision, Formal analysis, Writing—original draft, Writing—review and editing. All authors have read and agreed to the published version of the manuscript.

Funding: This research was funded by Czech Science Foundation (GAČR No. 22-07164S) and by the grants of specific university research (A2_FCHI_2022_002 and A1_FCHI_2022_002).

Institutional Review Board Statement: Not applicable.

Informed Consent Statement: Not applicable.

Data Availability Statement: All data relevant to the publication are included.

Acknowledgments: The authors would like to acknowledge the Zentiva Group (Prague, Czech Republic) and Corbion (Gorinchem, Netherlands) for the chemicals donated to conduct this research.

Conflicts of Interest: The authors declare no conflict of interest.

References

1. Savjani, K.T.; Gajjar, A.K.; Savjani, J.K. Drug Solubility: Importance and Enhancement Techniques. *ISRN Pharm.* **2012**, *2012*, 195727. [CrossRef] [PubMed]
2. He, Y.; Ho, C. Amorphous Solid Dispersions: Utilization and Challenges in Drug Discovery and Development. *J. Pharm. Sci.* **2015**, *104*, 3237–3258. [CrossRef] [PubMed]
3. Khan, K.U.; Minhas, M.U.; Badshah, S.F.; Suhail, M.; Ahmad, A.; Ijaz, S. Overview of nanoparticulate strategies for solubility enhancement of poorly soluble drugs. *Life Sci.* **2022**, *291*, 120301. [CrossRef] [PubMed]
4. Makadia, H.K.; Siegel, S.J. Poly Lactic-co-Glycolic Acid (PLGA) as Biodegradable Controlled Drug Delivery Carrier. *Polymers* **2011**, *3*, 1377–1397. [CrossRef] [PubMed]
5. Wang, Y.; Qu, W.; Choi, S.H. FDA's Regulatory Science Program for Generic PLA/PLGA-Based Drug Products. *Am. Pharm. Rev.* **2016**, *20*. Available online: https://www.americanpharmaceuticalreview.com/Featured-Articles/188841-FDA-s-Regulatory-Science-Program-for-Generic-PLA-PLGA-Based-Drug-Products/ (accessed on 19 December 2022).
6. Erbetta, C. Synthesis and Characterization of Poly(D,L-Lactide-co-Glycolide) Copolymer. *J. Biomater. Nanobiotechnol.* **2012**, *3*, 208–225. [CrossRef]
7. Nieto, K.; Mallery, S.; Schwendeman, S. Microencapsulation of Amorphous Solid Dispersions of Fenretinide Enhances Drug Solubility and Release from PLGA in vitro and in vivo. *Int. J. Pharm.* **2020**, *586*, 119475. [CrossRef]
8. Bohr, A.; Kristensen, J.; Stride, E.; Dyas, M.; Edirisinghe, M. Preparation of microspheres containing low solubility drug compound by electrohydrodynamic spraying. *Int. J. Pharm.* **2011**, *412*, 59–67. [CrossRef]
9. Bala, I.; Hariharan, S.; Kumar, M.N.V.R. PLGA Nanoparticles in Drug Delivery: The State of the Art. *Crit. Rev. Ther. Drug Carr. Syst.* **2004**, *21*, 387–422. [CrossRef] [PubMed]
10. Panyam, J.; Williams, D.; Dash, A.; Leslie-Pelecky, D.; Labhasetwar, V. Solid-state solubility influences encapsulation and release of hydrophobic drugs from PLGA/PLA nanoparticles. *J. Pharm. Sci.* **2004**, *93*, 1804–1814. [CrossRef] [PubMed]
11. Blasi, P. Poly(lactic acid)/poly(lactic-co-glycolic acid)-based microparticles: An overview. *J. Pharm. Investig.* **2019**, *49*, 337–346. [CrossRef]
12. Zumaya, A.L.V.; Ulbrich, P.; Vilčáková, J.; Dendisová, M.; Fulem, M.; Šoóš, M.; Hassouna, F. Comparison between two multicomponent drug delivery systems based on PEGylated-poly (l-lactide-co-glycolide) and superparamagnetic nanoparticles: Nanoparticulate versus nanocluster systems. *J. Drug Deliv. Sci. Technol.* **2021**, *64*, 102643. [CrossRef]
13. Zumaya, A.L.V.; Martynek, D.; Bautkinová, T.; Šoóš, M.; Ulbrich, P.; Raquez, J.-M.; Dendisová, M.; Merna, J.; Vilčáková, J.; Kopecký, D.; et al. Self-assembly of poly(L-lactide-co-glycolide) and magnetic nanoparticles into nanoclusters for controlled drug delivery. *Eur. Polym. J.* **2020**, *133*, 109795. [CrossRef]
14. Iemtsev, A.; Hassouna, F.; Klajmon, M.; Mathers, A.; Fulem, M. Compatibility of selected active pharmaceutical ingredients with poly(D, L-lactide-co-glycolide): Computational and experimental study. *Eur. J. Pharm. Biopharm.* **2022**, *179*, 232–245. [CrossRef]
15. Mahieu, A.; Willart, J.-F.; Dudognon, E.; Danède, F.; Descamps, M. A New Protocol To Determine the Solubility of Drugs into Polymer Matrixes. *Mol. Pharm.* **2013**, *10*, 560–566. [CrossRef]
16. Marsac, P.J.; Shamblin, S.L.; Taylor, L.S. Theoretical and Practical Approaches for Prediction of Drug–Polymer Miscibility and Solubility. *Pharm. Res.* **2006**, *23*, 2417. [CrossRef]
17. Sun, Y.; Tao, J.; Zhang, G.G.Z.; Yu, L. Solubilities of Crystalline Drugs in Polymers: An Improved Analytical Method and Comparison of Solubilities of Indomethacin and Nifedipine in PVP, PVP/VA, and PVAc. *J. Pharm. Sci.* **2010**, *99*, 4023–4031. [CrossRef] [PubMed]
18. Gill, P.; Moghadam, T.T.; Ranjbar, B. Differential scanning calorimetry techniques: Applications in biology and nanoscience. *J. Biomol. Tech.* **2010**, *21*, 167–193.
19. Mathers, A.; Hassouna, F.; Klajmon, M.; Fulem, M. Comparative Study of DSC-Based Protocols for API–Polymer Solubility Determination. *Mol. Pharm.* **2021**, *18*, 1742–1757. [CrossRef] [PubMed]
20. Höhne, G.W.H.; Hemminger, W.; Flammersheim, H.-J. *Differential Scanning Calorimetry*; Springer: Berlin/Heidelberg, Germany, 2003; Volume 2.
21. Kwei, T.K. The effect of hydrogen bonding on the glass transition temperatures of polymer mixtures. *J. Polym. Sci. Polym. Lett. Ed.* **1984**, *22*, 307–313. [CrossRef]
22. Simha, R.; Boyer, R.F. On a General Relation Involving the Glass Temperature and Coefficients of Expansion of Polymers. *J. Chem. Phys.* **1962**, *37*, 1003–1007. [CrossRef]
23. Iemtsev, A.; Hassouna, F.; Mathers, A.; Klajmon, M.; Dendisová, M.; Malinová, L.; Školáková, T.; Fulem, M. Physical stability of hydroxypropyl methylcellulose-based amorphous solid dispersions: Experimental and computational study. *Int. J. Pharm.* **2020**, *589*, 119845. [CrossRef]
24. Von Solms, N.; Kouskoumvekaki, I.A.; Lindvig, T.; Michelsen, M.L.; Kontogeorgis, G.M. A novel approach to liquid-liquid equilibrium in polymer systems with application to simplified PC-SAFT. *Fluid Phase Equilib.* **2004**, *222*, 87–93. [CrossRef]

25. Gross, J.; Sadowski, G. Perturbed-Chain SAFT: An Equation of State Based on a Perturbation Theory for Chain Molecules. *Ind. Eng. Chem. Res.* **2001**, *40*, 1244–1260. [CrossRef]
26. Gross, J.; Sadowski, G. Application of the Perturbed-Chain SAFT Equation of State to Associating Systems. *Ind. Eng. Chem. Res.* **2002**, *41*, 5510–5515. [CrossRef]
27. Gross, J.; Spuhl, O.; Tumakaka, F.; Sadowski, G. Modeling copolymer systems using the perturbed-chain SAFT equation of state. *Ind. Eng. Chem. Res.* **2003**, *42*, 1266–1274. [CrossRef]
28. Klajmon, M. Investigating Various Parametrization Strategies for Pharmaceuticals within the PC-SAFT Equation of State. *J. Chem. Eng. Data.* **2020**, *65*, 5753–5767. [CrossRef]
29. Prudic, A.; Lesniak, A.K.; Ji, Y.H.; Sadowski, G. Thermodynamic phase behaviour of indomethacin/PLGA formulations. *Eur. J. Pharm. Biopharm.* **2015**, *93*, 88–94. [CrossRef]
30. Cocchi, G.; Angelis, M.G.D.; Sadowski, G.; Doghieri, F. Modelling polylactide/water/dioxane systems for TIPS scaffold fabrication. *Fluid Phase Equilib.* **2014**, *374*, 1–8. [CrossRef]
31. Štejfa, V.; Pokorný, V.; Mathers, A.; Růžička, K.; Fulem, M. Heat capacities of selected active pharmaceutical ingredients. *J. Chem. Thermodyn.* **2021**, *163*, 106585. [CrossRef]
32. In Pyo Park, P.; Jonnalagadda, S. Predictors of glass transition in the biodegradable poly-lactide and poly-lactide-co-glycolide polymers. *J. Appl. Polym. Sci.* **2006**, *100*, 1983–1987. [CrossRef]
33. Lee, J.; Chae, G.; Khang, G.; Kim, M.; Cho, S.; Lee, H. The effect of gamma irradiation on PLGA and release behavior of BCNU from PLGA wafer. *Macromol. Res.* **2003**, *11*, 352–356. [CrossRef]
34. Luebbert, C.; Huxoll, F.; Sadowski, G. Amorphous-Amorphous Phase Separation in API/Polymer Formulations. *Molecules* **2017**, *22*, 296. [CrossRef] [PubMed]

pharmaceutics

MDPI

Article

Low-Density Lipoprotein Pathway Is a Ubiquitous Metabolic Vulnerability in High Grade Glioma Amenable for Nanotherapeutic Delivery

Adenike O. Adekeye [1], David Needham [2] and Ruman Rahman [1,*]

1 Biodiscovery Institute, School of Medicine, University of Nottingham, Nottingham NG7 2RD, UK
2 Department of Mechanical Engineering and Material Science, School of Engineering, Duke University, Durham, NC 27708, USA
* Correspondence: ruman.rahman@nottingham.ac.uk

Abstract: Metabolic reprogramming, through increased uptake of cholesterol in the form of low-density lipoproteins (LDL), is one way by which cancer cells, including high grade gliomas (HGG), maintain their rapid growth. In this study, we determined LDL receptor (LDLR) expression in HGGs using immunohistochemistry on tissue microarrays from intra- and inter tumour regions of 36 adult and 133 paediatric patients to confirm LDLR as a therapeutic target. Additionally, we analysed expression levels in three representative cell line models to confirm their future utility to test LDLR-targeted nanoparticle uptake, retention, and cytotoxicity. Our data show widespread LDLR expression in adult and paediatric cohorts, but with significant intra-tumour variation observed between the core and either rim or invasive regions of adult HGG. Expression was independent of paediatric tumour grade or identified clinicopathological factors. LDLR-expressing tumour cells localized preferentially within perivascular niches, also with significant adult intra-tumour variation. We demonstrated variable levels of LDLR expression in all cell lines, confirming their suitability as models to test LDLR-targeted nanotherapy delivery. Overall, our study reveals the LDLR pathway as a ubiquitous metabolic vulnerability in high grade gliomas across all ages, amenable to future consideration of LDL-mediated nanoparticle/drug delivery to potentially circumvent tumour heterogeneity.

Keywords: low-density lipoprotein; high grade glioma; perivascular; nanotherapy

1. Introduction

Gliomas are the most common primary brain tumours and originate from the glial cells of the brain. They are a heterogenous spectrum of slow-growing to highly aggressive infiltrating tumours and have been classified based on their histological features into four grades by the World Health Organization (WHO); grade 1 (pilocytic astrocytoma), grade 2 (diffuse astrocytoma), grade 3 (anaplastic astrocytoma), and grade 4 (glioblastoma, GBM) [1]. Grades 3 and 4 constitute nearly half of all malignant primary brain tumours and are collectively referred to as high grade gliomas (HGG) due to their highly aggressive nature and dismal prognosis. GBM in particular is one of the most aggressive and treatment-resistant cancers in adults and children alike with a median survival of 4.6 months in the absence of treatment and approximately 14 months with optimal multimodal treatment, with a 5-year survival rate of less than 5% and over 20 average years of life lost signifying the devastating burden it poses [2–7].

Since the mid-twentieth century, standard-of-care treatment has involved surgical resection of the tumour followed by external beam radiotherapy (except in children under 3 years) and more recently the use of adjuvant chemotherapeutic agents such as temozolomide, nitrosoureas and polymer-based carmustine (Gliadel® wafer) placed at the affected tumour site during surgery [3,8–10]. However, despite these advances in treatment methods, the near-impossibility of complete tumour removal, poor drug delivery to the tumour regions due to the blood–brain barrier (BBB), and the debilitating adverse effects of treatment especially in children, have all contributed to the failure in significant outcome improvement, as even with the best of treatment, HGGs recur in over 90% of cases [3,10].

Over the past decade, there has been increasing evidence to suggest that intra-tumour heterogeneity (different sub-clonal cell populations with different molecular profiles within the same tumour) exists in the majority of cancers including malignant brain tumours, and it is in fact fundamental to our understanding of the variation in patient response to treatment and treatment failure [3]. Studies have also shown that although histologically similar, paediatric, and adult brain tumours in-fact differ in their genetic alterations and pathologic features with paediatric tumours presenting fewer genetic mutations. This discovery throws in doubt the effectiveness of current standard-of-care drugs administered to children which were originally developed for use in their adult counterparts. Hence, there is increasing interest in finding a common, ubiquitous vulnerability that is prevalent across the various intra-tumour sub-clones, disregarding the heterogeneity and molecular differences between paediatric and adult tumour types towards which novel therapeutic strategies can be targeted [11–14].

Abnormal cellular metabolism is one of the hallmarks of cancer, with most cancer cells including HGGs relying on glycolysis rather than oxidative phosphorylation for energy supply, a phenomenon known as the Warburg effect. Additionally, the increased requirement for cholesterol to maintain tumour proliferation and spread is an observed feature in many cancers [15–17]. Cholesterol is made available to tumour cells either through; (i) increased cellular synthesis regulated by the enzymes fatty acid synthase (FASN), which catalyses the conversion of acetyl-CoA and malonyl-CoA to palmitate and 3-hydroxyl-3-methylglutaryl-coenzyme A (HMG-CoA) reductase or (ii) from cellular uptake of preformed small-sized (20–25 nm) low-density lipoproteins (LDL), which bind to the high affinity LDL receptors (LDLR) found in clathrin-coated pits on the cell surface, which are internalised into the cytoplasm via receptor-mediated endocytosis as lipid vesicles and then transported to lysosomes, where they are hydrolysed for subsequent use by the cells and the dissociated receptors recycled to the cell surface for more LDL uptake [15,16,18–21].

Overexpression of FASN is associated with various pathological conditions such as cardiovascular diseases, diabetes and several different cancers including gliomas, breast, prostate, lung and colon cancer [15]. Similarly, LDLR has been shown to be upregulated by a factor of hundreds in rapidly proliferating tumours such as leukaemias, gynaecological malignancies, lung cancer, and across numerous human glioma cell lines compared to their respective normal tissues [16,17,20]. Due to this pro-tumorigenic effect, LDLR expression has been shown to be a predictor of tumour progression and survival with studies of breast cancer and small-cell lung cancer patients exhibiting better survival rates in those whose tumours demonstrated low LDLR expression [22,23]. Although the extent to which tumour cells including malignant gliomas meet their cholesterol requirement is not fully understood, FASN and LDLR are potential targets for therapeutic intervention to circumvent tumour heterogeneity in HGGs [16,20]. One such intervention, described by Lacko et al., is the creation of synthetic nano-LDL particles within which the chemotherapeutic agent is encapsulated to serve as drug delivery vehicles, targeting the LDLRs, or where the drug is designed to be of similar size to LDL so that they are preferentially taken up by the cancer cells as they would for normal LDL- effectively the cancer cell's "food" [11,20,24,25]. A new improvement to this method is the synthesis of dual-targeting nanoparticles directed at both the tumour and BBB such that the nanoparticle can access the tumour irrespective of whether or not the BBB is intact [26]. The ability of LDL nanoparticle-delivery systems to escape immune detection, due to its properties as a natural molecule, has made it a better prospect compared to other delivery systems and also of potential use in other cancer types such as breast cancer [27].

Here, we test the hypothesis that invariably high levels of expression of LDLR protein across intra- and inter-tumour regions in adult and paediatric high-grade gliomas confirm

LDLR as a therapeutic target. This was determined through immunohistochemistry (IHC) using an LDL-receptor targeting antibody carried out on patient-derived tissue microarrays (TMA), which incorporated varying intra-tumour regions in the adult GBM cohort and single region samples for the paediatric HGGs. Additionally, representative cell lines of adult (GIN28) and paediatric (SF188, KNS42) glioblastoma were assessed for LDLR expression by immunofluorescence (IF) in vitro to confirm the future utility of these cell lines as models to test fluorescently labelled LDLR-targeted stealth nanoparticle uptake, retention, and cytotoxicity.

2. Materials and Methods

2.1. Tissue Samples and Cell Lines

Tissue microarrays (TMAs) were previously prepared in the Children's Brain Tumour Research Centre (CBTRC) from adult and paediatric patients with confirmed diagnosis of high-grade glioma (HGG) from multi-centre trials, retrospective studies and cases registered with the Children's Cancer and Leukaemia Group (CCLG). A total of 21 paediatric TMAs, each consisting of cores from different areas of a single biopsy specimen and 4 adult TMAs containing cores taken from three distinct intra-tumour regions were studied, with each TMA housing samples from 10–27 patients in triplicates. There was a total of 36 adult GBM and 133 paediatric high grade glioma cases, of which clinical diagnosis was available for 68 of paediatric cases. Of these, 13 were grade III and 55 grade IV.

Primary glioblastoma multiforme (GBM) cell line (GIN28) was generated from invasive tumour margin of adult brain tumour patients treated at the Nottingham University Hospitals (NUH) while, paediatric GBM cell lines (SF188, KNS42) were obtained from Professor Chris Jones of the Institute of Cancer Research (ICR), London, UK.

This project was approved by the National Research Ethics Committee (NRES Committee East Midlands) and the granted Ethics Reference Number is 11/EM/0076.

2.2. Immunohistochemistry (IHC)

Low density lipoprotein receptor (LDLR) protein in human HGG samples was detected using indirect IHC for better signal amplification. Formalin-fixed paraffin-embedded (FFPE) slide sections were deparaffinized in xylene for 15 min and rehydrated in decreasing concentrations of alcohol and running water. Slides were placed in a pre-heated Sodium Citrate buffer solution (pH 6.0) for 40 min in a steamer for antigen retrieval, cooled for 10–20 min, and washed in phosphate-buffered saline (PBS). All incubations were carried out in a humidified box at room temperature and washes between stages in PBS. Sections were covered in 20% normal goat serum (NGS) in PBS for 20 min to minimize non-specific antibody binding and endogenous peroxidase activity was blocked by incubating slides in DAKO peroxidase blocking solution for 5 min to prevent background staining. The primary antibody used was Abcam (Biomedical Campus, Discovery Drive, Trumpington, Cambridge, CB2 0AX, UK) anti-LDLR rabbit monoclonal antibody (EP1553Y) ab52818 and was optimized on normal liver tissue (positive control) at dilutions ranging from 1:250 to 1:1400 incubated for 1 h at room temperature, subsequent experiments were performed at 1:1400 dilution. Slides were incubated in DAKO REAL Envision HRP Rabbit/Mouse secondary antibody for 30 min. Detection of antibody binding was enzyme-based following incubation in DAKO diaminobenzindine (DAB) chromogen substrate for 10–15 min. Slides were counterstained in Gills Haematoxylin, differentiated in lithium carbonate solution, washed in running water, and dehydrated in increasing alcohol concentrations prior to immersion in xylene for mounting. Slides were mounted with coverslips using DePeX mounting medium. Slides were scanned at 40× magnification using a Nanozoomer at the Histopathology Department, NUH and expression levels analysed using the NDPview2 software. Scoring was based on proportion (absent, <25%, 25–50% or >50%) of cells stained. A double-blinded approach was implemented in scoring to produce fair and reproducible results.

2.3. Cell Culture and Reagents

SF188 and KNS42 paediatric GBM cell lines were cultured in Dulbecco Modified Eagle's Medium (DMEM)/F-12 nutrient mix (Gibco) supplemented with 10% Foetal Bovine Serum (FBS) and 2% L-glutamine. Adult GBM cell line, GIN28 was cultured in DMEM (Gibco; Milton Park, Oxford, Innovation Centre 99 Park Drive Milton Park, Oxford OX14 4RY, UK) containing 15% FBS, 1% L-glutamine and antibiotics (1% penicillin and streptomycin). HeLa cell line were used as positive control and grown in DMEM (Gibco) with 10% FBS and 2% L-glutamine. Cells were passaged at approximately 80% confluency by dissociating with trypsin-EDTA and seeded onto treated tissue culture plastic (polystyrene) and chamber slides. GBM cell lines were passaged twice a week to maintain exponential growth at varying seeding ratios of 1:3–1:5 and at cellular densities of 15,000–30,000 cells per well. HeLa cells were passaged daily due to their rapid doubling time and plated at 5000 cells per well. Optimal culture conditions of 37 °C, 21% Oxygen and 5% Carbon IV Oxide was maintained for all experiments.

2.4. Immunofluorescence (IF)

SF188, KNS42, GIN28 and HeLa cells were plated on 8-well chamber slides at optimised densities and incubation periods. Cells were washed in PBS, fixed in 0.4% paraformaldehyde for 20 min, permeabilised with PBS-T (0.1% tween, 0.25% triton-X in PBS) and blocked with 5% NGS in PBS-T for 1 h at room temperature. Primary anti-LDLR antibody was optimised on HeLa cells at dilutions ranging from 1:50–1:1000 and incubated overnight at 4 °C, subsequent experiments were performed at 1:250 dilution. Cells were incubated with goat anti-rabbit IgG secondary antibody labelled with Alexa Fluor 488 (Invitrogen, A11008) at 1:200 dilution for 1 h in the dark and washed in PBS-T. Slides were mounted with coverslips using DAPI-containing Vectastain mounting medium (Vector Labs Ltd., The Hauser Forum, 3 Charles Babbage Road, Cambridge, CB3 0GT, UK). Microscopy was carried out using a Nikon eclipse 90*i* fluorescent microscope and images were captured with Hamamatsu camera and Volocity 6.0 software at 40× magnification. ImageJ software was used to analyse images and correlate corrected total cell fluorescence (CTCF) with amount of LDLR protein expressed based on intensity of signal generated.

2.5. Statistical Analysis

All statistical analysis was performed using SPSS 24.0 software. Friedman ANOVA and Wilcoxon signed rank tests were used to analyse the significance of adult intra-tumour results. Mann–Whitney U or Kruskal–Wallis tests were used to determine the relationship between LDLR expression and clinicopathological factors. The overall survival time was compared using Kaplan–Meier and log-rank methods. Localisation of positive tumour cells within peri-vascular niches and LDLR expression in GBM cell lines were assessed with the One-way ANOVA test and Tukey's HSD post hoc test for multiple comparisons, and $p < 0.05$ was considered statistically significant.

3. Results

3.1. LDLR Protein Is Expressed in a Majority of Adult GBMs with Intra-Tumour Variation

Patient diagnosis and clinico-pathological details of TMA samples are shown in Table 1.

Table 1. Clinicopathological characteristics of adult and paediatric high grade glioma cases.

Variable	Adult [a] Frequency (%)	Paediatric [b] Frequency (%)
Age (years) [c]		
paediatric [d]		29 (39.2)
<6		30 (40.5)
6–12		15 (20.3)
>12		
Adult	7 (19.4)	
<50	16 (44.4)	
50–65	13 (36.1)	
>65		
Gender [d]		
Male	18 (50)	53 (64.6)
Female	18 (50)	29 (35.4)
Tumour grade [d]		
III (AA)	-	13 (19.1)
IV (GBM)	36 (100)	55 (80.9)
Tumour location [d]		
Supratentorial		
Infratentorial	36 (100)	57 (73.1)
Treatment (post-surgery) [d]	-	21 (26.9)
Radiotherapy		
Chemotherapy	2 (12.5)	9 (18.8)
Chemotherapy + Radiotherapy	1 (6.3)	15 (31.3)
None	12 (75)	23 (47.9)
Survival time (months) [d,e]	1 (6.3)	1 (2.1)
<12		
12–36	5 (100)	29 (63)
>36	-	15 (32.6)
Tumour type [d]	-	2 (4.3)
Primary		
Recurrent		
progressive	N/A	66 (84.6)
	N/A	10 (12.8)
	N/A	2 (2.6)

[a] Total number of cases—36. [b] Total number of cases—133. [c] Paediatric mean age 7.7 years (95% CI: 6.6–8.8), adult median age 59 years (interquartile range: 53.25, 67.75). [d] Missing details for some cases. [e] Median survival time; adults—4 months (interquartile range: 3, 6), paediatric—8 months (interquartile range: 3, 22). N/A—Not available.

LDLR protein expression was assessed in three distinct GBM intra-tumour regions representing core, rim, and invasive margin, from 36 patients. A universal expression was observed across all the regions, where 92% of samples from the tumour core had a frequency of positive cells above 25%. Similar results were also observed in the rim and invasive margins where more than 70% of cases expressed a frequency of LDLR positive cells greater than 25% Figure 1 and Table 2). Friedman ANOVA test showed a statistically significant difference among these intra-tumour regions ($p = 0.008$) and further pairwise analysis revealed that the observed significant differences were between the core and either rim or invasive margins ($p = 0.003$ and 0.014 respectively). No significant difference was detected in LDLR expression between the rim and invasive margin ($p = 0.683$, Wilcoxon's signed rank test).

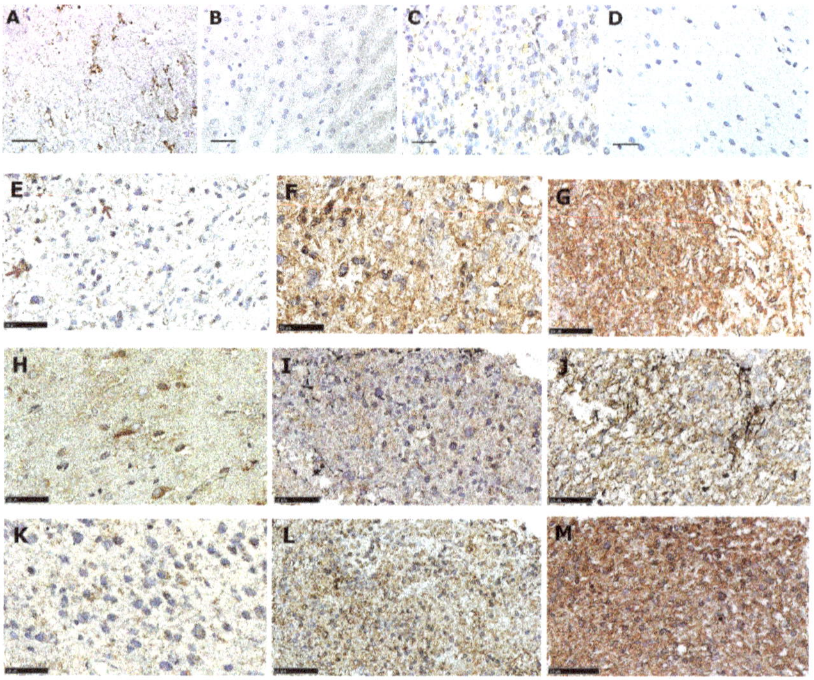

Figure 1. Anti-LDLR immunohistochemical analyses of adult GBM primary tissue. (**A**) Positive liver control. (**B**) Negative liver control. (**C**) Positive normal brain control. (**D**) Negative normal brain control. (**E**) Core with <25% expression. (**F**) Core with 25–50% expression. (**G**) Core with >50% expression. (**H**) Rim with <25% expression. (**I**) Rim with 25–50% expression. (**J**) Rim with >50% expression. (**K**) Invasive margin showing <25% expression. (**L**) Invasive margin with 25–50% expression. (**M**) Invasive margin with >50% expression based on proportion of positive cells. All images were captured at 40× magnification; scale bars correspond to 100 μm in (**A–D**) and 50 μm in (**E–M**).

Table 2. LDLR expression in distinct intra-tumour regions of adult GBM.

% Expression	Core	Frequency (%) Rim	Invasive
Negative	0	0	0
<25	3 (8.3)	8 (22.2)	8 (25.8)
25–50	14 (38.9)	19 (52.8)	15 (48.4)
>50	19 (52.8)	9 (25)	8 (25.8)
Total	36 (100)	36 (100)	31 (100) *

* Invasive margin samples not available for five patients.

3.2. LDLR Is Expressed in Paediatric Malignant Gliomas Irrespective of Tumour Grade

The expression of LDLR protein in grade 3 (anaplastic astrocytoma) and grade 4 (GBM) paediatric samples (Figure 2) shows an expression pattern similar to that observed in adult GBM cases with 112 of the 133 samples (84% of cases) exhibiting expression levels above 25% (Table 3). Statistical analysis showed no significant difference in expression between grades 3 and 4 ($p = 0.285$, Mann–Whitney U test). This suggests a ubiquitous role for LDLR in high grade glioma metabolism, which could be exploited in developing novel future therapeutic strategies for both adult and paediatric anaplastic astrocytoma and glioblastoma cases.

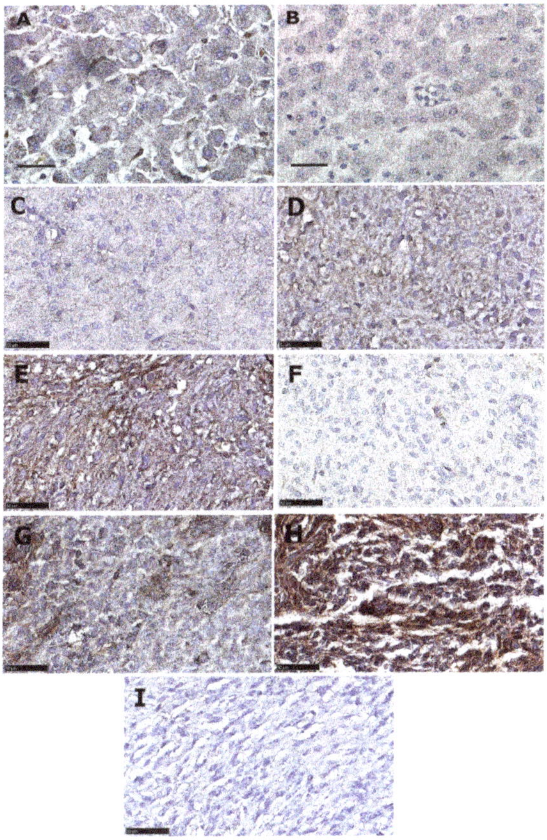

Figure 2. Anti-LDLR immunohistochemical analyses of paediatric high-grade glioma. (**A**) Positive liver control. (**B**) Negative liver control. (**C**) Anaplastic astrocytoma (AA, grade III) cells showing less than 25% expression. (**D**) AA showing 25–50% expression. (**E**) AA with >50% expression. (**F**) GBM (grade IV) cells with less than 25% expression. (**G**) GBM cells with 25–50% expression. (**H**) GBM cells with >50% expression. (**I**) GBM cells with negative protein expression. All images were captured at 40× magnification; scale bars correspond to 100µm in (**A**,**B**) and 50 µm in (**C**–**I**).

Table 3. LDLR expression in paediatric anaplastic astrocytoma and GBM.

% LDLR Expression	Frequency (%)
Negative	5 (3.8)
<25	16 (12)
25–50	77 (57.9)
>50	35 (26.3)
Total	133 (100%)

3.3. *Glioma LDLR Expression Is Localized within Peri-Vascular Niches*

Positively stained tumour cells were repeatedly observed to localize preferentially in varying proportions near blood vessels compared to other areas of the tumour. We therefore examined the distribution of positive tumour cells in respect to vasculature in our adult and paediatric TMAs (Figure 3A–C). The average number of positively stained tumour cells around each identified vessel in paediatric samples was found to be 4.74 (95% CI: 4.48–5.00), while in adults it was 5.19 (95% CI: 4.84–5.84). However, intra-tumour variation was observed in the three distinct adult regions with the core having on average, the greatest number of positively stained tumour cells per vessel, while the invasive margin had the least (Table 4). Additionally, statistical comparisons among these regions by One-way ANOVA and Tukey HSD tests confirmed significantly higher numbers of tumour cells localized around blood vessels ($p = 0.043$, Figure 3D). This therapeutically relevant finding suggests that LDLR-positive tumour cells are more likely to be located within the penetration distance of nanoparticles upon extravasation from the bloodstream and will therefore potentially benefit from such future therapeutic targeting.

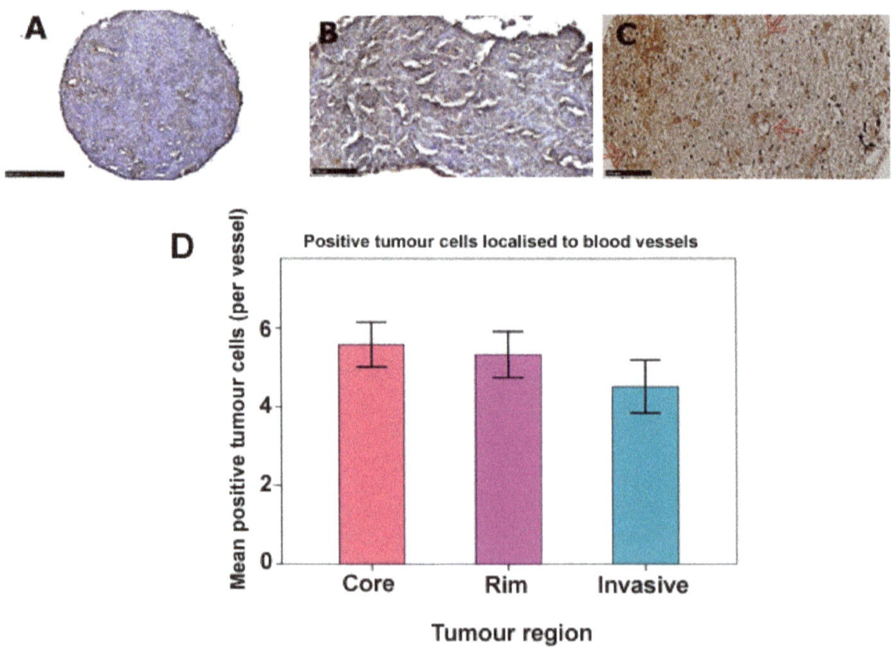

Figure 3. Localization of LDLR positive tumour cells in peri-vasculature regions. (**A**) Paediatric GBM whole tumour core showing preferential localisation of LDLR expression (brown staining) around blood vessels compared to other areas. (**B**) Highly predominant localisation of LDLR-expressing paediatric GBM cells around the vasculature. (**C**) Invasive region of adult GBM showing tumour cells with positive LDLR staining in perivascular areas (indicated by red arrows). Image magnification; A, 10×, B and C, 20×. scale bars correspond to 250 μm for A and 100 μm for B and C. (**D**) Variation in average number of positive tumour cells per vessel in adult intra-tumour regions ($p = 0.043$), where pair-wise analysis revealed a significant difference in core and invasive regions only ($p = 0.040$, Tukey HSD test).

Table 4. Categorical proportions of positive tumour cells per vessel in adult and paediatric cases and corresponding *p*-values.

LDLR Expression [a]	Core	Frequency (%) Adult Rim	Invasive	Paediatric [b]
Low	1 (2.8)	4 (11.1)	8 (27.6)	22 (18.0)
Moderate	24 (66.7)	23 (63.9)	17 (58.6)	86 (70.5)
High	11 (30.6)	9 (25.0)	4 (13.8)	14 (11.5)
Total	36 (100)	36 (100)	29 (100)	122 (100)

[a] Number of LDLR positive cells around each identified blood vessel. Low < 4, moderate 4–6, high > 6. [b] Comparison between anaplastic astrocytoma and GBM was limited by lack of clinical diagnosis for the majority of cases.

3.4. LDLR Expression Is Not Dependent on Clinicopathological Variables

We next assessed the relationship between LDLR expression and clinicopathological characteristics of known prognostic and predictive significance such as age, gender, tumour location, recurrence, and treatment. Statistical comparisons revealed that in both paediatric and adult cases, expression of LDLR generally did not correlate with these factors ($p > 0.05$, Mann–Whitney U and Kruskal–Wallis tests, with the exception of age in tumour core regions of in the adult cohort) (Table 5).

Table 5. Statistical analyses of LDLR expression and clinico-pathological characteristics.

Variable	Core	Adult Rim	Invasive	Paediatric
Age	0.047	0.653	0.578	0.445
Gender	0.106	0.223	0.623	0.270
Tumour site [a]	0.382	0.173	0.462	0.846
Treatment	0.186	0.173	0.362	0.431
Tumour type [b]	-	-	-	0.958

[a] Supra-vs. infra-tentorial location. In adults, where all cases were supratentorial, differences were examined amongst frontal, parietal, temporal and occipital tumour sites. [b] Primary or recurrent. No clinical information available for adult cases.

To determine the usefulness of LDLR expression as a predictor of survival as previous studies in breast and lung cancers have shown [22,23], we examined overall survival among our paediatric cohort (adult cases were excluded from analysis due to lack of information on survival for the majority of the cases). We observed that the median survival was lowest in cases with less than 25% expression and highest in those above 50% (5 and 11 months respectively); results were however not statistically significant ($p = 0.126$, Log-rank test) (Figure 4).

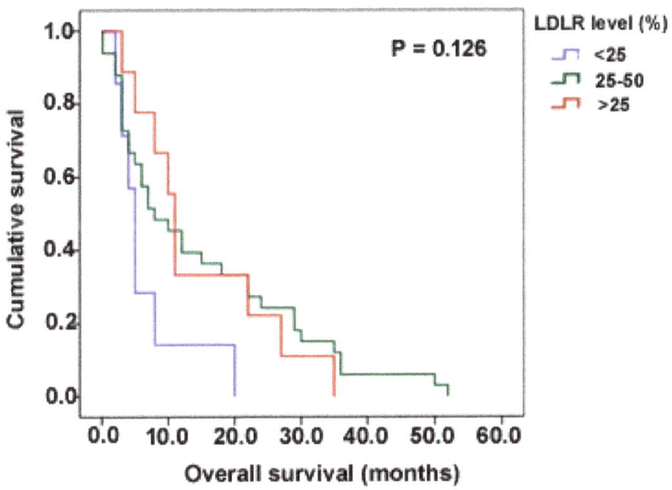

Figure 4. Association of LDLR expression with clinico-pathological characteristics. Kaplan–Meier plot showing median overall survival time of 5, 8, and 11 months with LDLR expression levels of <25%, 25–50% and >50% respectively in paediatric high-grade gliomas. Statistical analyses were not significant, indicating no correlation between overall survival and LDLR expression ($p = 0.126$).

3.5. Adult and Paediatric GBM Cell Lines Express Variable Levels of LDLR

To ascertain that our cell lines expressed LDLR and were representative of results obtained from IHC of adult and paediatric TMAs, and therefore amenable to utilize as future in vitro models to test LDLR-targeted nanoparticle uptake/retention, expression levels were examined in three GBM cell lines. SF188, KNS42 (both paediatric) and GIN28 (adult). This revealed widespread staining confirming LDLR expression (Figure 5A). The corresponding amount of expression in the different cell lines determined by correlating corrected total cell fluorescence (CTCF) with integrated density based on the intensity of signal generated, showed that the SF188 and KNS42 paediatric GBM lines have relatively higher LDLR expression compared to GIN28 GBM cells (Figure 5B). Further analysis confirmed these differences were statistically significant ($p = 0.001$ and $p = 0.008$ for SF188 vs. GIN28 and KNS42 vs. GIN28 respectively, One-way ANOVA test) and on pair-wise comparisons, no significant difference was observed between the paediatric cell lines ($p = 0.623$, Tukey HSD test). This was in harmony with previous studies showing variable LDLR expression between different cell lines [17].

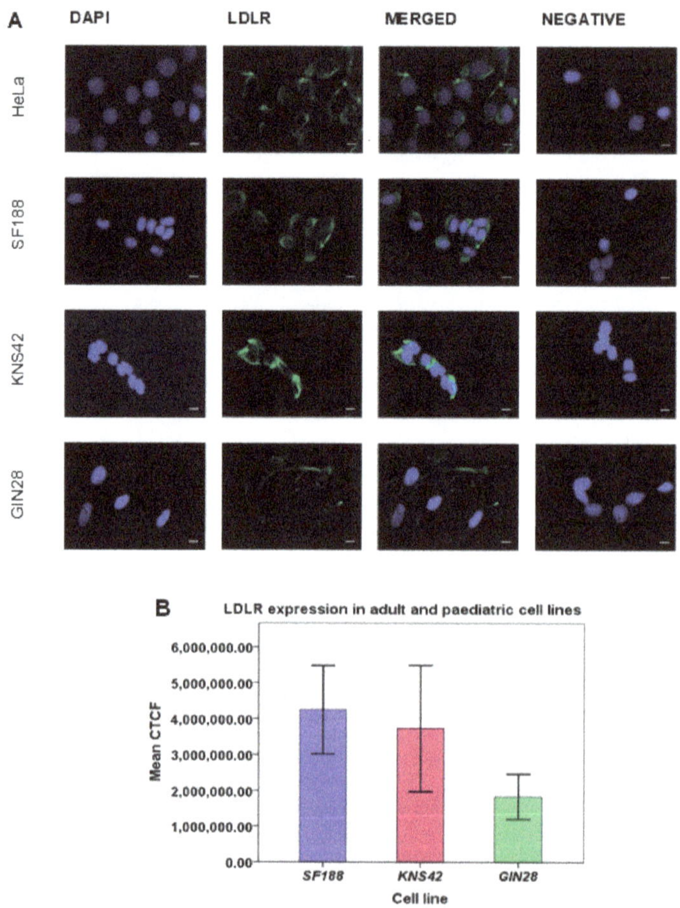

Figure 5. In vitro analysis of LDLR expression in adult and paediatric GBM cell lines. (**A**) Immunofluorescence analysis showing LDLR expressions in control HeLa cell line, paediatric cell lines SF188 and KNS42, and adult cell line GIN28. Blue—nuclear stain DAPI; Green—LDLR protein. Magnification 40×; scale bar: 100 µm. (**B**) Mean LDLR expression in all cell lines showing highest expression in SF188 and lowest in GIN28 ($p = 0.001$), where error bars correspond to 95% CI.

4. Discussion

As metabolic reprogramming is a well-established hallmark of cancer [28], research emphasis is shifting towards identifying altered metabolic pathways to serve as novel targets for anti-cancer drug delivery in order to maximize therapeutic benefits and efficiency [24]. Furthermore, increasing understanding of inter- and intra-patient genetic heterogeneity, renders common metabolic vulnerabilities particularly attractive for therapy, when considering the difficulties of applying personalized medicine to highly heterogeneous neoplasms. While the role of LDLR in maintaining tumour growth and proliferation has been well established in other cancers such as breast, lung, and gynaecological malignancies for decades, its role in malignant gliomas has not been extensively researched and evidence surrounding it is limited despite high grade gliomas being a leading cause of cancer related deaths worldwide [16,22,29]. LDLs, which are the principal carriers of cholesterol in circulation and required to support the growth of rapidly proliferating cancer cells, are the major ligands of LDLR, at least for canonical LDLR-based signalling. Abundance of LDLR thus signifies increased uptake and accumulation of cholesterol within the tumour and is an indicator of tumour aggressiveness and invasive potential as evident in high-grade gliomas [20,27,30,31].

We hypothesized that high-grade gliomas express invariably high levels of LDLR, despite the known genetic inter- and intra-tumour heterogeneity. Our results demonstrated widespread LDLR expression in the majority of the adult GBM cases with intra-tumour variation noted and a similar paediatric expression pattern which was independent of tumour grade, confirming LDLR as a valid therapeutic target across all ages for malignant glioma. Intriguingly, the localization of LDLR-expressing tumour cells within peri-vascular niches was observed in the vast majority of both adult and paediatric high-grade gliomas. Additionally, LDLR expression was discovered not to vary with clinicopathological factors, nor did it correlate with survival. However, a p-value of 0.126 (when comparing the <25% and >50% LDLR expression cohorts) certainly warrants re-examination in a larger cohort of high-grade glioma patients to refute a prognostic role more confidently for LDLR expression. Finally, we confirmed variable levels of detectable LDLR expression in adult (GIN28) and paediatric (SF188, KNS42) GBM in vitro cell lines, thus rendering these lines as future tools to test LDLR-targeted nanotherapy delivery.

The intra-tumour variation observed in adult GBM cases, where the core had the highest LDLR expression and the invasive margin the least, is consistent with current literature on intra-tumour heterogeneity [3,14]. The core is known to be the most cellular, proliferative and vascular region of the tumour compared to the periphery (which includes the rim and invasive areas) as demonstrated previously in our laboratory [32]. Previous studies have established mutually exclusive phenomena of proliferation and migration in tumours. The cells in the core are thus more proliferative, while those in the periphery/invasive margin are more migratory, invading into surrounding local tissue [33]. Although the invasive margin is the least proliferative, it is almost impossible to completely resect this infiltrative region during surgery. As a direct consequence, residual invasive cells left behind in this region are thought to be responsible for the high degree of tumour recurrence and high mortality [34]. Based on our findings, it is thus logical to assume that LDLR expression at the invasive margin represents a more therapeutically relevant population of cells amenable to LDL-mediated nanoparticle/drug targeting, which could potentially and significantly reduce recurrence rates and improve patient outcome. Furthermore, it also suggests that in cases where extensive tumour debulking is not possible, cells in the core expressing a higher percentage of LDLR could benefit from such targeting as an alternative form of intervention.

The non-significant difference in levels of LDLR expression between anaplastic astrocytoma and glioblastoma ($p = 0.285$) in our paediatric cohort points to a possible ubiquitous role of LDLR in paediatric high grade glioma metabolism. Based on this finding, we can predict that both anaplastic astrocytoma (grade 3) and glioblastoma (grade 4) may be amenable to anti-LDLR targeting to the same extent. However, re-examination in a larger paediatric cohort is required to validate a claim of LDLR as a broadly ubiquitous target. Considering that children are not able to tolerate the same levels of intense treatments as adults and that current standard-of-care treatment modalities, such as radiation therapy, are exempt in children under 3 years of age, paediatric patients might potentially benefit more from preferential LDLR targeting. Although our data did not show paediatric LDLR expression to be dependent on tumour grade, differences may be more subtle and require a larger sample size to interrogate.

LDLR expressing tumour cells were consistently observed to localize within perivascular areas, consistent with seminal studies by Calabrese et al., in 2007, across a host of primary brain tumours, notably glioblastoma, ependymoma, medulloblastoma and oligodendroglioma, where cancer stem cells (CSCs) were shown to associate in close proximity to tumour capillaries creating vascular niches. This tumour microenvironment is essential for the maintenance of tumour growth and proliferation and is one of the key rationales for several anti-angiogenic therapies [35]. The overall significant difference in vascularity detected across adult GBM intra-tumour regions was essentially due to the high vascularity of tumour cells in the core compared to the invasive margin ($p = 0.040$) as confirmed by Tukey's HSD test. This phenomenon can be explained by the higher cellular density and proliferation of the core compared to other regions which outstrips it of its blood supply, inducing a state of nutrient deficiency and hypoxia—the main driver of angiogenesis and neo-vascularization within the tumour. Similar comparison between paediatric anaplastic astrocytoma and glioblastoma was limited by disparity in sample sizes (13 and 55 respectively) which would have given a biased result. A larger sample size, particularly of grade 3 tumours, would have improved the power and validity of such analysis.

The association of positive tumour cells with the vasculature is of immense therapeutic significance. Hitherto, one of the greatest challenges of anti-cancer drug delivery to tumour sites is the difficulty in penetrating and accumulating to sufficient levels as to effect cell kill within the target organ, without incurring significant toxicity to surrounding normal tissues. In attempts to overcome this, novel drug delivery strategies such as macromolecular drug carriers have been developed [36,37]. A study by Dreher i.e., in 2006, showed that lower weight macromolecular drug carriers and liposomes have a higher vascular permeability and longer plasma half-life due to enhanced permeability and retention effects [38]. This in effect indicates that LDLR-expressing tumour cells within close proximity to the blood vessels as shown in our results, are potentially more susceptible to efficacious therapeutic targeting using LDL-mimicking stealth nanoparticles. In light of our data, and for future research into this aspect, we hypothesise that effective nanoparticle-mediated tumour-cell kill is a function of peri-vasculature localization. Furthermore, the use of statins has been reported to reduce the efflux activity of drug transporters Pgp/ABCB1 and BCRP/ABCG2, increase the number of LDLR exposed on the BBB surface, thereby increasing doxorubicin-loaded nanoparticle entry across the BBB [39]; polyester-based nanoparticles loaded with paclitaxel improved BBB permeability in vitro [40]; and an infiltrative GBM preclinical model has been recently shown to be sensitive to enhanced proton treatment with LDLR-ligand peptide-conjugated gold nanoparticles [41].

5. Conclusions

Taken together, our results have clearly shown the LDLR pathway to represent a metabolic vulnerability in both paediatric and adult high-grade gliomas, which potentially circumvents intra- and inter-tumour heterogeneity and is therefore potentially amenable to future lipoprotein based anti-LDLR nanotherapy, ubiquitously in patient cohorts. To maximize the benefits from this novel therapeutic strategy, there may be a need for selection of cases likely to benefit the most based on demonstrated levels of LDLR expression, tumour vascularity and/or further lipidomic studies.

Author Contributions: Conceptualization, D.N. and R.R.; methodology, A.O.A.; validation, A.O.A.; formal analysis, A.O.A.; resources, R.R.; data curation, A.O.A.; writing—original draft preparation, A.O.A.; writing—review and editing, R.R. and D.N.; supervision, R.R. and D.N. All authors have read and agreed to the published version of the manuscript.

Funding: This research was funded by a philanthropic donation from Mr. Michael Hodgkinson.

Institutional Review Board Statement: The study was conducted in accordance with the National Research Ethics Service Committee East Midlands, Nottingham (ethics committee reference 11/EM/0076).

Informed Consent Statement: Informed consent was obtained from all patients involved in the study.

Data Availability Statement: Data are contained within the article.

Acknowledgments: The authors acknowledge the University of Nottingham MSc Oncology course for provision of research funds. We thank Chris Jones (Institute of Cancer Research, London) for kindly providing the SF188 and KNS42 cell lines.

Conflicts of Interest: The authors declare no conflict of interest.

References

1. Louis, D.N.; Ohgaki, H.; Wiestler, O.D.; Cavenee, W.K.; Burger, P.C.; Jouvet, A.; Scheithauer, B.W.; Kleihues, P. The 2007 WHO Classification of Tumours of the Central Nervous System. *Acta Neuropathol.* **2007**, *114*, 97–109. [CrossRef] [PubMed]
2. Weller, M.; Fisher, B.; Taphoorn, M.J.B.; Belanger, K.; Brandes, A.A.; Marosi, C.; Bogdahn, U.; Curschmann, J.; Janzer, R. Radiotherapy plus Concomitant and Adjuvant Temozolomide for Newly Diagnosed Glioblastoma. *N. Engl. J. Med.* **2005**, *1*, 19–28. [CrossRef]
3. Sottoriva, A.; Spiteri, I.; Piccirillo, S.G.M.; Touloumis, A.; Collins, V.P.; Marioni, J.C.; Curtis, C.; Watts, C.; Tavare, S. Intratumor Heterogeneity in Human Glioblastoma Reflects Cancer Evolutionary Dynamics. *Proc. Natl. Acad. Sci. USA* **2013**, *110*, 4009–4014. [CrossRef]
4. Rao, J.S. Molecular Mechanisms of Glioma Invasiveness: The Role of Proteases. *Nat. Rev. Cancer* **2003**, *3*, 489–501. [CrossRef] [PubMed]
5. Masui, K.; Cloughesy, T.F.; Mischel, P.S. Molecular Pathology in Adult High-grade Gliomas: From Molecular Diagnostics to Target Therapies. *Neuropathol. Appl. Neurobiol.* **2012**, *38*, 271–291. [CrossRef]
6. Maher, E.A.; Furnari, F.B.; Bachoo, R.M.; Rowitch, D.H.; Louis, D.N.; Cavenee, W.K.; DePinho, R.A. Malignant Glioma: Genetics and Biology of a Grave Matter. *Genes Dev.* **2001**, *15*, 1311–1333. [CrossRef]
7. Burnet, N.G.; Jefferies, S.J.; Benson, R.J.; Hunt, D.P.; Treasure, F.P. Years of Life Lost (YLL) from Cancer Is an Important Measure of Population Burden—And Should Be Considered When Allocating Research Funds. *Br. J. Cancer* **2005**, *92*, 241–245. [CrossRef]
8. Walker, M.D.; Green, S.B.; Byar, D.P.; Alexander, E.; Batzdorf, U.; Brooks, W.H.; Hunt, W.E.; MacCarty, C.S.; Mahaley, M.S.; Mealey, J.; et al. Randomized Comparisons of Radiotherapy and Nitrosoureas for the Treatment of Malignant Glioma after Surgery. *N. Engl. J. Med.* **1980**, *303*, 1323–1329. [CrossRef]
9. Brada, M.; Hoang-Xuan, K.; Rampling, R.; Dietrich, P.-Y.; Dirix, L.Y.; Macdonald, D.; Heimans, J.J.; Zonnenberg, B.A.; Bravo-Marques, J.M.; Henriksson, R.; et al. Multicenter Phase II Trial of Temozolomide in Patients with Glioblastoma Multiforme at First Relapse. *Ann. Oncol.* **2001**, *12*, 259–266. [CrossRef]
10. Germano, I.; Swiss, V.; Casaccia, P. Primary Brain Tumors, Neural Stem Cell, and Brain Tumor Cancer Cells: Where Is the Link? *Neuropharmacology* **2010**, *58*, 903–910. [CrossRef]
11. Needham, D.; Arslanagic, A.; Glud, K.; Hervella, P.; Karimi, L.; Høeilund-Carlsen, P.-F.; Kinoshita, K.; Mollenhauer, J.; Parra, E.; Utoft, A.; et al. Bottom up Design of Nanoparticles for Anti-Cancer Diapeutics: "Put the Drug in the Cancer's Food". *J. Drug Target.* **2016**, *24*, 836–856. [CrossRef] [PubMed]
12. Paugh, B.S.; Qu, C.; Jones, C.; Liu, Z.; Adamowicz-Brice, M.; Zhang, J.; Bax, D.A.; Coyle, B.; Barrow, J.; Hargrave, D.; et al. Integrated Molecular Genetic Profiling of Pediatric High-Grade Gliomas Reveals Key Differences with the Adult Disease. *J. Clin. Oncol. Off. J. Am. Soc. Clin. Oncol.* **2010**, *28*, 3061–3068. [CrossRef] [PubMed]
13. Marusyk, A.; Polyak, K. Tumor Heterogeneity: Causes and Consequences. *Biochim. Biophys. Acta (BBA)—Rev. Cancer* **2010**, *1805*, 105–117. [CrossRef] [PubMed]
14. Smith, S.J.; Wilson, M.; Ward, J.H.; Rahman, C.V.; Peet, A.C.; Macarthur, D.C.; Rose, F.R.A.J.; Grundy, R.G.; Rahman, R. Recapitulation of Tumor Heterogeneity and Molecular Signatures in a 3D Brain Cancer Model with Decreased Sensitivity to Histone Deacetylase Inhibition. *PLoS ONE* **2012**, *7*, e52335. [CrossRef]
15. Yasumoto, Y.; Miyazaki, H.; Vaidyan, L.K.; Kagawa, Y.; Ebrahimi, M.; Yamamoto, Y.; Ogata, M.; Katsuyama, Y.; Sadahiro, H.; Suzuki, M.; et al. Inhibition of Fatty Acid Synthase Decreases Expression of Stemness Markers in Glioma Stem Cells. *PLoS ONE* **2016**, *11*, e0147717. [CrossRef]
16. Rudling, M.J.; Angelin, B.; Peterson, C.O.; Collins, V.P. Low Density Lipoprotein Receptor Activity in Human Intracranial Tumors and Its Relation to the Cholesterol Requirement. *Cancer Res.* **1990**, *50*, 483–487.
17. Maletínská, L.; Blakely, E.A.; Bjornstad, K.A.; Deen, D.F.; Knoff, L.J.; Forte, T.M. Human Glioblastoma Cell Lines: Levels of Low-Density Lipoprotein Receptor and Low-Density Lipoprotein Receptor-Related Protein. *Cancer Res.* **2000**, *60*, 2300–2303.
18. Havel, R.J. The Formation of LDL: Mechanisms and Regulation. *J. Lipid Res.* **1984**, *25*, 1570–1576. [CrossRef]
19. Brown, M.S.; Goldstein, J.L. M58 A Receptor-Mediated Pathway for Cholesterol Homeostasis. *Science* **1986**, *232*, 34–47. [CrossRef]
20. Nikanjam, M.; Gibbs, A.R.; Hunt, C.A.; Budinger, T.F.; Forte, T.M. Synthetic Nano-LDL with Paclitaxel Oleate as a Targeted Drug Delivery Vehicle for Glioblastoma Multiforme. *J. Control. Release* **2007**, *124*, 163–171. [CrossRef]

21. Teerlink, T.; Scheffer, P.G.; Bakker, S.J.L.; Heine, R.J. Combined Data from LDL Composition and Size Measurement Are Compatible with a Discoid Particle Shape. *J. Lipid Res.* **2004**, *45*, 954–966. [CrossRef] [PubMed]
22. Zhou, T.; Zhan, J.; Fang, W.; Zhao, Y.; Yang, Y.; Hou, X.; Zhang, Z.; He, X.; Zhang, Y.; Huang, Y.; et al. Serum Low-Density Lipoprotein and Low-Density Lipoprotein Expression Level at Diagnosis Are Favorable Prognostic Factors in Patients with Small-Cell Lung Cancer (SCLC). *BMC Cancer* **2017**, *17*, 269. [CrossRef] [PubMed]
23. Rodrigues dos Santos, C.; Fonseca, I.; Dias, S.; de Almeida, J.C.M. Plasma Level of LDL-Cholesterol at Diagnosis Is a Predictor Factor of Breast Tumor Progression. *BMC Cancer* **2014**, *14*, 132. [CrossRef] [PubMed]
24. Lacko, A.G.; Nair, M.; Prokai, L.; McConathy, W.J. Prospects and Challenges of the Development of Lipoprotein-Based Formulations for Anti-Cancer Drugs. *Expert Opin. Drug Deliv.* **2007**, *4*, 665–675. [CrossRef] [PubMed]
25. Sabnis, N.; Lacko, A.G. Drug Delivery via Lipoprotein-Based Carriers: Answering the Challenges in Systemic Therapeutics. *Ther. Deliv.* **2012**, *3*, 599–608. [CrossRef]
26. Zhang, B.; Sun, X.; Mei, H.; Wang, Y.; Liao, Z.; Chen, J.; Zhang, Q.; Hu, Y.; Pang, Z.; Jiang, X. LDLR-Mediated Peptide-22-Conjugated Nanoparticles for Dual-Targeting Therapy of Brain Glioma. *Biomaterials* **2013**, *34*, 9171–9182. [CrossRef]
27. Huntosova, V.; Buzova, D.; Petrovajova, D.; Kasak, P.; Nadova, Z.; Jancura, D.; Sureau, F.; Miskovsky, P. Development of a New LDL-Based Transport System for Hydrophobic/Amphiphilic Drug Delivery to Cancer Cells. *Int. J. Pharm.* **2012**, *436*, 463–471. [CrossRef]
28. Beloribi-Djefaflia, S.; Vasseur, S.; Guillaumond, F. Lipid Metabolic Reprogramming in Cancer Cells. *Oncogenesis* **2016**, *5*, 189. [CrossRef]
29. dos Santos, C.R.; Domingues, G.; Matias, I.; Matos, J.; Fonseca, I.; de Almeida, J.M.; Dias, S. LDL-Cholesterol Signaling Induces Breast Cancer Proliferation and Invasion. *Lipids Health Dis.* **2014**, *13*, 16. [CrossRef]
30. Gallagher, E.J.; Zelenko, Z.; Neel, B.A.; Antoniou, I.M.; Rajan, L.; Kase, N.; LeRoith, D. Elevated Tumor LDLR Expression Accelerates LDL Cholesterol-Mediated Breast Cancer Growth in Mouse Models of Hyperlipidemia. *Oncogene* **2017**, *36*, 6462–6471. [CrossRef]
31. Yue, S.; Li, J.; Lee, S.-Y.; Lee, H.J.; Shao, T.; Song, B.; Cheng, L.; Masterson, T.A.; Liu, X.; Ratliff, T.L.; et al. Cholesteryl Ester Accumulation Induced by PTEN Loss and PI3K/AKT Activation Underlies Human Prostate Cancer Aggressiveness. *Cell Metab.* **2014**, *19*, 393–406. [CrossRef] [PubMed]
32. Smith, S.J.; Diksin, M.; Chhaya, S.; Sairam, S.; Estevez-cebrero, M.A. The Invasive Region of Glioblastoma Defined by 5ALA Guided Surgery Has an Altered Cancer Stem Cell Marker Profile Compared to Central Tumour. *Int. J. Mol. Sci.* **2017**, *18*, 2452. [CrossRef] [PubMed]
33. Giese, A.; Loo, M.A.; Tran, N.; Haskett, D.; Coons, S.W.; Berens, M.E. Dichotomy of Astrocytoma Migration and Proliferation. *Int. J. Cancer* **1996**, *67*, 275–282. [CrossRef]
34. Xie, Q.; Mittal, S.; Berens, M.E. Targeting Adaptive Glioblastoma: An Overview of Proliferation and Invasion. *Neuro-Oncology* **2014**, *16*, 1575–1584. [CrossRef] [PubMed]
35. Calabrese, C.; Poppleton, H.; Kocak, M.; Hogg, T.L.; Fuller, C.; Hamner, B.; Oh, E.Y.; Gaber, M.W.; Finklestein, D.; Allen, M.; et al. A Perivascular Niche for Brain Tumor Stem Cells. *Cancer Cell* **2007**, *11*, 69–82. [CrossRef]
36. Kopeček, J.; Kopečková, P.; Minko, T.; Lu, Z.-R. HPMA Copolymer–Anticancer Drug Conjugates: Design, Activity, and Mechanism of Action. *Eur. J. Pharm. Biopharm.* **2000**, *50*, 61–81. [CrossRef]
37. Duncan, R. The Dawning Era of Polymer Therapeutics. *Nat. Rev. Drug Discov.* **2003**, *2*, 347–360. [CrossRef]
38. Dreher, M.R.; Liu, W.; Michelich, C.R.; Dewhirst, M.W.; Yuan, F.; Chilkoti, A. Tumor Vascular Permeability, Accumulation, and Penetration of Macromolecular Drug Carriers. *J. Natl. Cancer Inst.* **2006**, *98*, 335–344. [CrossRef]
39. Pinzõn-Daza, M.L.; Garzõn, R.; Couraud, P.O.; Romero, I.; Weksler, B.; Ghigo, D.; Bosia, A.; Riganti, C. The Association of Statins plus LDL Receptor-Targeted Liposome-Encapsulated Doxorubicin Increases in Vitro Drug Delivery across Blood-Brain Barrier Cells. *Br. J. Pharmacol.* **2012**, *167*, 1431–1447. [CrossRef] [PubMed]
40. Di Mauro, P.P.; Cascante, A.; Vilà, P.B.; Vallejo, V.G.; Llop, J.; Borrós, S. Peptide-Functionalized and High Drug Loaded Novel Nanoparticles as Dual-Targeting Drug Delivery System for Modulated and Controlled Release of Paclitaxel to Brain Glioma. *Int. J. Pharm.* **2018**, *553*, 169–185. [CrossRef]
41. Seo, S.; Kim, E.H.; Chang, W.-S.; Lee, W.-S.; Kim, K.-H.; Kim, J.-K. Enhanced Proton Treatment with a LDLR-Ligand Peptide-Conjugated Gold Nanoparticles Targeting the Tumor Microenvironment in an Infiltrative Brain Tumor Model. *Am. J. Cancer Res.* **2022**, *12*, 198–209. [PubMed]

pharmaceutics

Article

Selective Targeting and Eradication of Various Human Non-Small Cell Lung Cancer Cell Lines Using Self-Assembled Aptamer-Decorated Nanoparticles

Daniel Barak [1,†], Shira Engelberg [1,†], Yehuda G. Assaraf [2,*] and Yoav D. Livney [1,*]

1 Lab of Biopolymers for Food & Health, Department of Biotechnology & Food Engineering, Technion, Israel Institute of Technology, Haifa 3200003, Israel
2 The Fred Wyszkowski Cancer Research Lab, Department of Biology, Technion-Israel Institute of Technology, Haifa 3200003, Israel
* Correspondence: assaraf@technion.ac.il (Y.G.A.); livney@technion.ac.il (Y.D.L.)
† These authors contributed equally to this work.

Abstract: The leading cause of cancer mortality remains lung cancer (LC), of which non-small cell lung cancer (NSCLC) is the predominant type. Chemotherapy achieves only low response rates while inflicting serious untoward toxicity. Herein, we studied the binding and internalization of S15-aptamer (S15-APT)-decorated polyethylene glycol-polycaprolactone (PEG-PCL) nanoparticles (NPs) by various human NSCLC cell lines. All the NSCLC cell lines were targeted by S15-APT-decorated NPs. Confocal microscopy revealed variable levels of NP binding and uptake amongst these NSCLC cell lines, decreasing in the following order: Adenocarcinoma (AC) A549 cells > H2228 (AC) > H1299 (large cell carcinoma) > H522 (AC) > H1975 (AC). Flow cytometry analysis showed a consistent variation between these NSCLC cell lines in the internalization of S15-APT-decorated quantum dots. We obtained a temperature-dependent NP uptake, characteristic of active internalization. Furthermore, cytotoxicity assays with APT-NPs entrapping paclitaxel, revealed that A549 cells had the lowest IC_{50} value of 0.03 μM PTX (determined previously), whereas H2228, H1299, H522 and H1975 exhibited higher IC_{50} values of 0.38 μM, 0.92 μM, 2.31 μM and 2.59 μM, respectively (determined herein). Cytotoxicity was correlated with the binding and internalization of APT-NPs in the various NSCLC cells, suggesting variable expression of the putative S15 target receptor. These findings support the development of APT-targeted NPs in precision nanomedicine for individual NSCLC patient treatment.

Keywords: lung cancer; anticancer drugs; paclitaxel; targeted delivery; aptamers; nanoparticles

1. Introduction

Lung cancer (LC) is the most common form of cancer worldwide amounting 12.3% of all cancers. The main cause of LC is tobacco smoking, with 80–90% of cases arising in cigarette smokers [1]. Over half of LC patients die within one year of diagnosis, and the 5-year survival rate is ~17.8%, based on the currently available treatments [2]. The LC death toll in 2020 attained nearly 1.8 million people and is currently the leading cause of cancer mortality worldwide [3,4]. In general, there are four major histological types which comprise the majority of LC; they are roughly split into small cell LC, and non-small cell lung cancer (NSCLC), the latter of which accounts for ~85% of all LC cases [5]. NSCLC can be further divided into several histological subtypes: adenocarcinoma (AC), squamous cell carcinoma, adenosquamous cell carcinoma, large cell carcinoma, and sarcomatoid carcinoma, with AC accounting over 40% of NSCLC cases [6,7]. Current treatment options include surgery, immunotherapy, chemotherapy, radiotherapy, and adjuvant therapy [2]. Non-targeted chemotherapy is administered in combination with surgery in most patients, albeit inflicting serious side effects related to the cytotoxicity of anticancer drugs to rapidly dividing cells (e.g., bone marrow, epithelial cells of the gastrointestinal tract, hair follicles, and sperm cells). These deleterious effects damage the host's defense mechanisms, which are often already compromised by the underlying cancer. Cancer patients are therefore highly susceptible to infections, especially bacterial viral and fungal infections. Furthermore, all types of infections are associated with higher rates of morbidity and mortality in cancer patients receiving non-specific chemotherapy treatment [8].

Despite its many side effects, paclitaxel (PTX) is currently one of the most effective chemotherapeutic drugs used for the treatment of various cancers including breast, ovarian, lung, head and neck cancer and NSCLC [9]. PTX promotes intracellular polymerization of tubulin, thereby causing cell death by disrupting the normal microtubule dynamics required for cell division and vital interphase processes [10–12]. Nonetheless, administration of PTX has been associated with multiple toxicities [13]. Formulation aids such as Cremophor EL are currently used in the clinic for various poorly water soluble chemotherapeutic drugs such as PTX, since the hydrophobic nature of many antitumor agents limits their administration at high doses [14]. Cremophor EL is not an inert agent by itself, exerting a range of adverse biological effects, some of which have serious clinical implications [15]. Common short- and long-term side effects of these treatments include severe anaphylactoid hypersensitivity reactions, hyperlipidemia, abnormal lipoprotein patterns, aggregation of erythrocytes and peripheral neuropathy [16].

The untoward toxicity of PTX, its high lipophilicity and the toxic side effects, call for the development of novel, selectively targeted, and non-immunogenic modalities to eliminate cancer cells in patients, using a minimally invasive approach and without harming normal cells. For example, micellar nanoparticles made of amphiphilic polymers can be used for drug entrapment and solubilization, and can be targeted by specific tumor-targeting ligands [17].

Recent nanomedicine studies have shown it to be an avenue of incredible potential for the development of specific tumor-targeted drug delivery systems [18–21]. This continuously developing therapeutic approach is expected to revolutionize cancer treatment and management [14]. More precisely, nanoscale drug delivery systems hold great promise in successfully formulating and enhancing the therapeutic efficacy of a large number of anticancer agents [14]. Nanoparticles (NPs) are known to enhance biodistribution, increase therapeutic efficacy and reduce untoward toxicity of potent anticancer drugs. Their superior biocompatibility, ability to protect encapsulated compounds from metabolism and degradation, ability to deliver therapeutics to cancer cells in vivo, and/or to passively target tumors via the enhanced permeability and retention (EPR) effect [22–24], render NPs highly promising delivery vehicles [14]. Currently, several types of NP-based drug delivery systems are being developed to address the problems raised above and to allow for a more facile, precise, and efficacious LC treatment. For example, several types of nanocarriers that are currently studied for drug and gene delivery applications include liposomes, micellar NPs, nanostructured lipid carriers, polymeric NPs and more [20,21,25,26]. Recently, NPs made of polymers such as polyethylene glycol (PEG), poly (lactic acid) (PLA), poly (lactic-co-glycolic) acid (PLGA), gelatin, albumin, chitosan, polycaprolactone (PCL), polyalkyl-cyanoacrylates, and their block copolymers have gained popularity in use, because of their controlled and sustained release properties, subcellular size, and biocompatibility [27]. Polymeric NPs have also been shown to enhance the chemo- and radio-therapeutic efficacy of anticancer agents [28]. The amphiphilic nature of some of the polymers constructing the NPs, particularly block copolymers, allows them to self-assemble into micelles and thus entrap and carry hydrophobic drugs such as PTX through the blood circulation [29,30].

To allow for a more efficacious drug delivery and pharmacological activity against the tumor cells, specific active targeting ligands are used by decorating the NPs. These targeting ligands include, e.g., vitamin E derivatives, biotinylated epidermal growth factors, folic acid, antibodies, and aptamers. They allow obtaining high targeting precision and selectivity for the tumor cells, thereby markedly reducing the side effects of the chemotherapeutic drug cargo [31].

Aptamers (APTs), which are usually single-stranded DNA or RNA molecules, are emerging as promising targeting ligands for both cancer diagnostics and therapy [32]. APTs are comparable to antibodies regarding specificity and affinity for their target, which is typically a cell-surface protein or receptor [33]. APTs are much smaller than antibodies and are comparable in their molecular weight to peptides (1–5 kDa). APTs can be chemically produced for any cell-surface target, e.g., using a methodology called "SELEX" (Systematic Evolution of Ligands by EXponential Enrichment) [34]. APTs possess several desirable features over other targeting strategies, mainly their small size, low immunogenicity, and high specificity [35]. APTs can be coupled to diagnostic or therapeutic agents or to polymers, such as PEG, to enable selective targeting [33]. However, aptamers, as oligonucleotides, are susceptible to nuclease degradation, which may affect in vivo function. Thus, aptamer modifications are being added to thwart and protect them from nuclease degradation, leading to an improvement in their stability in biological environments [36,37].

We have recently demonstrated the selective targeting and eradication of human A549 NSCLC cells with PEG-PCL (poly (ethylene glycol)-poly (ε-caprolactone)) block co-polymer NPs, without harming healthy cells [38]. These NPs were conjugated to the APT S15, an 85-base-long single-stranded DNA, as the targeting ligand. To create NPs suitable for endocytosis by target NSCLC cells, we aimed at forming NPs with a size < 50 nm. At 70 μM PEG-PCL, NPs formed with a mean size of 30 ± 1 nm. The drug loading capacity and encapsulation efficiency of PTX with increasing drug: polymer molar ratios were recently determined [38]. The APT-NPs morphology was studied with cryogenic transmission electron microscopy (Cryo-TEM), NPs entrapping PTX were observed, with a clear core and shell structure. The in vitro release of PTX from APT-NPs was assessed [38]. APT-NPs loaded with PTX were shown to have selective cytotoxicity to target A549 cells, entering the cells via clathrin-mediated endocytosis [38,39].

In the current study, we aimed at characterizing the S15 APT-NP delivery system, in terms of selectivity and cytotoxicity to various model human NSCLC cell lines. After reviewing the relevant literature, to the best of our knowledge, no previous study had investigated the sensitivity of various NSCLC subtypes to APT-decorated NPs. Our findings can serve as the initial basis for the identification of the putative surface receptor that specifically binds and internalizes the S15-decorated NPs in NSCLC cells. Further studies can also pave the way toward tailoring NPs for a personalized medicine treatment of individual cancer patients.

2. Materials and Methods

2.1. Materials

The tumor cell lines used in the current study were generously provided by the following labs: AC H2228 [40] NSCLC cell line was provided by Prof. Arie Admon (Faculty of Biology, Technion, Haifa, Israel), lung carcinoma H1299 [41] and AC H1975 [42] cell lines were provided by Dr. Iris Kamer (Shiba Medical Center, Ramat-Gan, Israel) and the AC H522 [43] cell line was provided by Prof. Yosef Yarden (Dept. of Biological regulation, Weizmann Institute of Science, Rehovot, Israel).

All chemicals and cell culture reagents were obtained from Sigma-Aldrich (Merck) unless otherwise stated. PEG-PCL (5:2.5) KDa block-copolymer containing an addition of a carboxylic group at the PEG terminus, was custom synthesized (Creative PEGWorks, Durham, NC. Item number: DCD-5k2500, lot number: ZXM07050). Qdot™ 545 ITK™ Carboxyl QDs were purchased from Thermo Fisher Scientific (Waltham, MA, USA), lot number: 2273708. The S15 aptamer with the following nucleotide sequence: 5′-NH2-ACG CTC GGA TGC CAC TAC AGG CTA TCT TAT GGA AAT TTC GTG TAG GGT TTG GTG TGG CGG GGC TAC TCA TGG ACG TGC TGG TGA CidT-3′ was purchased from BioSpring Gesellschaft für Biotechnologie mbH, (Frankfurt, Germany), lot number:

257225_E. idT = inverted deoxythymidine incorporated at the 3′-end leading to a 3′-3′ linkage for protection against exonuclease degradation, see Supplementary Information [39].

2.2. Cell Cultures

Human NSCLC cell lines: A549, H2228, H1299, H1975, and H522 were maintained in RPMI-1640 medium, supplemented with 10% fetal bovine serum (FBS), 2 mM glutamine, 100 µg/mL penicillin and streptomycin (Biological Industries, Beit Ha-Emek, Israel). Cells were incubated under a controlled temperature of 37 °C in a humidified atmosphere of 5% CO_2, and their monolayer status was followed using standard light microscopy. Freezing and thawing of cells was carried out using a previously published protocol [44].

2.3. Methods

2.3.1. Preparation of Cy7-Labeled APT-NPs

Self-assembled PEG-PCL micelles were prepared by surfactant-free nanoprecipitation [45,46] with minor modifications of a previous protocol [38]. PEG-PCL was dissolved in acetonitrile (ACN), then added dropwise into a 0.8 mL–0.9% saline solution (Biological Industries, Kibbutz Beit-HaEmek, Israel) to a final concentration of 70 µM, and stirred at 161× *g* until complete evaporation of ACN was achieved. Thereafter, 1.3 µM EDC (1-Ethyl-3-(3-dimethylaminopropyl) carbodiimide) and 0.2 µM NHS (N-Hydroxy succinimide) freshly dissolved in saline, were added to the mixture for 10 min to create the NHS ester group at the ends of the PEG chains, in the well-established carbodiimide conjugation reaction [47,48]. Then, 0.03 µM of amine-modified S15-APTs, which were previously filtered through a 4 µm syringe filter, were added and left under stirring for 2 h to create the amide linkage with the PEG-NHS tail. The APT-conjugated polymeric NPs were ultra-filtered twice (1700× *g*, 4 min, 4 °C) through a 100 kDa MWCO membrane (Merck, Darmstadt, Germany) to remove unbound free aptamers. The carbodiimide EDC-NHS conjugation reaction was repeated using 0.5 µM Cy7 (Cyanine7)-amine fluorescent dye, to attach it to the carboxylic end of yet undecorated PEG blocks, and was followed by a similar filtration step. NPs were stored in the dark at 4 °C until the incubation with cells. A schematic structure of the Cy7-labeled APT-NPs and their preparation steps are depicted in Figure 1.

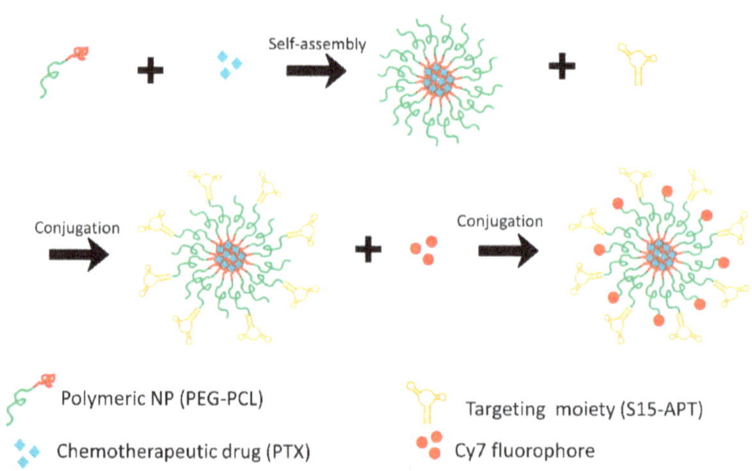

Figure 1. Schematic structure of the Cy7-labeled APT-NPs and their preparation steps.

2.3.2. Particle Size Distribution Analysis and Cryo-TEM Imaging

Particle size distribution of the APT-NPs was determined by dynamic light scattering (DLS), using a DLS analyzer (NICOMP™ 380, Particle Sizing System (PSS), Inc., Santa Barbara, CA, USA), as previously described [38]. APT-decorated NPs were observed by Cryo-TEM imaging using FEI Talos 200 C electron microscope (accelerating voltage 200 kV) at a low dose imaging mode. Specimen preparation was conducted as previously described [49]. Images were acquired by a FEI Falcon III direct-imaging camera, using "phase-plates" (FEI), to enhance image contrast [50].

2.3.3. Characterization and Comparison of the Specificity of APT-NPs to Various NSCLC Cell Lines Using Confocal Laser Microscopy

Specific internalization of APT-NPs by different NSCLC cell lines including H2228, H1299, H1975 and H2228 was determined using confocal laser microscopy imaging. 120 μL of each cell suspension were seeded on μ-slides VI 0.1 (Ibidi, Martinsried, Germany) at 50% confluency (15×10^4 cells/mL) and incubated overnight to allow for cell attachment. Cells were then washed with phosphate-buffered saline (PBS) pH 7.4. For internalization studies, APT-NPs were diluted 1:5 (v/v) in FBS-free medium and incubated with the cells for 2 h at 37 °C. Cells were washed three times with PBS to remove unbound APT-NPs. Cells were then incubated with 10 μg/mL Hoechst 33342 in growth medium for 20 min to achieve nuclear DNA staining. Fluorescence was studied using an inverted laser scanning confocal microscope (Zeiss LSM 710). Two fluorescence channels were used during all image capturing: 405 nm excitation laser for the blue viable DNA dye Hoechst 33342 (461 nm max emission) and 639 nm excitation laser for the red Cy7 dye (767 nm max emission). Confocal microscopy and data acquisition were performed using a ×40 magnification water-immersion lens. Further analysis and image processing was performed using the ZEN 2011 microscope imaging software by ZEISS.

Selective binding of S15 APTs to the studied NSCLC cell lines was performed similarly to the internalization procedure described above with minor modifications. Cells were exposed to 10 μg/mL Hoechst 33342 in growth medium for 20 min before incubation with NPs to allow for nuclear staining. Following NPs dilution in FBS-free medium, they were added to the cells and kept at 4 °C for 50 min. The cells were then washed three times with PBS and kept on ice until fluorescence microscopy analysis was performed.

2.3.4. Image Analysis

To obtain quantitative measurements from the captured photos, IMARIS 9.8 software for analysis of image data was used (Oxford instruments, Oxon, UK). The red fluorescence channel detection threshold was set in the range of 10–100 relative fluorescence units for all presented images, which allowed us to acquire the values of intracellular mean fluorescence intensity (MFI) and the mean value for the number of fluorescent endolysosomes per cell (FEPC).

2.3.5. Characterization of Active S15 APT-NPs Internalization by NSCLC Cell Lines

Conjugation of QDs

S15 aptamers were conjugated to carboxy terminals of QDs, as previously described [51]. Qdot® 545 ITK™ carboxyl QDs, at a final concentration of 1 μM, were added to 0.75 nmol S15 filtered APTs in a final volume of 100 μL of freshly prepared 10 mM sodium borate buffer pH 7.4—and kept in the dark. Following 5 min incubation at room temperature, 3 μL of a 10 mg/mL EDC solution (freshly prepared in 10 mM sodium borate buffer at pH 7.4) were added to the mixture and kept for 18 h at room temperature with continuous stirring at 40× g. After the reaction took place, the vial was centrifuged and the APT-QDs solution was ultra-filtered (1700× g, 4 min, at 4 °C) via centrifugal concentrators (100 kDa MWCO PES, Vivaspin, Sartorius Stedim Biotech GmbH, Göttingen, Germany) to remove free unconjugated aptamers. The purification step was repeated 4 times using 50 mM sodium borate buffer at pH 8.3 to increase pH and maintain the APT-QDs concentration.

Flow Cytometric Analysis

H2228, H1299, H1975 and H522 cells were plated in 24-well plates at approximately 50% confluence, depending on their respective growth rates. Following 18 h of incubation at 37 °C, cells were first washed with 1 mL PBS to remove serum remnants and dead cells. Next, the previously prepared APT-QD's solution was diluted 1:4 (*v*/*v*) in FBS-free medium and incubated with the cells for 1 h at 37 °C or 4 °C. An unstained control group, which was not exposed to the APT-QDs, was maintained simultaneously at 37 °C for each of the tumor cell lines. Following incubation, cells were washed 3 times with 1 mL PBS and incubated with 0.3 mL HBSS buffer containing 0.05% trypsin and 0.53 mM EDTA for 15 min. After incubation, 0.3 mL of FBS-containing protease inhibitors, including α1-antitrypsin (Biological Industries, Beit-HaEmek, Israel) [52], were added, and the cells were then transferred to 2 mL Eppendorf microfuge tubes and centrifuged at 800× *g* for 3 min, at room temperature. The supernatant was aspirated, and the sedimented cells were then resuspended in 0.5 mL growth medium and transferred into flow cytometry tubes through a 35 µm strainer cap. Flow cytometry analysis was performed using the BD LSR II flow cytometer (BD Biosciences). The QDs were excited at 355 nm, while their emission was detected using a 530/30 nm filter (green). Raw data processing and further downstream analysis was carried out using the flow cytometry analysis tool FCS Express™ 7.12.0005.

2.3.6. Cytotoxicity Assays

Preparation of Aptamer-Decorated Nanoparticles Harboring a Cytotoxic Drug

S15 APT polymeric NPs were prepared similarly to the above protocol, with minor modifications, to allow for the efficient entrapment of the hydrophobic drug PTX in the core of the NPs. To this end, final concentrations of 70 µM PEG-PCL and 14 µM PTX were dissolved simultaneously in ACN. The ACN solution was then added dropwise into 1 mL of saline and stirred at 161× *g* until complete ACN evaporation was achieved. Exonuclease protected S15 APTs (modified with a 3′ deoxy thymidine cap [53]) were conjugated to the PEG carboxy terminal as described above, using EDC and NHS freshly dissolved in saline. The conjugated NPs were ultra-filtered and repeatedly concentrated (2200× *g*, 4 min, 4 °C) until a concentration of 60 µM was achieved.

Cytotoxicity Assay

The selective cytotoxicity of PTX entrapped within ATP-NPs was studied in the various tumor cell lines. First, a confluency calibration was performed: cell lines were seeded and grown on a 96-well plate at different cell densities for 72 h, to calibrate the minimal necessary cell density for full confluency at the exposure time. Seeding cell densities for each tumor cell line were determined to be 4×10^4, 1.7×10^4, 1.8×10^4, and 2.5×10^4 cells/mL for H2228, H1299, H1975 and H522, respectively (cell counting was performed using a hemocytometer). Cell lines were seeded in their appropriate calibrated density in 96-well plates- 100 µL of cell suspension in each well. After an incubation period of 24 h, with the same nanoparticle concentration in all cell lines, the growth medium was replaced by FBS-free medium and the APT-NPs entrapping PTX were serially diluted in the wells to obtain 11 increasing concentrations of 0.3–30 µM. Cells in the 12th column were left drug free and were used as the 100% cell viability control reference. The procedure was performed in 9 replicates for each of the cell lines and the results presented were obtained from five independent experiments. Cell exposure to the drug-loaded NPs was for 2 h at 37 °C, followed by three washes with growth medium to remove the remaining non-internalized APT-NPs/PTX. The cells were then incubated for additional 72 h in growth medium, to allow for PTX to elicit its cytotoxic effect. Following this incubation period, cell viability was assessed using a colorimetric XTT-based cell proliferation assay [54] (Biological Industries, Beit-HaEmek, Israel). The XTT reagent (2,3-Bis-(2-Methoxy-4-Nitro-5-Sulfophenyl)-2H-Tetrazolium-5-Carboxanilide) and activation phenazine methosulfate reagent (N-methyl dibenzopyrazine methyl sulfate) were thoroughly thawed. The reagent was mixed with the intended activator at a 1:50 (*v*:*v*) ratio. Then, 33 µL of the mixture were added to 67 µL of growth medium in each well, and the cells were incubated at 37 °C for 20 min. Blank samples were performed in triplicates by adding the same volumes specified above to cell-free wells. Following 20 min of incubation, gentle orbital shaking was applied to the plates to evenly distribute the orange color of the obtained formazan in each well. Thereafter, the absorbance of the formazan dye formed by the metabolically active cells was measured at 450 nm, using a scanning multiwell spectrophotometer (Biotek Microplate Reader, Winooski, VT, USA). Data collection and its initial handling was performed using the BioTek Gen5 v3.11 software. The collected data was further analyzed using a nonlinear curve fitting model (Hill1) [55] with the OriginPro 9.0 graphing and analysis software for dose–response curve represented according to Equation (1) [56]:

$$P = P_\infty + (P_0 - P_\infty)\cdot([D]^n/(IC_{50})^n + [D]^n) \quad (1)$$

In Equation (1), P is the percentage of viable cells; P_∞ is the minimal percent of live cells at theoretical infinite drug concentration which results in zero cell viability; P_0 is the maximal percent of surviving viable cells in the absence of drug (when its concentration is zero, as in the control column, cell viability is 100%); [D] is a the drug concentration; IC_{50} is the drug concentration at which half-maximal inhibitory effect in cell growth is achieved; n represents the Hill slope coefficient, indicating the steepness of the dose–response curve. Curve fitting was performed at a significance level of $\alpha = 0.05$.

3. Results and Discussion

3.1. Size Distribution of NPs and Cryogenic Transmission Electron Microscopy (Cryo-TEM) Analysis

The average diameter of the APT-NPs was determined to be 30 ± 1 nm using DLS. The polydispersity index (PDI) was calculated using Equation (8) in reference [57] and was found to be 0.055 ± 0.005 (Figure 2A). A PDI value < 0.1 indicates a stable and monodisperse NP population [58]. The ideal size, at which NPs escape renal exclusion and RES uptake, allowing for enhanced cell permeability into solid tumors via passive diffusion, and supporting endocytosis into malignant cells is ~10–50 nm [59,60]. Cryo-TEM analysis further supported these data and revealed the morphology of APT-decorated PEG-PCL NPs encapsulating 14 μM PTX (Figure 2B); the APT-NPs appeared as small stable spheres.

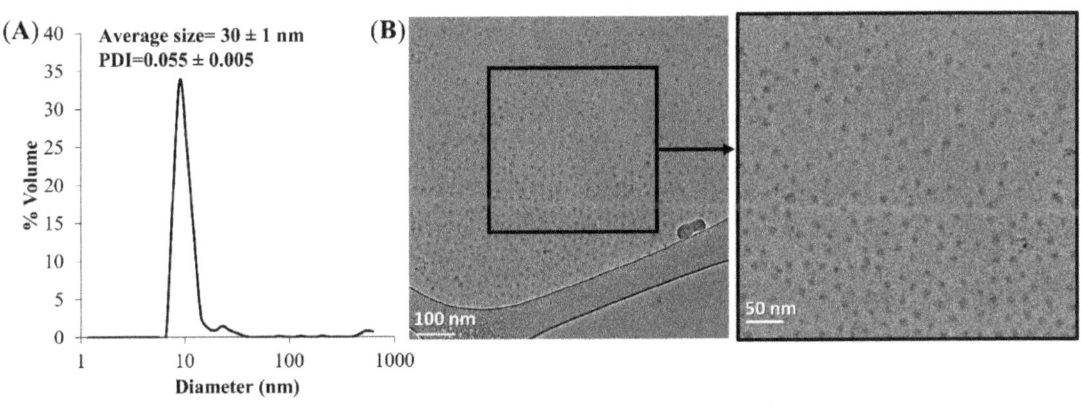

Figure 2. (**A**) Particle size distribution of APT functionalized PEG-PCL NPs. Average NP size = 30 ± 1 nm; PDI = 0.055 ± 0.005 (the measurements are based on duplicates). (**B**) Cryo-TEM image of APT-functionalized PEG-PCL NPs.

3.2. Specificity of Binding and Internalization of APT-NPs in Various Human NSCLC Cell Lines

The specificity of binding and internalization of APT-NPs in various human NSCLC cell lines was studied using confocal laser microscopy. Following an incubation with 14 μM S15 APT-NPs for 2 h at 37 °C, a variability in the red fluorescent intracellular vesicles was observed among the NSCLC cell lines H2228, H1299, H522 and H1975 (Figure 3A). This was reflected in the numbers of accumulated FEPC, containing the Cy7-labeled NPs (Figure 3B) [38,61]. H2228, H1299, H522 and H1975 cells displayed FEPC values of 24.1,

21.7, 18.7 and 9.5, respectively, whereas human NSCLC A549 cells, which were previously found to be well targeted by these NPs [38], were found to have a remarkable mean FEPC value of 30.2. The observed difference in the internalization capacity of the APT-NPs by these NSCLC cells, suggests a variable surface expression of a putative receptor recognized by the S15 APT-NPs, which accordingly impacts their uptake abilities. Indeed, physiological differences between subtypes of NSCLC cell lines have been previously studied and characterized. Molina-Romero Camil et al., found differential gene expression patterns amongst several subtypes of AC [62]. They were able to identify a gene expression signature consisting of 13 genes which affected different functional networks related to six distinct biological categories: lipid metabolism; biochemistry of small molecules; vitamin and mineral metabolism; DNA replication; recombination and repair; and cell cycle. Additional studies have concluded that different molecular subtypes of NSCLC exhibit diverse gene mutation rates in key genes, including EGFR, KRAS, STK11, and TP53, chromosomal instability, regional copy number, and genome-wide DNA methylation [63]. In addition, lung cancer subtypes, patient behavior such as smoking, and/or patient germline sequence may further modulate gene expression patterns. The intrinsic properties of cancer cell types may promote specific genomic alterations due to a hyper-mutator phenotype, along with the selective advantage that a specific mutation confers upon a specific cell type [63]. Despite the expected molecular subtype variance, even H522 and H1975 cells, which displayed the lowest level of internalization of S15 APT-NPs, still appeared to be substantially targeted by these S15 NPs. In contrast, in similar experiments we recently conducted with the same S15 APT-NPs, we found that neither normal lung epithelial BEAS2B cells, nor cervical carcinoma HeLa cells, colorectal adenocarcinoma CaCo-2 cells, neonatal foreskin FSE cells and embryonic kidney HEK-293 cells, showed any cellular fluorescence when incubated with these S15 APT-NPs [38]. These cumulative findings suggest that S15 APT-NPs demonstrate affinity and selectivity toward various NSCLC subtypes, but at the same time, they have no affinity toward healthy tissues, which can be highly important in the clinical setting. These findings are also supported by previous studies, which demonstrated that mucin 1 and S15 APTs have the ability to target molecular abnormalities, which are specific for NSCLC adenocarcinoma, while displaying a slightly lower targeting ability for large cell carcinoma, and almost no targeting ability to other cancerous lung cells nor to other types of cancer cells [39,64]. Powell et al., reported that the A6-APT NP system for the selective delivery of P-gp-targeted siRNA (aimed at P-gp knockdown) into breast cancer cells had been evaluated using seven different cell lines. Using fluorescence microscopy, they found that different expression levels of the A6-APT target receptor Her-2 led to distinct internalization levels in the cells [65]. Therefore, these data support our findings shown in Figure 3.

Consistently, a previous theranostic study by Kim et al., reported that hydrophobic QDs were first effectively incorporated into lipid bilayers, and then therapeutic siRNAs were complexed with QD-lipid nanocarriers (QLs) [66]. Thereafter, anti-EGFR aptamer-lipid conjugates were inserted into the QLs for triple negative breast cancer (TNBC) targeting (aptamo-QLs). TNBC-targeted aptamo-QLs were compared to anti-EGFR antibody-coupled immuno-QLs. The in vitro delivery of therapeutic siRNAs and QDs to target TNBC cells was verified by flow cytometry and confocal microscopy. The in vivo tumor targeting of siRNAs and their therapeutic efficacy were evaluated in mice harboring TNBC MDA-MB-231 tumors. Both types of EGFR-targeting QLs displayed enhanced delivery to target TNBC cells, yielding more effective gene silencing and enhanced tumor imaging compared to non-targeted control QLs. Furthermore, combination therapy with Bcl-2 and PKC-ι siRNAs loaded into the anti-EGFR QLs was efficacious in inhibiting both tumor growth and metastasis.

(A)

(B)

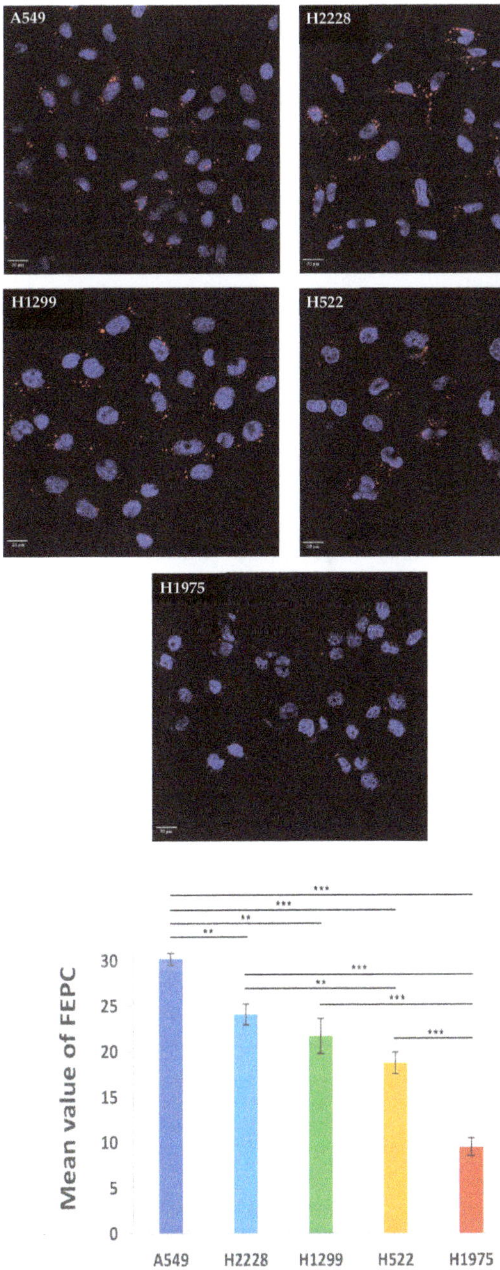

Figure 3. (**A**) Variable internalization of APT-NPs by various NSCLC cells at 37 °C. Confocal laser microscopy images of NSCLC subtypes exposed to PEG-PCL NPs decorated with S15 APTs. The tumor cell lines were H2228, H1299, H522 and H1975. Cells were incubated at 37 °C, for 2 h, in FBS-free medium, containing APT-NPs labeled with Cy7 (red). Nuclear DNA was labeled with Hoechst 33342 (2 μg/mL) (blue). A 20 μm scale bar is shown. (**B**) Mean FEPC values ± SE of Cy7-labeled S15 APT-NPs. Values were determined using IMARIS software in the images shown in (**A**). ** $p < 0.01$, *** $p < 0.001$, unpaired, two-tailed t-test.

Specific binding of the S15 APT-NPs to the different cell lines was further explored, to confirm that their uptake by the cells is an active internalization process. Cells were incubated with Cy7-labeled APT-NPs for 50 min, at 4 °C, a temperature at which surface binding is retained, whereas endocytosis is prevented. Cells were then gently washed with ice-cold PBS to remove free unbound NPs. Using confocal laser microscopy, a fluorescent layer was observed colocalized with the plasma membrane, whereas fluorescent endosomes were essentially absent from the cytoplasm (Figure 4A). H2228 displayed the highest mean fluorescence intensity (MFI) of attached NPs, followed by H522 and lastly H1299 and H1975 cells (Figure 4B), in accord with their respective uptake ability shown above (Figure 3), excluding H522 cells which displayed a higher apparent specific binding of NPs than H1299. The higher MFI observed with H522 cells might be attributed to the partial internalization of NPs, which can be seen as cytoplasmic fluorescent dots near the cell nuclei (Figure 4A). A possible explanation is that the cells were transiently exposed to a temperature slightly higher than 4 °C, while being transferred to the confocal microscope, and therefore, some internalization might have occurred.

These findings suggest that the mechanism of entry of the NPs into the cells is based upon a putative receptor expressed on the cell surface, which specifically binds S15-decorated NPs, since unbound NPs were completely washed out [67]. Similarly, previous studies have also shown the ability of aptamers to bind to specific plasma membrane cell receptors, such as the receptor tyrosine kinases ErbB2, epidermal growth factor (HER)-3, as well as CD4 [68–70]. Thus, the possibility of a nonspecific adsorptive uptake route, such as micropinocytosis, or diffusion through the lipid bilayer, is highly unlikely, suggesting that an active internalization process was taking place, as was similarly shown for ligand conjugated polymeric NPs, targeting a specific entity on the plasma membrane of NSCLC and prostate cancer cells [71,72]. Thus, the entry mechanism into these cells is suggested to be a clathrin-mediated endocytosis, as we have previously described for A549 cells [73,74].

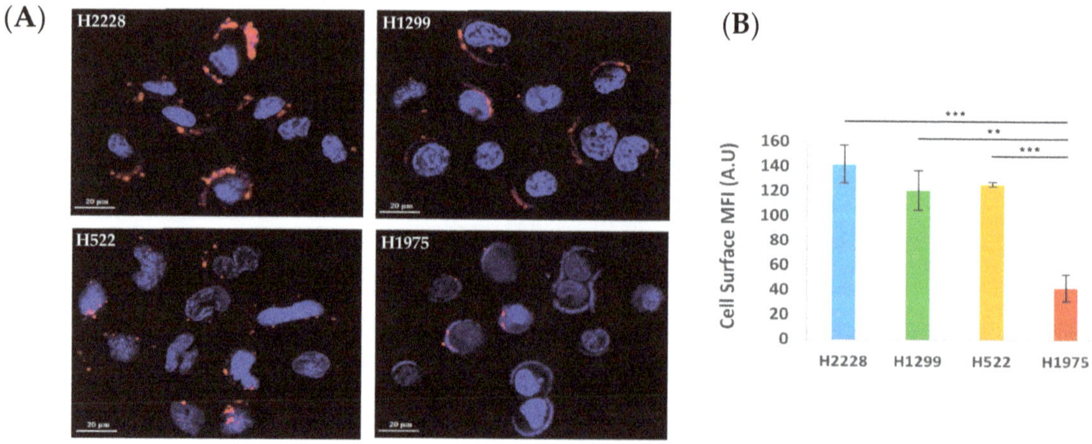

Figure 4. Cell surface binding of APT-NPs at 4 °C differs among NSCLC cell lines. (**A**) Confocal laser microscopy images of NSCLC cell lines incubated with PEG-PCL NPs decorated with S15 APTs. The studied tumor cell lines were H2228, H1299, H522 and H1975. Cells were incubated at 4 °C, for 50 min, in FBS-free medium, containing APT-NPs labeled with Cy7 (red). Nuclear DNA was labeled with Hoechst 33342 (2 μg/mL) (blue). (**B**) Mean fluorescence intensity (MFI) values of Cy7-labeled NPs bound to cells in panel A were determined using IMARIS software. Error bars represent SE. ** $p < 0.01$, *** $p < 0.001$, unpaired, two-tailed t-test.

3.3. Further Exploration of the Selective Binding and Internalization of the S15-Decorated NPs

Flow cytometry analysis was performed to verify the observed differences in the binding ability and internalization of S15 NPs into NSCLC cell lines. The four NSCLC cell lines were pre-incubated at 37 °C, a temperature at which the cells were shown to actively internalize the NPs to different extents. In contrast, at 4 °C (Figure 4A), active internalization is known to be abolished [75], as we have previously shown using the same NPs with A549 NSCLC cells [51]. Following 1 h incubation with S15-APT QDs, cells were detached by trypsinization, and cellular MFI was determined using flow cytometry. We found a minor shift in cellular fluorescence intensity [76] in the profile of all examined cells incubated at 4 °C, compared to their autofluorescence level Figure 5A). This shift is also consistently much smaller compared to the large shift in the histograms from the experiments carried out at 37 °C, compared to the autofluorescence of unstained cells. These results reveal the specific binding of the S15 APTs to the cell surface of NSCLC cells, as was previously demonstrated [39]. Since trypsin is well known to remove large amounts of both glycoproteins and glycosaminoglycans from the cell surface [77], this trypsin treatment, routinely used for cultured cell detachment, most likely resulted in the removal of the putative surface receptor protein which specifically interacted with the S15-APT QDs, thus eliminating their specific binding to the target cells. These findings are also consistent with the results obtained with confocal microscopy analysis (Figures 3 and 4), where the internalization of S15 NPs at 4 °C, was markedly blocked and the NPs exhibited cell surface binding, while being readily internalized at 37 °C. This supports the conclusion that the mode of entry of NPs into the cells, is a temperature-dependent process, which is characteristic of an active (ATP-dependent) endocytosis process, similarly to the results of our previous study with A549 cells, targeted by the same NPs [51]. The prominent up-shift in cellular fluorescence intensity observed at 37 °C is a result of the APT-QDs being actively internalized by the cells prior to the trypsin treatment. The net cellular fluorescence of cells incubated at 37 °C implies a difference in their active internalization ability (Figure 5B). Although the net cellular fluorescence values do not significantly differ from each other, they demonstrate a similar trend of internalization ability to what was observed in Figure 3A. H1975 cells showed the lowest net cellular fluorescence intensity which was 2.5-fold lower than that of H1299 cells, which exhibited the highest fluorescence. The same general correlation of MFI and FEPC values was observed for the four tumor cell lines, both in the confocal laser microscopy (Figures 3 and 4) and flow cytometry analyses (Figure 5), further strengthening our observations [39,51]. Consistently, cell-type specific binding (at 4 °C) and internalization (at 37 °C) was previously observed by Ferreira et al., for three types of APTs directed against MUC1. Using flow cytometry, cell surface binding revealed different binding affinities to the distinct target cells (MCF-7, T47D, PANC-1) and no binding to non-target Chinese hamster ovary cells. The internalization of APTs at 37 °C was determined using confocal laser microscopy and is consistent with our results as different fluorescence intensities could be observed for the four cell lines we studied [78].

3.4. Cytotoxicity of APT-NPs to NSCLC Cell Lines

Prior to investigating cytotoxicity of drug-loaded APT-NPs, we first verified that the drug-free APT-NP are devoid of cytotoxic activity. For this purpose, we chose A549 cells, which, according to Figure 3, displayed the highest uptake of the APT-NPs, and hence, they are expected to be most sensitive to these APT-NPs.

Figure 6 shows A549 cells incubated with PTX-lacking or PTX-containing APT-NPs. Drug-free NPs did not show any cytotoxic activity against A549 target cells as opposed to PTX-loaded APT-NPs, which displayed a low IC_{50} value of 30 nM against these target A549 cells. These findings demonstrate that the drug-free APT-decorated NPs are non-toxic.

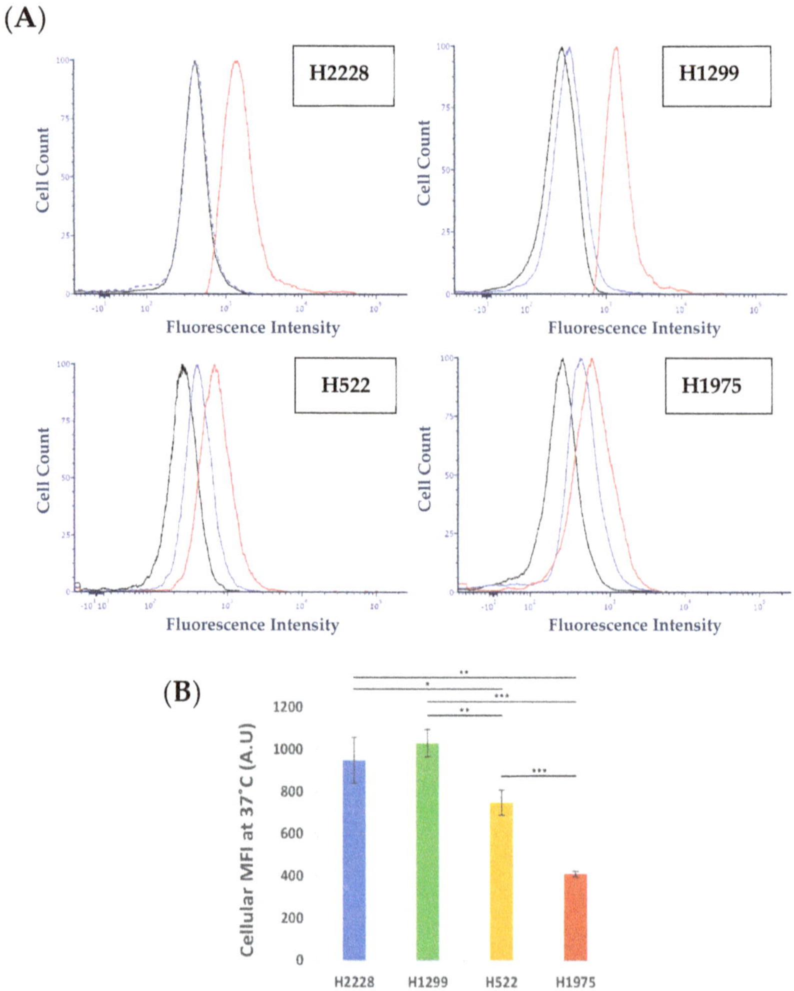

Figure 5. Internalization of APT-QDs in various NSCLC cell lines. (**A**) Cellular fluorescence profiles: APT-QDs exposed to the NSCLC cell lines H2228, H1299, H522 and H522 at 37 °C and 4 °C for 1 h, followed by a 15 min trypsin treatment to remove membrane bound APT-QDs. Black line: control, unstained cells. Blue line: cells incubated at 4 °C. Red line: cells incubated at 37 °C. (**B**) Mean intracellular fluorescence intensity values ± SE after APT-QD exposure at 37 °C. Autofluorescence was subtracted. * $p < 0.05$, ** $p < 0.01$, *** $p < 0.001$, unpaired, two-tailed *t*-test.

The selectivity and degree of cytotoxic activity of S15-APT-NPs entrapping the chemotherapeutic drug PTX, was determined in the four NSCLC cell lines: H2228, H1299, H1975 and H522. Cell viability was plotted against PTX concentration, accompanied by a fitted dose–response curve (Figure 7A). A similar sigmoidal curve was observed for H1975 and H522 cells, while the curves for H1299 and H2228 were steeper, showing that H2228 displays the steepest cytotoxicity curve. This observation is accompanied by the half maximal inhibitory concentrations (IC_{50}) of the encapsulated drug PTX for each NSCLC cell line (Figure 7B). H2228 displayed a remarkable IC_{50} value of 0.38 μM PTX, whereas higher IC_{50} values of 0.92 μM, 2.31 μM and 2.59 μM were obtained with H1299, H522 and H1975 cells, respectively. Although these results reveal some variability in the cytotoxic activity toward the different NSCLC cell lines, these PTX-loaded S15-APT NPs exhibit a relatively potent cytotoxic activity at the sub-micromolar to low micromolar range, toward all the studied NSCLC cell lines.

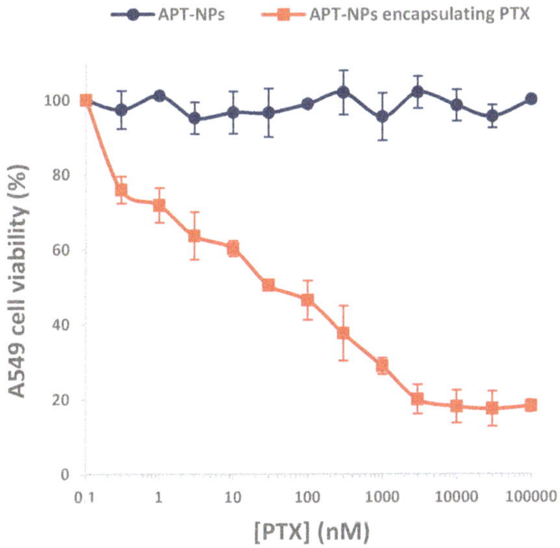

Figure 6. The impact of drug-free APT-NPs vs. APT-NPs encapsulating PTX on the viability of target A549 cells.

This difference in the cytotoxic activity between the tumor cell lines also correlated with the internalization findings above, where H2228 cells demonstrated the highest binding and internalization of NPs, hence being the most sensitive to PTX-loaded NPs in this cytotoxicity assay. The H1299 cell line exhibited the second highest IC_{50} value, while showing less binding and internalization ability (Figures 3–5). The IC_{50} values of H1975 and H522 were 6–7 fold higher than those of H2228 cells (H1975 cells having the highest IC_{50} value although not statistically different from that of H522). In contrast, a previous study performed in our lab revealed very high IC_{50} values of 87 and 980 μM PTX, entrapped within the same NP delivery system, toward neonatal foreskin fibroblast FSE cells, and human embryonic kidney HEK-293 cells, respectively [38]. These results demonstrate that the PTX-loaded NPs were 34- and 378-fold more cytotoxic toward the least drug-sensitive NSCLC cell line H1975, compared to these healthy cells, respectively.

This comparison of cytotoxicity between the examined NSCLC cell lines, highlights the promising therapeutic potential of the S15-APT NPs drug delivery system (IC_{50} (H1229) = 0.92 μM = 0.79 μg/mL), as it even markedly outperformed an existing EGFR specific small peptide AR-PEG-modified PTX-loaded nanostructured lipid carriers (A-P-NLC) by ~10 fold (IC_{50}(H1229) = 7.57 μg/mL), and compared to free PTX reported in that paper, by ~57-fold (IC_{50}(H1229) = 45.13 μg/mL) [79]. In this respect, previous studies obtained similar results: the Sgc8c aptamer, which can specifically recognize tyrosine kinase protein 7 (PTK7) on the surface of cancer cells, exhibited high cytotoxicity toward PTK7 receptor-positive CCRF-CEM and HCT116 cell lines, in contrast to receptor-negative K562 and HepG2 cells [80]. Li et al., demonstrated specific recognition of the Sgc8c APT against its target receptor PTK7; therefore, different cytotoxicity levels were observed [80].

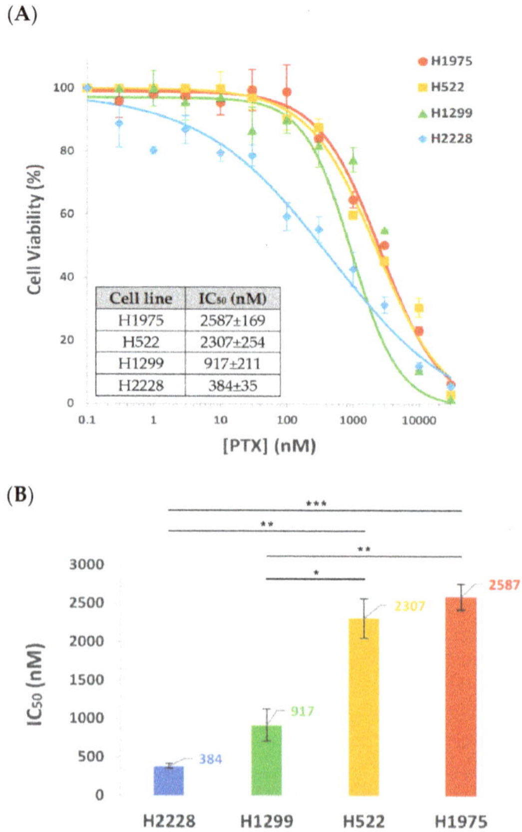

Figure 7. Cytotoxicity of PTX-loaded APT-NPs to various NSCLC cells. (**A**) APT-NPs entrapping PTX were incubated with the NSCLC cell lines H2228, H1299, H522 and H522. Cell viability as a function of the PTX concentration is presented as mean ± SE. (**B**) IC_{50} values derived from each appropriate dose-response curve ± SE. P-values obtained were < 0.05 for all cell lines. * $p < 0.05$, ** $p < 0.01$, *** $p < 0.001$, unpaired, two-tailed *t*-test.

The use of a protected S15-APT, with PEG at the 5′ and dT at the 3′ is the same as was applied for Pegaptanib (Macugen), the only aptamer which has been approved by the FDA for clinical use [81]. Pegaptanib is a pegylated anti-vascular endothelial growth factor (VEGF) aptamer for the treatment of age-related macular degeneration (AMD). Therefore, our S15-APT NPs should proceed to in vivo studies, hopefully followed by clinical trials.

Considering the fact that the selective internalization mechanism for target NSCLC A549 cells occurs via classical clathrin-dependent, receptor-meditated endocytosis, as we have previously shown [51], it is herein suggested that the difference observed in the tumor cell lines cytotoxicity was due to different expression levels of the putative S15 APT-specific receptor. The correlation between NPs binding, internalization and cytotoxic activity observed in all NSCLC cell lines in the above experiments suggests that these cell lines express different levels of the putative S15 APT-specific receptor. The higher the levels of this surface receptor expressed by the NSCLC cell line, the higher the binding and the internalization, and consequently the more enhanced the cytotoxicity. These findings imply that the high affinity binding and internalization of the S15-NPs could readily overcome pre-existing mechanisms of anticancer drug resistance such as P-glycoprotein overexpression at the plasma membrane. Thus, these targeted S15-decorated NPs bear important therapeutic implications for the surmounting of cancer MDR by evading plasma membrane-localized MDR efflux pumps that extrude a multitude of structurally and functionally distinct antitumor agents.

These results suggest that the therapeutic efficacy of the developed APT-NPs drug delivery system can be pre-determined depending on the NSCLC tumor that the individual patient is harboring. For such future precision medicine to be applied, further studies should be conducted to identify the APT-specific receptor, which can be used as a strong biomarker for both diagnosis and selectively targeted personalized medicine. The comparison between the different NSCLC cell lines performed in the current research may facilitate finding a correlation between internalization capability of the NPs, to the expression of a specific protein receptor on the plasma membrane of tumor cells, which could allow for a more precisely targeted drug delivery to NSCLC cells and possibly to additional subtypes of LC.

3.5. Relation between the Internalization and Cytotoxic Effect of PTX-Loaded NPs on NSCLC Cells

Results obtained using confocal microscopy and cytotoxicity assays were combined and plotted to explore their correlation (Figure 8). A strong inverse correlation was observed between the number of FEPC and the respective IC_{50} values obtained with the PTX-loaded NPs (R^2 = 0.826). The higher the number of observed FEPC in the various NSCLC cells, the lower their respective IC_{50} values, which means a better eradication of cancer cells exerted by the PTX-loaded NPs. This observation supports the abovementioned conclusion: the higher the levels of the S15-targeted receptor on the surface of these NSCLC cells, the better is the performance of our S15 APT-NPs drug delivery system in specifically eliminating these target tumor cells.

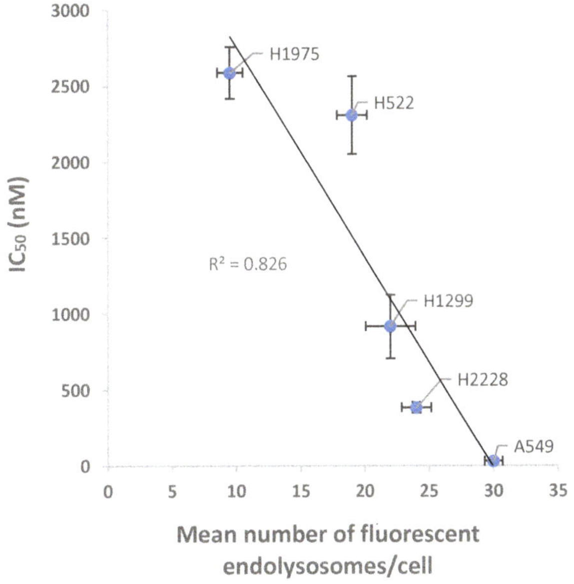

Figure 8. Correlation between the cytotoxic effect to various target NSCLC cells and the internalization ability of PTX-loaded APT-NPs. The IC_{50} value (30 nM) of A549 cells exposed to the same drug delivery system, was determined in our previous study, Engelberg et al. [38].

4. Conclusions

To the best of our knowledge, this study is the first to investigate the selectivity and cytotoxicity of APT-decorated NPs to various NSCLC subtypes. Such studies exploring the variability among cancer subtypes are scarce in the literature but are crucial for verifying that targeted NP-based treatments are effective against various subtypes of a given cancer. Our finding of the variable sensitivity of the various NSCLC cell lines to the APT-decorated PTX-containing NPs, supports the hypothesis that the internalization mechanism is based upon a specific interaction of the NPs with a putative plasma membrane receptor expressed on the tumor cell surface. Thus, elevated levels of the receptor in these tumor cells, enhance the specific targeting and cytotoxicity of the APT-NPs. Indeed, the micromolar IC_{50} values we obtained suggest that this receptor is ubiquitously expressed on the surface of lung cancer cells. This implies that our targeted NPs would be efficacious against a large variety of lung cancer types.

Supplementary Materials: The following supporting information can be downloaded at: https://www.mdpi.com/article/10.3390/pharmaceutics14081650/s1. Figure S1: Agarose gel electrophoresis of protected vs. unprotected APTs. Figure S2: Selective internalization of protected S15-APTs (with inverted thymidine) into human non-small cell lung A549 target cells. Figure S3: Drug release kinetics.

Author Contributions: The experiments were designed by D.B., S.E., Y.G.A. and Y.D.L. and conducted by D.B. and S.E. The results were evaluated by D.B., S.E., Y.G.A. and Y.D.L. The writing of this paper was carried out by D.B. and reviewed by S.E., Y.G.A. and Y.D.L. All authors have read and agreed to the published version of the manuscript.

Funding: This research was partly funded by the Israel Cancer Association, ICA USA Board of Directors, grant number 20190102 and by the state of Lower-Saxony and the Volkswagen Foundation, Hannover, Germany, grant number ZN3169.

Institutional Review Board Statement: Not applicable.

Informed Consent Statement: Not applicable.

Data Availability Statement: The data presented in this study are available on request from the corresponding authors.

Acknowledgments: We wish to thank Ravit Edelman for her valuable expert support.

Conflicts of Interest: We declare no conflict of interest.

References

1. Schabath, M.B.; Cote, M.L. Cancer Progress and Priorities: Lung Cancer. *Cancer Epidemiol. Biomarkers Prev.* **2019**, *28*, 1563–1579. [CrossRef] [PubMed]
2. Zappa, C.; Mousa, S.A. Non-Small Cell Lung Cancer: Current Treatment and Future Advances. *Transl. Lung Cancer Res.* **2016**, *5*, 288–300. [CrossRef] [PubMed]
3. de Groot, P.M.; Wu, C.C.; Carter, B.W.; Munden, R.F. The Epidemiology of Lung Cancer. *Transl. Lung Cancer Res.* **2018**, *7*, 220–233. [CrossRef]
4. Sung, H.; Ferlay, J.; Siegel, R.L.; Laversanne, M.; Soerjomataram, I.; Jemal, A.; Bray, F. Global Cancer Statistics 2020: GLOBOCAN Estimates of Incidence and Mortality Worldwide for 36 Cancers in 185 Countries. CA. *Cancer J. Clin.* **2021**, *71*, 209–249. [CrossRef]
5. Friedlaender, A.; Addeo, A.; Russo, A.; Gregorc, V.; Cortinovis, D.; Rolfo, C.D. Targeted Therapies in Early Stage NSCLC: Hype or Hope? *Int. J. Mol. Sci.* **2020**, *21*, 6329. [CrossRef]
6. Chen, F.; Zhang, Y.; Parra, E.; Rodriguez, J.; Behrens, C.; Akbani, R.; Lu, Y.; Kurie, J.M.; Gibbons, D.L.; Mills, G.B.; et al. Multiplatform-Based Molecular Subtypes of Non-Small-Cell Lung Cancer. *Oncogene* **2017**, *36*, 1384–1393. [CrossRef] [PubMed]
7. Hyun, S.H.; Ahn, M.S.; Koh, Y.W.; Lee, S.J. A Machine-Learning Approach Using PET-Based Radiomics to Predict the Histological Subtypes of Lung Cancer. *Clin. Nucl. Med.* **2019**, *44*, 956–960. [CrossRef]
8. Islas-Muñoz, B.; Volkow-Fernández, P.; Ibanes-Gutiérrez, C.; Villamar-Ramírez, A.; Vilar-Compte, D.; Cornejo-Juárez, P. Bloodstream Infections in Cancer Patients. Risk Factors Associated with Mortality. *Int. J. Infect. Dis.* **2018**, *71*, 59–64. [CrossRef]
9. Bernabeu, E.; Cagel, M.; Lagomarsino, E.; Moretton, M.; Chiappetta, D.A. Paclitaxel: What Has Been Done and the Challenges Remain Ahead. *Int. J. Pharm.* **2017**, *526*, 474–495. [CrossRef]
10. Mosca, L.; Ilari, A.; Fazi, F.; Assaraf, Y.G.; Colotti, G. Taxanes in Cancer Treatment: Activity, Chemoresistance and Its Overcoming. *Drug Resist. Updates* **2021**, *54*, 100742. [CrossRef]
11. Das, T.; Anand, U.; Pandey, S.K.; Ashby, C.R.; Assaraf, Y.G.; Chen, Z.S.; Dey, A. Therapeutic Strategies to Overcome Taxane Resistance in Cancer. *Drug Resist. Updates* **2021**, *55*, 100754. [CrossRef] [PubMed]

12. Hong, X.; Li, S.; Li, W.; Xie, M.; Wei, Z.; Guo, H.; Wei, W.; Zhang, S. Disruption of Protein Neddylation with MLN4924 Attenuates Paclitaxel-Induced Apoptosis and Microtubule Polymerization in Ovarian Cancer Cells. *Biochem. Biophys. Res. Commun.* **2019**, *508*, 986–990. [CrossRef] [PubMed]
13. Vermeer, C.J.C.; Hiensch, A.E.; Cleenewerk, L.; May, A.M.; Eijkelkamp, N. Neuro-Immune Interactions in Paclitaxel-Induced Peripheral Neuropathy. *Acta Oncol.* **2021**, *60*, 1369–1382. [CrossRef] [PubMed]
14. Babu, A.; Templeton, A.K.; Munshi, A.; Ramesh, R. Nanoparticle-Based Drug Delivery for Therapy of Lung Cancer: Progress and Challenges. *J. Nanomater.* **2013**, *2013*, 14. [CrossRef]
15. Jadhav, S.R.; Bryant, G.; Mata, J.P.; Eldridge, D.S.; Palombo, E.A.; Harding, I.H.; Shah, R.M. Structural Aspects of a Self-Emulsifying Multifunctional Amphiphilic Excipient: Part II. The Case of Cremophor EL. *J. Mol. Liq.* **2021**, *344*, 117881. [CrossRef]
16. Zeng, L.; Xin, X.; Zhang, Y. Development and Characterization of Promising Cremophor EL-Stabilized o/w Nanoemulsions Containing Short-Chain Alcohols as a Cosurfactant. *RSC Adv.* **2017**, *7*, 19815–19827. [CrossRef]
17. Calzoni, E.; Cesaretti, A.; Polchi, A.; Di Michele, A.; Tancini, B.; Emiliani, C. Biocompatible Polymer Nanoparticles for Drug Delivery Applications in Cancer and Neurodegenerative Disorder Therapies. *J. Funct. Biomater.* **2019**, *10*, 4. [CrossRef]
18. Su, Z.; Dong, S.; Zhao, S.C.; Liu, K.; Tan, Y.; Jiang, X.; Assaraf, Y.G.; Qin, B.; Chen, Z.S.; Zou, C. Novel Nanomedicines to Overcome Cancer Multidrug Resistance. *Drug Resist. Updates* **2021**, *58*, 100777. [CrossRef]
19. Zhao, C.Y.; Cheng, R.; Yang, Z.; Tian, Z.M. Nanotechnology for Cancer Therapy Based on Chemotherapy. *Molecules* **2018**, *23*, 826. [CrossRef]
20. Bar-Zeev, M.; Livney, Y.D.; Assaraf, Y.G. Targeted Nanomedicine for Cancer Therapeutics: Towards Precision Medicine Overcoming Drug Resistance. *Drug Resist. Updates* **2017**, *31*, 15–30. [CrossRef]
21. Cohen, L.; Livney, Y.D.; Assaraf, Y.G. Targeted Nanomedicine Modalities for Prostate Cancer Treatment. *Drug Resist. Updates* **2021**, *56*, 100762. [CrossRef] [PubMed]
22. Stylianopoulos, T. EPR-Effect: Utilizing Size-Dependent Nanoparticle Delivery to Solid Tumors. *Ther. Deliv.* **2013**, *4*, 421–423. [CrossRef] [PubMed]
23. Kang, H.; Rho, S.; Stiles, W.R.; Hu, S.; Baek, Y.; Hwang, D.W.; Kashiwagi, S.; Kim, M.S.; Choi, H.S. Size-Dependent EPR Effect of Polymeric Nanoparticles on Tumor Targeting. *Adv. Healthc. Mater.* **2020**, *9*, 8–15. [CrossRef] [PubMed]
24. Caro, C.; Pozo, D. Polysaccharide Colloids as Smart Vehicles in Cancer Therapy. *Curr. Pharm. Des.* **2015**, *21*, 4822–4836. [CrossRef]
25. Wilczewska, A.Z.; Niemirowicz, K.; Markiewicz, K.H.; Car2, H. Nanoparticles as Drug Delivery Systems. *Pharmacol. Rep.* **2012**, *64*, 1020–1037. [CrossRef]
26. Livney, Y.D.; Assaraf, Y.G. Rationally Designed Nanovehicles to Overcome Cancer Chemoresistance. *Adv. Drug Deliv. Rev.* **2013**, *65*, 1716–1730. [CrossRef]
27. El-Say, K.M.; El-Sawy, H.S. Polymeric Nanoparticles: Promising Platform for Drug Delivery. *Int. J. Pharm.* **2017**, *528*, 675–691. [CrossRef]
28. Liu, T.I.; Yang, Y.C.; Chiang, W.H.; Hung, C.K.; Tsai, Y.C.; Chiang, C.S.; Lo, C.L.; Chiu, H.C. Radiotherapy-Controllable Chemotherapy from Reactive Oxygen Species-Responsive Polymeric Nanoparticles for Effective Local Dual Modality Treatment of Malignant Tumors. *Biomacromolecules* **2018**, *19*, 3825–3839. [CrossRef]
29. Rios-Doria, J.; Carie, A.; Costich, T.; Burke, B.; Skaff, H.; Panicucci, R.; Sill, K. A Versatile Polymer Micelle Drug Delivery System for Encapsulation and In Vivo Stabilization of Hydrophobic Anticancer Drugs. *J. Drug Deliv.* **2012**, *2012*, 951741. [CrossRef]
30. Begines, B.; Ortiz, T.; Pérez-Aranda, M.; Martínez, G.; Merinero, M.; Argüelles-Arias, F.; Alcudia, A. Polymeric Nanoparticles for Drug Delivery: Recent Developments and Future Prospects. *Nanomaterials* **2020**, *10*, 1403. [CrossRef]
31. Aravind, A.; Jeyamohan, P.; Nair, R.; Veeranarayanan, S.; Nagaoka, Y.; Yoshida, Y.; Maekawa, T.; Kumar, D.S. AS1411 Aptamer Tagged PLGA-Lecithin-PEG Nanoparticles for Tumor Cell Targeting and Drug Delivery. *Biotechnol. Bioeng.* **2012**, *109*, 2920–2931. [CrossRef] [PubMed]
32. Hori, S.I.; Herrera, A.; Rossi, J.J.; Zhou, J. Current Advances in Aptamers for Cancer Diagnosis and Therapy. *Cancers* **2018**, *10*, 9. [CrossRef] [PubMed]
33. Zhu, G.; Chen, X. Aptamer-Based Targeted Therapy. *Adv. Drug Deliv. Rev.* **2018**, *134*, 65–78. [CrossRef]
34. Liu, Q.; Zhang, W.; Chen, S.; Zhuang, Z.; Zhang, Y.; Jiang, L.; Lin, J.S. SELEX Tool: A Novel and Convenient Gel-Based Diffusion Method for Monitoring of Aptamer-Target Binding. *J. Biol. Eng.* **2020**, *14*, 1. [CrossRef] [PubMed]
35. Engelberg, S.; Lin, Y.; Assaraf, Y.G.; Livney, Y.D. Targeted Nanoparticles Harboring Jasmine-Oil-Entrapped Paclitaxel for Elimination of Lung Cancer Cells. *Int. J. Mol. Sci.* **2021**, *22*, 1019. [CrossRef] [PubMed]
36. Ni, S.; Yao, H.; Wang, L.; Lu, J.; Jiang, F.; Lu, A.; Zhang, G. Chemical Modifications of Nucleic Acid Aptamers for Therapeutic Purposes. *Int. J. Mol. Sci.* **2017**, *18*, 1683. [CrossRef]
37. Kratschmer, C.; Levy, M. Effect of Chemical Modifications on Aptamer Stability in Serum. *Nucleic Acid Ther.* **2017**, *27*, 335–344. [CrossRef]
38. Engelberg, S.; Netzer, E.; Assaraf, Y.G.; Livney, Y.D. Selective Eradication of Human Non-Small Cell Lung Cancer Cells Using Aptamer-Decorated Nanoparticles Harboring a Cytotoxic Drug Cargo. *Cell Death Dis.* **2019**, *10*, 702. [CrossRef]
39. Zhao, Z.; Xu, L.; Shi, X.; Tan, W.; Fang, X.; Shangguan, D. Recognition of Subtype Non-Small Cell Lung Cancer by DNA Aptamers Selected from Living Cells. *Analyst* **2009**, *134*, 1808–1814. [CrossRef]
40. Kim, H.R.; Kim, W.S.; Choi, Y.J.; Choi, C.M.; Rho, J.K.; Lee, J.C. Epithelial-Mesenchymal Transition Leads to Crizotinib Resistance in H2228 Lung Cancer Cells with EML4-ALK Translocation. *Mol. Oncol.* **2013**, *7*, 1093–1102. [CrossRef]

41. Pawlik, A.; Nowak, J.M.; Grzanka, D.; Gackowska, L.; Michalkiewicz, J.; Grzanka, A. Hyperthermia Induces Cytoskeletal Alterations and Mitotic Catastrophe in P53-Deficient H1299 Lung Cancer Cells. *Acta Histochem.* **2013**, *115*, 8–15. [CrossRef] [PubMed]
42. Hong, Y.S.; Jang, W.J.; Chun, K.S.; Jeong, C.H.O. Hsp90 Inhibition by WK88-1 Potently Suppresses the Growth of Gefitinib-Resistant H1975 Cells Harboring the T790M Mutation in EGFR. *Oncol. Rep.* **2014**, *31*, 2619–2624. [CrossRef] [PubMed]
43. Elango, R.; Athinarayanan, J.; Subbarayan, V.P.; Lei, D.K.Y.; Alshatwi, A.A. Hesperetin Induces an Apoptosis-Triggered Extrinsic Pathway and a P53- Independent Pathway in Human Lung Cancer H522 Cells. *J. Asian Nat. Prod. Res.* **2018**, *20*, 559–569. [CrossRef] [PubMed]
44. Baust, J.M.; Campbell, L.H.; Harbell, J.W. Best Practices for Cryopreserving, Thawing, Recovering, and Assessing Cells. *Vitr. Cell. Dev. Biol.-Anim.* **2017**, *53*, 855–871. [CrossRef] [PubMed]
45. Martínez Rivas, C.J.; Tarhini, M.; Badri, W.; Miladi, K.; Greige-Gerges, H.; Nazari, Q.A.; Galindo Rodríguez, S.A.; Román, R.Á.; Fessi, H.; Elaissari, A. Nanoprecipitation Process: From Encapsulation to Drug Delivery. *Int. J. Pharm.* **2017**, *532*, 66–81. [CrossRef]
46. Liu, Y.; Yang, G.; Baby, T.; Tengjisi; Chen, D.; Weitz, D.A.; Zhao, C. Stable Polymer Nanoparticles with Exceptionally High Drug Loading by Sequential Nanoprecipitation. *Angew. Chem.* **2020**, *132*, 4750–4758. [CrossRef]
47. Zhang, B.; Jayalath, I.M.; Ke, J.; Sparks, J.L.; Hartley, C.S.; Konkolewicz, D. Chemically Fueled Covalent Crosslinking of Polymer Materials. *Chem. Commun.* **2019**, *55*, 2086–2089. [CrossRef]
48. Aravind, A.; Varghese, S.H.; Veeranarayanan, S.; Mathew, A.; Nagaoka, Y.; Iwai, S.; Fukuda, T.; Hasumura, T.; Yoshida, Y.; Maekawa, T.; et al. Aptamer-Labeled PLGA Nanoparticles for Targeting Cancer Cells. *Cancer Nanotechnol.* **2012**, *3*, 1–12. [CrossRef]
49. Israeli-Lev, G.; Pitchkhadze, M.; Nevo, S.; Fahoum, L.; Meyron-Holtz, E.; Livney, Y.D. Harnessing Proteins to Control Crystal Size and Morphology, for Improved Delivery Performance of Hydrophobic Bioactives, Using Genistein as a Model. *Food Hydrocoll.* **2017**, *63*, 97–107. [CrossRef]
50. Engelberg, Y.; Landau, M. The Human LL-37(17-29) Antimicrobial Peptide Reveals a Functional Supramolecular Structure. *Nat. Commun.* **2020**, *11*, 1894. [CrossRef]
51. Engelberg, S.; Modrejewski, J.; Walter, J.G.; Livney, Y.D.; Assaraf, Y.G. Cancer Cell-Selective, Clathrin-Mediated Endocytosis of Aptamer- Decorated Nanoparticles. *Oncotarget* **2018**, *9*, 20993–21006. [CrossRef]
52. Tang, W.H.; Wang, C.F.; Liao, Y. Di Fetal Bovine Serum Albumin Inhibits Antimicrobial Peptide Activity and Binds Drug Only in Complex with A1-Antitrypsin. *Sci. Rep.* **2021**, *11*, 1267. [CrossRef]
53. Ochoa, S.; Milam, V.T. Modified Nucleic Acids: Expanding the Capabilities of Functional Oligonucleotides. *Molecules* **2020**, *25*, 4659. [CrossRef]
54. Nowak, E.; Kammerer, S.; Küpper, J.H. ATP-Based Cell Viability Assay Is Superior to Trypan Blue Exclusion and XTT Assay in Measuring Cytotoxicity of Anticancer Drugs Taxol and Imatinib, and Proteasome Inhibitor MG-132 on Human Hepatoma Cell Line HepG2. *Clin. Hemorheol. Microcirc.* **2018**, *69*, 327–336. [CrossRef]
55. Yang, L.; Wang, J.; Cheke, R.A.; Tang, S. A Universal Delayed Difference Model Fitting Dose-Response Curves. *Dose-Response* **2021**, *19*, 15593258211062785. [CrossRef]
56. Focke, W.W.; Van Der Westhuizen, I.; Musee, N.; Loots, M.T. Kinetic Interpretation of Log-Logistic Dose-Time Response Curves. *Sci. Rep.* **2017**, *7*, 2234. [CrossRef]
57. Wilson, B.K.; Prud'homme, R.K. Nanoparticle Size Distribution Quantification from Transmission Electron Microscopy (TEM) of Ruthenium Tetroxide Stained Polymeric Nanoparticles. *J. Colloid Interface Sci.* **2021**, *604*, 208–220. [CrossRef]
58. D'Angelo, N.A.; Câmara, M.C.C.; Noronha, M.A.; Grotto, D.; Chorilli, M.; Lourenço, F.R.; Rangel-Yagui, C.d.O.; Lopes, A.M. Development of PEG-PCL-Based Polymersomes through Design of Experiments for Co-Encapsulation of Vemurafenib and Doxorubicin as Chemotherapeutic Drugs. *J. Mol. Liq.* **2022**, *349*, 118166. [CrossRef]
59. Feng, S.S.; Chien, S. Chemotherapeutic Engineering: Application and Further Development of Chemical Engineering Principles for Chemotherapy of Cancer and Other Diseases. *Chem. Eng. Sci.* **2003**, *58*, 4087–4114. [CrossRef]
60. Jiang, W.; Kim, B.Y.S.; Rutka, J.T.; Chan, W.C.W. Nanoparticle-Mediated Cellular Response Is Size-Dependent. *Nat. Nanotechnol.* **2008**, *3*, 145–150. [CrossRef]
61. Kaksonen, M.; Sun, Y.; Drubin, D.G. A Pathway for Association of Receptors, Adaptors, and Actin during Endocytic Internalization. *Cell* **2003**, *115*, 475–487. [CrossRef]
62. Molina-Romero, C.; Rangel-Escareño, C.; Ortega-Gómez, A.; Alanis-Funes, G.J.; Avilés-Salas, A.; Avila-Moreno, F.; Mercado, G.E.; Cardona, A.F.; Hidalgo-Miranda, A.; Arrieta, O. Differential Gene Expression Profiles According to the Association for the Study of Lung Cancer/American Thoracic Society/European Respiratory Society Histopathological Classification in Lung Adenocarcinoma Subtypes. *Hum. Pathol.* **2017**, *66*, 188–199. [CrossRef]
63. Wilkerson, M.D.; Yin, X.; Walter, V.; Zhao, N.; Cabanski, C.R.; Hayward, M.C.; Miller, C.R.; Socinski, M.A.; Parsons, A.M.; Thorne, L.B.; et al. Differential Pathogenesis of Lung Adenocarcinoma Subtypes Involving Sequence Mutations, Copy Number, Chromosomal Instability, and Methylation. *PLoS ONE* **2012**, *7*, e36530. [CrossRef]
64. Press, D. Aptamer-Hybrid Nanoparticle Bioconjugate Efficiently Delivers MiRNA-29b to Non-Small-Cell Lung Cancer Cells and Inhibits Growth by Downregulating Essential Oncoproteins. *Int. J. Nanomed.* **2016**, *11*, 3533–3544.
65. Powell, D.; Chandra, S.; Dodson, K.; Shaheen, F.; Wiltz, K.; Ireland, S.; Syed, M.; Dash, S.; Wiese, T.; Mandal, T.; et al. Aptamer-Functionalized Hybrid Nanoparticle for the Treatment of Breast Cancer. *Eur. J. Pharm. Biopharm.* **2017**, *114*, 108–118. [CrossRef]

66. Kim, M.W.; Jeong, H.Y.; Kang, S.J.; Jeong, I.H.; Choi, M.J.; You, Y.M.; Im, C.S.; Song, I.H.; Lee, T.S.; Lee, J.S.; et al. Anti-EGF Receptor Aptamer-Guided Co-Delivery of Anti-Cancer SiRNAs and Quantum Dots for Theranostics of Triple-Negative Breast Cancer. *Theranostics* **2019**, *9*, 837–852. [CrossRef]
67. Wan, L.Y.; Yuan, W.F.; Ai, W.B.; Ai, Y.W.; Wang, J.J.; Chu, L.Y.; Zhang, Y.Q.; Wu, J.F. An Exploration of Aptamer Internalization Mechanisms and Their Applications in Drug Delivery. *Expert Opin. Drug Deliv.* **2019**, *16*, 207–218. [CrossRef]
68. Kim, M.Y.; Jeong, S. In Vitro Selection of RNA Aptamer and Specific Targeting of ErbB2 in Breast Cancer Cells. *Nucleic Acid Ther.* **2011**, *21*, 173–178. [CrossRef]
69. Farokhzad, O.C.; Karp, J.M.; Langer, R. Nanoparticle-Aptamer Bioconjugates for Cancer Targeting. *Expert Opin. Drug Deliv.* **2006**, *3*, 311–324. [CrossRef]
70. Zhou, J.; Rossi, J.J. Cell-Specific Aptamer-Mediated Targeted Drug Delivery. *Oligonucleotides* **2011**, *21*, 1–10. [CrossRef]
71. Dhar, S.; Gu, F.X.; Langer, R.; Farokhza, O.C.; Lippard, S.J. Targeted Delivery of Cisplatin to Prostate Cancer Cells by Aptamer Functionalized Pt(IV) Prodrug-PLGA-PEG Nanoparticles. *Proc. Natl. Acad. Sci. USA* **2008**, *105*, 17356–17361. [CrossRef]
72. Jiang, Z.M.; Dai, S.P.; Xu, Y.Q.; Li, T.; Xie, J.; Li, C.; Zhang, Z.H. Crizotinib-Loaded Polymeric Nanoparticles in Lung Cancer Chemotherapy. *Med. Oncol.* **2015**, *32*, 193. [CrossRef]
73. Kettler, K.; Veltman, K.; van de Meent, D.; van Wezel, A.; Hendriks, A.J. Cellular Uptake of Nanoparticles as Determined by Particle Properties, Experimental Conditions, and Cell Type. *Environ. Toxicol. Chem.* **2014**, *33*, 481–492. [CrossRef]
74. Behzadi, S.; Serpooshan, V.; Tao, W.; Hamaly, M.A.; Alkawareek, M.Y.; Dreaden, E.C.; Brown, D.; Alkilany, A.M.; Farokhzad, O.C.; Mahmoudi, M. Cellular Uptake of Nanoparticles: Journey inside the Cell. *Chem. Soc. Rev.* **2017**, *46*, 4218–4244. [CrossRef]
75. Xiao, Z.; Shangguan, D.; Cao, Z.; Fang, X.; Tan, W. Cell-Specific Internalization Study of an Aptamer from Whole Cell Selection. *Chem.-A Eur. J.* **2008**, *14*, 1769–1775. [CrossRef]
76. Adan, A.; Alizada, G.; Kiraz, Y.; Baran, Y.; Nalbant, A. Flow Cytometry: Basic Principles and Applications. *Crit. Rev. Biotechnol.* **2017**, *37*, 163–176. [CrossRef]
77. Tsuji, K.; Ojima, M.; Otabe, K.; Horie, M.; Koga, H.; Sekiya, I.; Muneta, T. Effects of Different Cell-Detaching Methods on the Viability and Cell Surface Antigen Expression of Synovial Mesenchymal Stem Cells. *Cell Transplant.* **2017**, *26*, 1089–1102. [CrossRef]
78. Ferreira, C.S.M.; Cheung, M.C.; Missailidis, S.; Bisland, S.; Gariépy, J. Phototoxic Aptamers Selectively Enter and Kill Epithelial Cancer Cells. *Nucleic Acids Res.* **2009**, *37*, 866–876. [CrossRef]
79. Zhou, J.; Sun, M.; Jin, S.; Fan, L.; Zhu, W.; Sui, X.; Cao, L.; Yang, C.; Han, C. Combined Using of Paclitaxel and Salinomycin Active Targeting Nanostructured Lipid Carriers against Non-Small Cell Lung Cancer and Cancer Stem Cells. *Drug Deliv.* **2019**, *26*, 281–289. [CrossRef]
80. Li, Y.; Peng, Y.; Tan, Y.; Xuan, W.; Fu, T.; Wang, X.Q.; Tan, W. A New Paradigm for Artesunate Anticancer Function: Considerably Enhancing the Cytotoxicity via Conjugating Artesunate with Aptamer. *Signal Transduct. Target. Ther.* **2021**, *6*, 2020–2022. [CrossRef]
81. Ng, E.W.M.; Shima, D.T.; Calias, P.; Cunningham, E.T.; Guyer, D.R.; Adamis, A.P. Pegaptanib, a Targeted Anti-VEGF Aptamer for Ocular Vascular Disease. *Nat. Rev. Drug Discov.* **2006**, *5*, 123–132. [CrossRef]

pharmaceutics MDPI

Review

Business Risk Mitigation in the Development Process of New Monoclonal Antibody Drug Conjugates for Cancer Treatment

Balázs Kiss [1,2] and János Borbély [3,4,*]

1 Faculty of Economics, University of Debrecen, 4032 Debrecen, Hungary; kiss.balazs@econ.unideb.hu
2 BBS Dominus LLC, 4225 Debrecen, Hungary
3 Doctoral School of Clinical Medicine, University of Debrecen, 4032 Debrecen, Hungary
4 BBS Biochemicals LLC, 4225 Debrecen, Hungary
* Correspondence: jborbely@med.unideb.hu

Abstract: Recent developments aim to extend the cytotoxic effect and therapeutic window of mAbs by constructing antibody–drug conjugates (ADCs), in which the targeting moiety is the mAb that is linked to a highly toxic drug. According to a report from mid of last year, the global ADCs market accounted for USD 1387 million in 2016 and was worth USD 7.82 billion in 2022. It is estimated to increase in value to USD 13.15 billion by 2030. One of the critical points is the linkage of any substituent to the functional group of the mAb. Increasing the efficacy against cancer cells' highly cytotoxic molecules (warheads) are connected biologically. The connections are completed by different types of linkers, or there are efforts to add biopolymer-based nanoparticles, including chemotherapeutic agents. Recently, a combination of ADC technology and nanomedicine opened a new pathway. To fulfill the scientific knowledge for this complex development, our aim is to write an overview article that provides a basic introduction to ADC which describes the current and future opportunities in therapeutic areas and markets. Through this approach, we show which development directions are relevant both in terms of therapeutic area and market potential. Opportunities to reduce business risks are presented as new development principles.

Keywords: antibody–drug conjugate (ADC); nanomedicine; development; nanoparticle drug conjugate (NPDC); linkage; payload; warhead; business risk; open innovation; collaboration

1. Introduction

Medical (red) biotechnology is currently one of the most rapidly developing industries, with disruptive innovations emerging from time to time. Each decision, such as the development of a drug, has the potential to be worth billions of USD. The pharmaceutical industry is among the most research-intensive industries, with an average new product development (NPD) cycle of 11.9 years [1]. The uncertainty in the industry is determined primarily by its nature, which leads to very long development timelines (averaging approximately 10 years from research to market), very large investments (approximately USD 1–2 billion for new molecules), and extremely stringent regulations for product licensing. For these reasons, it is important to minimize business risks during the early stages of drug development. As antibody–drug conjugates (ADCs) are currently one of the most heavily researched areas in biotechnology, it is certainly worthwhile to consider this area from that perspective.

An example from 2017 illustrates the importance of prudent R&D where the pivotal trial of Bayer's ADC anetumab ravtansine did not meet its primary endpoint, as the drug failed to improve progression-free survival in mesothelioma patients. This resulted in a decline in the stock prices of Bayer's collaborators, ImmunoGen and MorphoSys [2].

ADC is a type of targeted therapy that combines the specificity of monoclonal antibodies with the cytotoxicity of a small molecule drug. ADCs are designed to deliver a toxic payload specifically to cancer cells, minimizing damage to healthy cells.

The market for ADC is expected to grow significantly in the coming years, driven by the increasing incidence of cancer, the growing demand for targeted therapies, and the development of new and improved ADC technologies. The cancer type that is most targeted by ADCs is breast cancer, followed by lung and ovarian cancers. The market is divided into segments on the basis of the type of drug, with the largest portion of the market held by microtubule inhibitors, followed by DNA-damaging agents and alkylating agents. Additionally, the market is segmented on the basis of its applications, with the highest market share held by solid tumors, followed by hematological malignancies. In terms of geography, North America holds the largest market share, due to the presence of a large number of biotechnology and pharmaceutical companies in the region, followed by Europe and Asia Pacific.

Companies such as Seattle Genetics, ImmunoGen, and Roche are some of the key players in the ADC market. The activity of the market is shown by the fact that continual acquisitions and mergers are also characteristic of the present times. An excellent example of this is the recently announced news that Pfizer and Seagen have entered into a definitive merger agreement under which Pfizer will acquire Seagen, a global biotechnology company that discovers, develops, and commercializes transformative cancer medicines, for USD 229 in cash per Seagen share for a total enterprise value of USD 43 billion [3]. It is the largest acquisition in biopharma since June of 2019, when AbbVie acquired Allergan for USD 63 billion. Early in 2019, Bristol Myers Squibb executed the largest transaction in industry history with its USD 74 billion purchase of Celgene [4].

It is worth mentioning that ADCs are still a relatively new class of therapeutics, and they are still facing a number of challenges. For example, many ADCs that have been developed so far have not shown significant efficacy in clinical trials [2,5], and many have experienced setbacks during development. Additionally, the cost of ADC therapy is high (e.g., yearly ADC treatment regimen costs~USD 100–500 k [6,7]); production costs also are high [8], which could be a barrier for some patients.

In order to overcome these weaknesses, recent developments are creating new constructions to improve efficacy. A new development introduces the advantages of peptide-drug-conjugated constructions [9] vs. ADCs.

In order to increase the payload, drug molecules are located in a nanoparticle, which is conjugated with the mAb. Those new structures are referred to in the literature as antibody-conjugated nanoparticles (ACNP). It is expected that through this structure, the selectivity and the efficacy are improved. Nanoparticles, e.g., liposomes, that are used as chemotherapeutic agent nanocarriers for doxorubicin (Doxil), are successfully conjugated with mAbs, forming the so-called immunoliposomes [10–14]. Targeting immunoliposomes are in clinical trials [15,16]. Other polymeric nanoparticles based on natural or synthetic biopolymers are conjugated with mAb and are carriers of anticancer drugs, such as as doxorubicin (DOX), paclitaxel (PTX), epirubicine, and cisplatin [17–21]. Many other nanocarriers have been conjugated with mAb, e.g., dendrimers [22], gold nanoparticles [23], and magnetic nanoparticles [24].

2. The Market of ADCs

The mAbs, e.g., Herceptin (Trastuzumab (TZM)), are a class of anticancer agents. Kadcyla represents the gold standard for the treatment of HER2+ breast cancer patients [25]. Recent developments aim to extend the cytotoxic effect and therapeutic window of these type of mAbs by constructing ADCs in which the targeting moiety is the mAb that is linked to a highly toxic drug. As of the end of 2022, there were 14 ADCs marketed, and there are approximately 100 others in clinical trials for different indications. According to a report from the middle of last year, the global ADC market accounted for USD 1387 million in 2016 and was worth USD 3.51 billion in 2020, registering a compound annual growth rate (CAGR) of 14.12%. It is estimated to grow at a rate of 17.08% and will increase in value to USD 13.15–16.4 billion by 2030 [26,27] as Figure 1 shows.

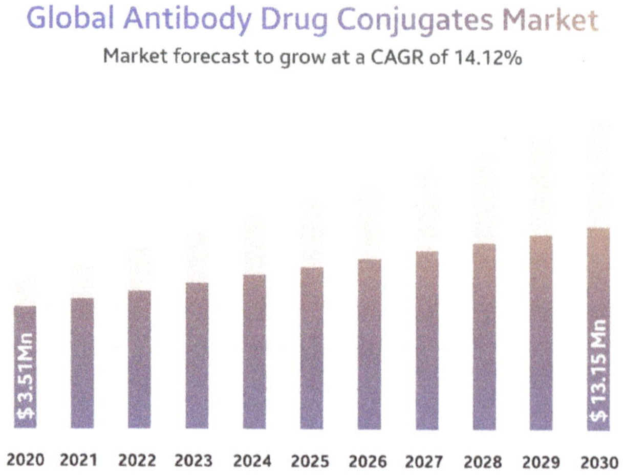

Figure 1. Global ADC market [26]. Quoted with permission from Strategic Market Research. Available online: https://www.globenewswire.com/en/news-release/2022/06/21/2465821/0/en/Antibody-Drug-Conjugate-Market-a-13-15-billion-Industry-by-2030-with-a-CAGR-of-14-12.html (accessed on 21 June 2022).

One might inquire as to the significance of ADCs and the reasons for their substantial market growth rate. Let us provide a concise summary of the principal determinants that underlie this phenomenon. Primary drivers stimulating the global market expansion are the following: ADCs are developed to target the cancer cells alone, bypassing the healthy cells. Using this phenomenon, ADCs are used primarily in cancer treatment. According to Eurostat, cancer was responsible for the deaths of approximately 1.2 million people in Europe, which was 26% of the total deaths in the region. According to the CDC, breast cancer is the second most common type of cancer in women in the US and is responsible for nearly 42,000 deaths each year. Thus, with rising cancer cases globally, the need for ADCs will boost the overall market growth. As per UN statistics, globally, the number of people over the age of 65 will rise from 9.3% of the total population in 2020 to 16.0% of the population by 2050. To ensure a good quality of life at this increased average age, more targeted, effective cancer therapies, such as ADCs, will be needed. The increase in the average age of individuals can be explained by their adoption of healthy habits and lifestyles, which has led to greater awareness of these practices [26].

The current market status is as follows: Since the first ADC, Mylotarg® (gemtuzumab ozogamicin), was approved in 2000 by the US Food and Drug Administration (FDA), there have been 14 ADCs that received market approval so far worldwide. As of November 2022, the FDA has approved 13 different ADCs, including Lumoxiti (moxetumomab pasudotox-tdfk), which we consider as an immunotoxin.

The market for ADCs is expected to continue to grow beyond 2026, as novel agents are introduced into clinical practice for several oncology indications [26,27].

Table 1 summarizes the trade name, maker, payload design, and approved indications of these drugs [28]. Currently, there are more than 100 ADCs being developed for clinical use, with the majority intended for the treatment of cancer.

Table 1. FDA-approved ADCs as of November of 2022 [28].

ADC Drug	Maker	Disease Indication	Payload/Payload Class	Target	mAb	Linker	Approval Year
Mirvetuximab soravtansine	ImmunoGen	Platinum-resistant ovarian cancer	Maytansinoid DM4	FRα	IgG1	/	2022
Tisotumab vedotin-tftv	Seagen Inc	Recurrent or metastatic cervical cancer	MMAE/auristatin	Tissue factor	IgG1	Enzyme-cleavable	2021
Loncastuximab tesirine-lpyl	ADC Therapeutics	Large B-cell lymphoma	SG3199/PBD dimer	CD19	IgG1	Enzyme-cleavable	2021
Belantamab mafodotin-blmf	GlaxoSmithKline (GSK)	Adult patients with relapsed or refractory multiple myeloma	MMAF/auristatin	BCMA	IgG1	Non-cleavable	2020, withdrawn on 22 November 2022
Sacituzumab govitecan	Immunomedics	Adult patients with metastatic triple-negative breast cancer (mTNBC) who have received at least two prior therapies for patients with relapsed or refractory metastatic disease	SN-38/camptothecin	TROP2	IgG1	Acid-cleavable	2020
Trastuzumab deruxtecan	AstraZeneca/Daiichi Sankyo	Adult patients with unresectable or metastatic HER2-positive breast cancer who have received two or more prior anti-HER2 based regimens	DXd/camptothecin	HER2	IgG1	Enzyme-cleavable	2019
Enfortumab vedotin	Astellas/Seagen Genetics	Adult patients with locally advanced or metastatic urothelial cancer who have received a PD-1 or PD-L1 inhibitor and a Pt-containing therapy	MMAE/auristatin	Nectin4	IgG1	Enzyme-cleavable	2019
Polatuzumab vedotin-piiq	Genentech, Roche	Relapsed or refractory (R/R) diffuse large B-cell lymphoma (DLBCL)	MMAE/auristatin	CD79	IgG1	Enzyme-cleavable	2019
Moxetumomab pasudotox	Astrazeneca	Adults with relapsed or refractory hairy cell leukemia (HCL)	PE38 (Pseudotox)	CD22	IgG1	Cleavable	2018
Inotuzumab ozogamicin	Pfizer/Wyeth	Relapsed or refractory CD22-positive B-cell precursor acute lymphoblastic leukemia	Ozogamicin/calicheamicin	CD22	IgG4	Acid-cleavable	2017
Trastuzumab emtansine	Genentech, Roche	HER2-positive metastatic breast cancer (mBC) following treatment with trastuzumab and a maytansinoid	DM1/maytansinoid	HER2	IgG1	Nnon-cleavable	2013

Table 1. *Cont.*

ADC Drug	Maker	Disease Indication	Payload/Payload Class	Target	mAb	Linker	Approval Year
Brentuximab vedotin	Seagen Genetics, Millennium/Takeda	Relapsed HL and relapsed sALCL	MMAE/auristatin	CD30	IgG1	Enzyme-cleavable	2011
Gemtuzumab ozogamicin	Pfizer/Wyeth	Relapsed acute myelogenous leukemia (AML)	Ozogamicin/calicheamicin	CD33	IgG4	Acid-cleavable	2017; 2000

3. Nanotechnology in Medical Biotechnology

New developments are in progress to increase the selectivity as well as the efficacy of ADCs. One of hot areas of these types of developments is nanomedicines. Recognizing the importance of a comprehensive understanding of advances in pharmaceutical R&D, we felt it valuable to provide a concise overview of current developments in nanomedicines within the field of medical biotechnology when discussing the development of ADCs. Some examples of the different types of nanomedicines that are being developed and studied for various therapeutic applications can be found below. Each type of nanomedicine has unique properties and advantages that make them suitable for specific therapeutic or diagnostic purposes.

3.1. Nanoparticle-Based Drug Delivery

Nanoparticles can be used to encapsulate the drugs and deliver them to specific target sites, such as tumors or inflamed tissues. This approach can improve the efficacy and reduce the side effects of the drugs [29,30].

3.2. Gene Therapy

Nanoparticles can be used to deliver genetic material, such as DNA or RNA, into cells for therapeutic purposes. This approach has the potential to treat genetic disorders and some types of cancer [31–33].

3.3. Photodynamic Therapy

This involves using photosensitizing agents that are activated by light to kill cancer cells or bacteria. Nanoparticles can be used to deliver photosensitizers to target sites and improve the specificity of the treatment [32,34,35].

3.4. Immunotherapy

Nanoparticles can be used to deliver immunotherapeutic agents, such as antibodies or cytokines, to target sites to boost the immune system's response against cancer or other diseases [32,36,37].

3.5. Tissue Engineering

Nanofibers and other nanomaterials can be used to create scaffolds for tissue engineering applications, such as repairing damaged tissues or organs [38,39].

3.6. Diagnostic Imaging

Nanoparticles can be used as contrast agents for various imaging modalities, such as positron-emission tomography (PET) [40,41], magnetic resonance imaging (MRI) [42,43], or computed tomography (CT) [44] scans, to detect and diagnose diseases.

The nanomedicine market is a rapidly growing sector that includes the development and application of nanotechnology in medicine. According to a market research report by Grand View Research, the global nanomedicine market size was valued at USD 215.0 billion in 2020 and is expected to grow at a CAGR of 13.6% from 2021 to 2028. The report suggests that the increasing prevalence of chronic diseases, such as cancer, diabetes, and cardiovascular diseases, along with the growing demand for personalized medicine, is driving the growth of the nanomedicine market. The report also highlights the rising investments in the research and development of nanomedicines, along with the increasing adoption of nanotechnology in drug delivery and diagnostic applications, as key factors contributing to the growth of the market [45].

When a drug or medical device has a therapeutic effect due to the use of nanomaterials, it is important to understand how the pharmacodynamic and pharmacokinetic responses have been affected by the size of the nanomaterial. The EUON (European Union Observatory for Nanomaterials) has a database with more than 1000 products that use nanomaterials, including 91 healthcare products, such as wound dressings, implants, and liposomal drugs for indications such as cancer, cardiovascular disease, diabetes, and infection. The US Nanomaterial Consumer Products Inventory lists 762 nanotechnology-related products in the fitness and health category, but information on the composition of the nanomaterial is missing for almost half of them.

In the pharmaceutical industry, only a small number of platform technologies are commonly used, according to submissions to the Center for Drug Evaluation and Research (CDER) of the US FDA. Some nanocrystal formulations and liposomes, considered non-biological complex drugs (NBCD), are challenging to manufacture under good manufacturing practice (GMP) conditions and require thorough physicochemical and biopharmaceutical characterization [46].

The nanomedicine market includes several segments, such as nanoscale therapeutic agents, nanodiagnostics, and nanotechnology-based medical devices. The report suggests that the nanoscale therapeutic agents segment held the largest share of the market in 2020, driven by the increasing demand for targeted drug delivery and the development of novel drug delivery systems [45].

Overall, the nanomedicine market is expected to continue its growth trajectory in the coming years, driven by advancements in nanotechnology, increasing demand for personalized medicine, and rising investments in research and development.

4. R&D of ADC Technologies

In the new class of drugs, mAbs and mimetics [47,48] can play a dual role. ADCs contain antibodies directly linked to a limited number of highly toxic drug warheads through a linker (Figure 2). However, in the class of ACNP drugs [14,49], a large number of drug molecules are enclosed in a special nanocarrier. The NPDCs serve as a cargo to improve the drug concentration at the tumor site, by passive targeting mechanisms due to their enhanced permeability and retention ability. In order to overcome the limitations due to low drug concentration of the ADCs, the nanoparticle–drug combination is conjugated with the antibody (Figure 3). Thus, the specificity of ACNP is improved by mAb or mimetics, and the higher cargo is provided by the NPDC. In the development process, researchers must take into account the strengths of each of the mentioned drug classes.

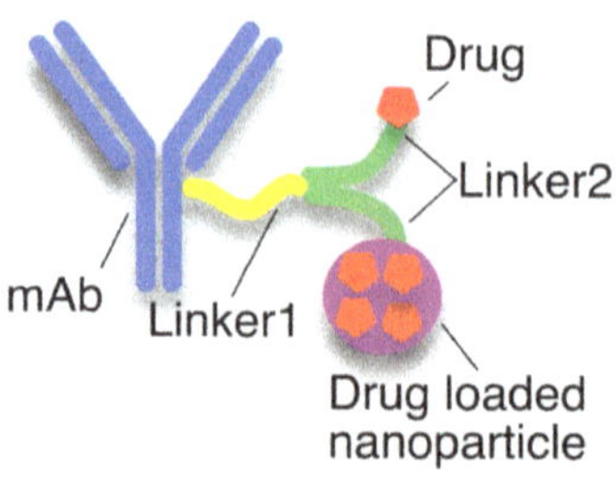

Figure 2. Schematic structure of antibody–drug conjugate (ADC). The functional group (e.g., amino group of Lys/Cys moiety) of mAb (monoclonal antibody). Drug molecules are linked directly to the mAb or are loaded into a nanoparticle linked to the mAb.

The mAbs that are created for the treatment of cancer use various mechanisms to fight the disease, such as antibody-dependent cell-mediated cytotoxicity (ADCC), complement-dependent cytotoxicity (CDC), and alterations in cell signaling. One well-designed drug for targeting HER2 in cancer therapy is TZM. However, clinical data show that many patients with HER2-overexpressing breast cancer do not respond to TZM-based therapies. To address this issue, a new type of biopharmaceutical drug, called ADC, is being developed. ADCs consist of an mAb as a targeting molecule, and a highly cytotoxic drug that is attached to a linker with a functional group, which is then connected to the Lys amine group and Cys thiol group of the mAb [50]. The primary advantage of ADCs is that the cytotoxic drug is specifically delivered to the tumor cell, increasing the efficacy and reducing the side effects due to their limited off-target toxicity [51].

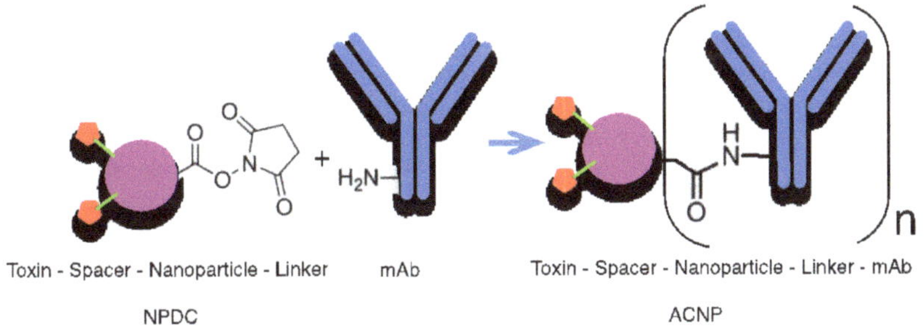

Figure 3. Schematic structure of NPDC and ACNP.

Recently, several drug-loaded nanoparticles have been in development for conjugation with mAb, e.g., functionalized magnetic nanoparticles [52], including trastuzumab-modified gold nanoparticles [53], as potential multimodal agents. It was found that these nanoparticles bind with high specificity to the HER2+ cell lines but not to the HER2- cell lines. Biopolymer-based as well as poly(lactide-co-glycolide) (PLGA) nanocarriers were loaded with the oxalilplatin chemotherapeutic drug and coated with mAb [54]. There is a study which demonstrated that the anti-cancer drug loaded in polymeric micelles or cyclodextrins improved the stability and the solubility of the prodrug, which was then connected to mAb, resulting in high efficacy [55].

Recent strategies for developing new ADCs involve selecting new components for targeting.

- mAb selection: For targeting, biologics, such as mAbs, fragments, and other backbones (e.g., single-chain variable fragment (scFv), affibody, Pentarin, and antibody–cytokine fusion proteins), are used to target HER2 or other antigens. Generally, lysines with free amines are more common than cysteines with disulfides and are not evenly distributed in the antibody [56,57]. However, Genentech is currently testing modified antibodies with engineered cysteines. Antibody mimetics are also the focus of recent research due to their importance. The site specificity/efficacy of the composed drug can be improved by adding a new building block as an antibody mimetic, e.g., single domain antibody [58], nanobody [59], or affibody [60].
- Toxins: There are numerous cytotoxic drugs that can be used as payloads, such as Maytansine (DM1, DM4), Auristatine (MMAE), SN-38, Doxorubicin, and Duocarmycin analogues [61].
- Linker selection: Linkers have a crucial role in the ADC and ACNP constructions, respectively. This part of the construction is responsible for the stability of cargo. The linker must be stable during the circulation in the bloodstream to avoid the leakage of drug molecules. A class of linkers is designed according to bio-orthogonal chemistry [62], which allows cleavage in the microenvironment of cancer cells [63]. Linkers can be either cleavable or non-cleavable, with various types of cleavable linkers, such as chemically labile linkers and enzyme-cleavable linkers (e.g., pH-sensitive linkers, disulfide linkers, peptide linkers, β-glucuronide linkers, and aldehyde tags [64]). Examples of linker platforms include the ImmunoGen Platform, Val-Cit, Disulphide, and Hydrazon [65]. For the nondegradable linkers, the connection of the cytotoxic and the antibody is non-sensitive to proteolytic degradation [66].

In the construction of ADCs, the biomolecule antibody requires precise chemical modification under specific conditions, and the payload capacity is relatively limited. In contrast, the formation of the nanoparticle drug constructs can occur in a less sensitive environment. In the final step of the ACNP formation, the activity of the antibody can be better protected.

There are many developments focusing on how to modify the TZM antibody. In this section, we describe a few steps of a synthesis project for creating a modified version of the trastuzumab antibody as an example. As a first step, lysine residues on the antibody are chosen as the conjugation sites to avoid disturbing the antibody's structure and function and to maintain its favorable pharmacokinetic properties. Secondly, toxins are selected on the basis of their functional groups, and linkers are chosen for their ability to bind the toxin to the biopolymer and the NPDC to the mAb. Finally, biopolymers are selected on the basis of their solubility in water, ability to self-assemble into nanoparticles of a desired size and distribution, and their functional and steric properties for connection to the mAb.

A new approach to combining mAbs with toxic drugs involves using a biopolymer-based nanoparticle. This type of ADC is composed of a biodegradable biopolymer nanoparticle that contains a highly cytotoxic payload and is decorated with the targeting TZM antibody on the surface. The synthetic route for creating this ADC begins with the creation of a drug/toxin biopolymer conjugate that is coupled through a spacer (PDC). The biopolymer (Poly_1) is capable of self-assembling, and by adding Biopolymer_2 (Poly_2), a nanoparticle containing the drug molecule is obtained (NPDC). The functional groups of NPDC are then modified with a maleimide-containing linker (e.g., SMCC) that can react with lysine groups of the TZM antibody.

This proposed method has potential advantages over other ADCs because the NPDC can be precisely prefabricated [67,68], and in the final step, it can be conjugated to the mAb, resulting in the appropriate ACNP (Figure 3). It is suspected that these constructions have the benefits of highly targeting cancer cells due to their antibody content, as well as a high efficacy due to the high payload carried by the conjugated nanoparticles.

This new structure will allow for a combination of immune and chemotherapeutic nanomedicine, resulting in overall higher anticancer efficacy.

For the class of antibody-functionalized lipid-based nanoparticles or solid–liquid particles, in the first step, different terminal groups, such as amino, carboxyl, maleimide, or NHS, are formed [69], then they are conjugated with the appropriate mAb.

New elements, such as nanoparticles (e.g., dendrimers, PLGA, and polymer-based ADCs), are being explored in ADC development. The Mersena (MA) technology uses a polyacetal polymer-based platform for creating ADCs.

In the field of ADCs, the technology of "cleavable linker" currently has the largest and most robust market share. Cleavable linkers are advantageous because they offer more varied applications compared with non-cleavable linkers due to their ability to use different mechanisms to act on disease sites. In May 2022, Aptamer and PinotBio entered into a collaborative effort to develop Optimer-drug conjugates as an alternative format for ADCs to target four specific non-blood-based cancer targets, namely Nectin-4, Tissue Factor, CEACAM5, and CD73. These biomarkers have the potential to target solid tumors, which have a poor response to chemotherapy. This could lead to the development of therapeutics with a smaller size that would allow for greater penetration of the tumor compared with standard antibody-based ADCs [70,71].

In terms of applications, "Breast Cancer" has the highest market share for ADCs. The number of breast cancer cases has increased dramatically over the years, with the World Health Organization reporting nearly 19.3 million cases in 2020, almost doubled from the 10 million cases reported in 2000. This increase in breast cancer cases has made it the most prevalent form of cancer, surpassing lung cancer. Consequently, ADCs are being used as a viable treatment option, which has increased the market size for ADCs.

Geographically, "North America" holds the largest market share for ADCs. Most of the development in the field of ADCs is conducted in North America, including research and development and clinical trials. The presence of major pharmaceutical companies based in the United States, such as Johnson & Johnson, Pfizer, Gilead Sciences, and Abbott International, has significantly contributed to the revenue share of the region. The American

Cancer Society has predicted 1.9 million cancer cases and 609,000 deaths in the region in 2022, which is likely to boost the market for ADCs [26].

Some recent development examples in the market are mentioned below. In June 2022, ADC Therapeutics was in the process of adding its second ADC to the company's product portfolio, with the introduction of camidanlumab tesirine. This drug has shown promising results in Phase 2 of clinical trials, and the company is likely to seek approval from the US Food and Drug Administration (FDA) to use it as a treatment option for Hodgkin's lymphoma [72].

Also in June 2022, Spirea Limited announced that it had received GBP 2.4 million in funding from investors to develop highly specialized ADCs for treating solid tumors. Their technology allows for a higher drug-to-antibody ratio, which means that more of the drug can be delivered to the affected area to eliminate cancer cells. This could lead to more effective cancer treatments [73].

In May 2022, the FDA approved the use of fam-trastuzumab deruxtecan-nxki (Enhertu®) for the treatment of metastatic HER2-positive breast cancer. The drug was developed by Daiichi Sankyo and AstraZeneca [74].

5. Challenges and Business Risks in R&D of ADCs

5.1. Risks for R&D, Limits, and Failures

For more than 100 years, the concept of ADCs has existed, but there are still only a limited number of varieties available on the market. Most of the varieties under research are still in the early stages of development. The main reason for this is that the development of ADC drugs is a challenging process, with high technical barriers. ADC drugs require multiple steps to become effective after entering the body, and each step presents significant technical difficulties that must be overcome.

The primary ethical concern regarding medical science and technology is often the potential risk involved, especially for new and unfamiliar technologies. While risk and risk–benefit analyses are only one aspect of ethical oversight, they are often used interchangeably in ethical review and risk assessment. This is because both the Common Rule and FDA emphasize the importance of minimizing risk for human subjects and require the local Institutional Review Boards (IRBs) to consider the risk–benefit analysis when making decisions about proposed research. In the case of ADCs, toxicological analyses are typically the first thing that comes to mind when assessing risk, as they are critical for evaluating the safety of these molecules. There is a significant body of literature on toxicological risk analysis for ADCs, and many new methodologies have been developed and published [75–77]. However, business risk methodologies for the development of these molecules have been less discussed. Developing an ADC involves significant technological challenges and a complex development process that can be very expensive. Therefore, the business risk associated with developing an ADC is very high.

In general, the high rates of drug development failures can be attributed to various reasons, including:

- Unreliable published data;
- Biopharmaceutical issues, such as suboptimal pharmacokinetics;
- Poorly predictive preclinical models used in discovery research and preclinical testing;
- The concept of target-based drug discovery, which involves complex target selection, competition for proprietary targets, and the validation process;
- Complexities of clinical trials, particularly in treating chronic diseases, along with increasing demands from regulatory authorities and payers.

Smaller organizations lacking the know-how and resources of larger organizations, leading to lower probability of technical and regulatory success from Phase I to submission [78,79].

While ADCs hold great promise for the treatment of cancer and other diseases, there are several challenges and business risks associated with their development. Some of these challenges and risks include:

- Complex Manufacturing: ADCs are complex molecules that require precise conjugation of the antibody and the cytotoxic agent. The manufacturing process can be challenging and time-consuming, and any variability in the manufacturing process can affect the quality and efficacy of the final product.
- Regulatory Challenges: ADCs are subject to strict regulatory oversight, and the approval process can be lengthy and expensive. Regulators require extensive data on the safety and efficacy of ADCs, including data on the pharmacokinetics, pharmacodynamics, and toxicology of the drug.
- Target Selection: Choosing the right target for an ADC is critical for its success. If the target is not expressed on the tumor cells, or if it is expressed on normal cells, the ADC may not be effective or may cause off-target toxicity [80–82].
- Resistance: As with any cancer therapy, the development of resistance is a significant challenge for ADCs. Cancer cells can develop resistance to the antibody, the cytotoxic agent, or both, rendering the ADC ineffective.
- Intellectual Property: ADC development involves complex intellectual property issues, including patenting of the antibody, the linker, and the cytotoxic agent. Companies must navigate these issues carefully to avoid infringement and protect their intellectual property.
- Cost: Developing ADCs can be extremely expensive, with high costs associated with manufacturing, clinical trials, and regulatory approval. There is also significant competition in the market, which can drive down prices and limit profitability.

The characterization of nanotechnology-related products requires a combination of different techniques to understand their physicochemical features and how these impact efficacy and product safety. From a regulatory standpoint, assessment of the environmental toxicity and effects on occupational health are required for the raw materials (excipients), while drug products and medical devices follow their own framework, with a greater emphasis on the therapeutic applications [83].

Pharmaceutical companies are currently facing significant barriers to entering the emerging nanomedicine market, with recent trends in the European regulatory landscape indicating increased restrictions and a narrower field of competitors. However, a growing knowledge base and a rising number of drug products and medical devices in the market offer new opportunities for the industry [83].

The European Medicines Agency (EMA) has established a framework for evaluating nanomedicines, which includes the following principles:

- The evaluation of any nanomedicine should be based on established principles of benefit/risk analysis, rather than solely on the basis of the technology itself (including the use of Risk Management Plans and Environmental Risk Assessment) [77,83].
- Specialized multidisciplinary expertise is required, with a group of mixed academia and regulatory experts pooling their knowledge of quality, safety, and kinetics to support evaluation and formulate guidelines [83].
- Close cooperation with other scientific committees (such as the Scientific Committee on Emerging and Newly Identified Health Risks and the European Food Safety Authority), networks (such as the Nanotechnology Knowledge Base and the European Technology Platform for Nanomedicine), and the European Commission [83].
- International cooperation, with EMA chairing an international expert group that includes the US FDA, Japan MHLW, Health Canada, and TGA Australia.
- Transparent dialogue with stakeholders [83,84].

5.2. Case Studies

In 2012, the FDA approvals were reviewed, and the most probable reasons for failures in Phase II and Phase III clinical development were found to be lack of efficacy (56%), safety issues (28%), changing strategies (7%), commercial reasons (5%), and operational challenges (5%). These results were confirmed by a second analysis of 142 drug R&D projects of AstraZeneca, which found that preclinical and Phase I projects failed primarily failed due to safety reasons, whereas projects failing in Phases II and III commonly lacked efficacy [78].

In the pharmaceutical research and development (R&D) process, discovery research takes 4.5 years, preclinical testing lasts for 1 year, and the three clinical development phases require 1.5, 2.5, and 2.5 years, respectively, with an additional 18 months from submission to launch. Basic research and post-approval Phase IV trials must also be considered in the overall R&D time. There are two additional findings when reviewing drug R&D timelines. First, clinical development takes longer than in the past. Second, the average time for FDA review and approval has decreased significantly since the Prescription Drug User Fee Act (PDUFA) was enacted, potentially due to fast-track status or accelerated approvals.

The long overall time of pharmaceutical R&D impacts on total R&D costs increases the risk of industry competition and raises the uncertainty of generic competition. This is due to the capitalization of R&D costs, which increases overall expenditures, the risk of competition, which reduces the chance to be first-to-market, and the commercial success of a drug candidate. Additionally, the effective date of generic competition can influence the ROI of a new drug by reducing the commercially usable patent term.

At present, the duration of the clinical trials is longer due to the COVID-19 pandemic because there are many vaccine candidates in development which are requiring many resources [78].

The biotechnology industry faces the challenge of covering the entire R&D process and clinical costs within a limited timeframe, typically no more than 10 years and sometimes as little as 5 years. To address this challenge, many companies are exploring ways to shorten the development process. One approach is to use open innovation, which involves collaborating with external partners and outsourcing R&D activities instead of relying solely on in-house resources. By leveraging the expertise of external partners, companies can avoid the time and cost of developing new skills internally. This strategy can help accelerate the development process and potentially reduce costs. All of the following factors can contribute to a decrease in R&D efficiency in the pharmaceutical industry. An inadequate number of projects in early R&D phases can lead to a lack of diversity in a company's pipeline and limit the potential for new drug discoveries. More technically complex research for new drug targets and subsequent preclinical and clinical studies can increase the time and resources required for R&D, which can negatively impact efficiency.

The burden for approval and reimbursement of new molecular entities (NMEs) can also be high, particularly in view of the already-approved drugs in the market. This can lead to a more risk-averse approach from both regulators and society, which may result in longer approval timelines and a lower likelihood of success for new drugs.

Mergers and acquisitions (M&As) can have a negative effect on R&D efficiency as well, particularly if the integration of the two companies results in the duplication of efforts and resources. Similarly, the decreasing number of research-based pharmaceutical companies taking the financial risk of drug R&D can limit competition and innovation in the industry.

Licensing, co-development, or joint ventures can have a negative effect on clinical development and approval durations if there are disagreements or delays in decision-making and resource allocation between the partnering companies. Overall, a combination of these factors can contribute to the challenges facing the pharmaceutical industry in maintaining R&D efficiency and productivity [78].

5.3. New Forms for R&D

It is encouraging to see that more pharmaceutical companies are recognizing the need to make changes to their R&D ecosystems in order to improve their R&D efficiencies. The process changes that these companies are making can help them to manage better their R&D activities and reduce costs while also expanding their competence and technology base and strengthening their innovation potential.

One way that companies are creating growth options is through M&As, which can provide access to new technologies and pipelines and expand their market reach [85].

Restructuring R&D into smaller, more manageable units, similar to those found in biotechnology companies, can also help to improve the efficiency and focus on specific therapeutic areas or technologies.

Outsourcing and virtual R&D are increasingly being used by pharmaceutical companies to reduce R&D costs, particularly in areas such as preclinical and clinical development, where third-party service providers can provide cost-efficient solutions. Companies are also widening their competence field by expanding collaborations and research partnerships, which can provide access to new technologies, scientific expertise, and drug candidates in all phases of development.

Venture capital investments are being used to strengthen innovation potential, particularly in the area of early-stage drug discovery, where startups and emerging companies are developing new technologies and drug candidates. Lastly, some companies are using the power of the crowd to broaden their knowledge base, for example, by crowdsourcing ideas and solutions through open innovation platforms.

Overall, the process changes being made by pharmaceutical companies reflect a recognition of the need to adapt to the changing landscape of R&D and to adopt more agile, innovative approaches to improve R&D efficiencies and productivity [78].

The concept of Open Innovation has garnered increasing interest in the past 20 years, particularly since Henry Chesbrough introduced the term in 2003, which led to the development of a new field of knowledge. However, despite the extensive research, there is no standardized, all-encompassing theory of Open Innovation. Instead, there exists a range of models that address different aspects and are applicable to specific contexts and industries. One of the industries that has adopted Open Innovation is the pharmaceutical industry. The pharmaceutical industry's move towards Open Innovation has unique features, such as the need to address the current productivity crisis as a driver for change and the R&D-intensive nature of the industry [86].

Due to the high complexity of the biopharmaceutical industry, outsourcing, "slicing up" of tasks, and collaboration with other companies are more likely to occur. Another characteristic of the sector is that the development and production of a biologically active ingredient involves thousands of steps, resulting in a long value chain. At almost every step, management must decide whether to keep the given activity in-house or to outsource it.

In modern biotechnology projects and programs, there is a trend towards project network organizations (PNOs) as opposed to project-based firms (PBFs), which have become increasingly important contexts for interorganizational project cooperation. Due to organizational specialization, PNOs have become general organizational forms that combine the coordination ability of PBFs with the resource richness of networks. PNOs connect legally independent but often operationally interdependent individuals and organizations in the form of strategically coordinated project teams and flexible partner groups that persist beyond individual projects. On the basis of empirical review, PNOs have outstanding characteristics in complex product and system development, research, open innovation, and international development. Differences between the two structures can be related to the variety and connectivity of the projects, the degree of specialization, and the geographical concentration of the resources [87].

6. Conclusions

A review of social research in the healthcare industry suggests that there will be a significant demand for effective oncology solutions in the near future, indicating a vast potential market and high risks for medical biotechnology, including ADCs. Market analyses, forecasts, and acquisitions clearly demonstrate this potential. We are on the cusp of a major breakthrough, with only 14 approved ADC products and more than 100 in clinical trials, a number that is set to increase rapidly. Regulatory and clinical acceptance for ADCs is also improving. However, there are technological challenges that make the situation difficult, which could be addressed by the NPDC route as a technological option.

Given the recent excitement surrounding the potential of artificial intelligence (AI) in drug discovery and development, particularly driven by advancements in machine learning and its role in the competitive race for the next blockbuster drug, it is expected that pharmaceutical companies will leverage these computer-based platforms for the development of the next-generation ADCs [88].

Moreover, the potential of ADCs as a therapeutic strategy extends beyond cancer and is substantial. Research is already being conducted on the use of ADCs for the treatment of non-oncological conditions, such as autoimmune and cardiovascular diseases, diabetes, and antimicrobial infections [89].

Since biopharma R&D is highly capital-intensive, it is crucial to identify business risks as early as possible and minimize them where possible. New development principles, such as open innovation, new assembly structures, and new structures for collaboration, offer an opportunity to achieve this.

Author Contributions: Conceptualization, B.K. and J.B.; methodology, B.K. and J.B.; software, B.K.; validation, B.K.; formal analysis, B.K.; investigation, B.K. and J.B.; resources, B.K. and J.B.; data curation, B.K. and J.B.; writing—original draft preparation, B.K. and J.B.; writing—review and editing, B.K.; visualization, B.K.; supervision, B.K. and J.B.; project administration, J.B. All authors have read and agreed to the published version of the manuscript.

Funding: This research received no external funding.

Institutional Review Board Statement: Not applicable.

Informed Consent Statement: Not applicable.

Data Availability Statement: Not applicable.

Conflicts of Interest: The authors declare no conflict of interest.

Abbreviations

ADC	Antibody–Drug Conjugate
mAb	Monoclonal Antibody
HER	Human Epidermal Growth Factor
TZM	Trastuzumab
CAGR	Compound Annual Growth Rate
ADCC	Antibody-Dependent Cell-Mediated Cytotoxicity
CDC	Complement-Dependent Cytotoxicity
MRI	Magnetic Resonance Imaging
CT	Computer Tomography
ACNP	Antibody-Conjugated Nanoparticle
NPDC	Nanoparticle–Drug Conjugate
PDUFA	Prescription Drug User Fee Act
ROI	Return of Investment
NME	New Molecular Entity
M&A	Mergers and Acquisitions
PBF	Project-Based Firm
PNO	Project Network Organization

References

1. Fernald, K.D.S.; Pennings, H.P.G.; Bosch, J.F.V.D.; Commandeur, H.R.; Claassen, E. The moderating role of absorptive capacity and the differential effects of acquisitions and alliances on Big Pharma firms' innovation performance. *PLoS ONE* **2017**, *12*, e0172488. [CrossRef]
2. Bayer ADC Fails Pivotal Mesothelioma Trial. Available online: https://www.fiercebiotech.com/biotech/bayer-adc-fails-pivotal-mesothelioma-trial (accessed on 24 July 2017).
3. Pfizer Invests $43 Billion to Battle Cancer. Available online: https://www.pfizer.com/news/press-release/press-release-detail/pfizer-invests-43-billion-battle-cancer (accessed on 13 March 2023).
4. With $43B Buyout, Pfizer Sees Cancer Specialist Seagen as a 'Goose' Laying 'Golden Eggs'. Available online: https://www.fiercepharma.com/pharma/43b-buyout-pfizer-sees-seagen-its-golden-goose (accessed on 13 March 2023).

5. After Dose De-Escalation, Death Drives Magenta to Pause Antibody-Drug Conjugate Leukemia Trial. Available online: https://www.fiercebiotech.com/biotech/after-dose-de-escalation-death-drives-magenta-pause-antibody-drug-conjugate-leukemia-trial (accessed on 26 January 2023).
6. Zolot, R.S.; Basu, S.; Million, R.P. Antibody–drug conjugates. *Nat. Rev. Drug Discov.* **2013**, *12*, 259–260. [CrossRef] [PubMed]
7. Kapinos, K.A.; Hu, E.; Trivedi, J.; Geethakumari, P.R.; Kansagra, A. Cost-Effectiveness Analysis of CAR T-Cell Therapies vs Antibody Drug Conjugates for Patients with Advanced Multiple Myeloma. *Cancer Control.* **2023**, *30*, 10732748221142945. [CrossRef]
8. Mckertish, C.M.; Kayser, V. Advances and Limitations of Antibody Drug Conjugates for Cancer. *Biomedicines* **2021**, *9*, 872. [CrossRef]
9. Wu, M.; Huang, W.; Yang, N.; Liu, Y. Learn from antibody–drug conjugates: Consideration in the future construction of peptide-drug conjugates for cancer therapy. *Exp. Hematol. Oncol.* **2022**, *11*, 93. [CrossRef]
10. Espelin, C.W.; Leonard, S.C.; Geretti, E.; Wickham, T.J.; Hendriks, B.S. Dual HER2 Targeting with Trastuzumab and Liposomal-Encapsulated Doxorubicin (MM-302) Demonstrates Synergistic Antitumor Activity in Breast and Gastric Cancer. *Cancer Res.* **2016**, *76*, 1517–1527. [CrossRef]
11. Hu, X.; Kwon, N.; Yan, K.; Sedgwick, A.C.; Chen, G.; He, X.; James, T.D.; Yoon, J. Bio-Conjugated Advanced Materials for Targeted Disease Theranostics. *Adv. Funct. Mater.* **2020**, *30*, 1907906. [CrossRef]
12. Rodallec, A.; Franco, C.; Robert, S.; Sicard, G.; Giacometti, S.; Lacarelle, B.; Bouquet, F.; Savina, A.; Lacroix, R.; Dignat-George, F.; et al. Prototyping Trastuzumab Docetaxel Immunoliposomes with a New FCM-Based Method to Quantify Optimal Antibody Density on Nanoparticles. *Sci. Rep.* **2020**, *10*, 4147. [CrossRef] [PubMed]
13. Matusewicz, L.; Filip-Psurska, B.; Psurski, M.; Tabaczar, S.; Podkalicka, J.; Wietrzyk, J.; Ziółkowski, P.; Czogalla, A.; Sikorski, A.F. EGFR-targeted immunoliposomes as a selective delivery system of simvastatin, with potential use in treatment of triple-negative breast cancers. *Int. J. Pharm.* **2019**, *569*, 118605. [CrossRef]
14. Juan, A.; Cimas, F.J.; Bravo, I.; Pandiella, A.; Ocaña, A.; Alonso-Moreno, C. Antibody Conjugation of Nanoparticles as Therapeutics for Breast Cancer Treatment. *Int. J. Mol. Sci.* **2020**, *21*, 6018. [CrossRef] [PubMed]
15. Kasenda, B.; König, D.; Manni, M.; Ritschard, R.; Duthaler, U.; Bartoszek, E.; Bärenwaldt, A.; Deuster, S.; Hutter, G.; Cordier, D.; et al. Targeting immunoliposomes to EGFR-positive glioblastoma. *ESMO Open* **2022**, *7*, 100365. [CrossRef]
16. Mamot, C.; Wicki, A.; Hasler-Strub, U.; Riniker, S.; Li, Q.; Holer, L.; Bärtschi, D.; Zaman, K.; von Moos, R.; Dedes, K.J.; et al. A multicenter phase II trial of anti-EGFR-immunoliposomes loaded with doxorubicin in patients with advanced triple negative breast cancer. *Sci. Rep.* **2023**, *13*, 3705. [CrossRef]
17. Kumar, A.; Lale, S.V.; Alex, M.A.; Choudhary, V.; Koul, V. Folic Acid and Trastuzumab Conjugated Redox Responsive Random Multiblock Copolymeric Nanocarriers for Breast Cancer Therapy: In-Vitro and in-Vivo Studies. *Colloids Surf. B Biointerfaces* **2017**, *149*, 369–378. [CrossRef]
18. Peng, J.; Chen, J.; Xie, F.; Bao, W.; Xu, H.; Wang, H.; Xu, Y.; Du, Z. Herceptin-Conjugated Paclitaxel Loaded PCL-PEG Worm-like Nanocrystal Micelles for the Combinatorial Treatment of HER2-Positive Breast Cancer. *Biomaterials* **2019**, *222*, 119420. [CrossRef] [PubMed]
19. Kolahkaj, F.F.; Derakhshandeh, K.; Khaleseh, F.; Azandaryani, A.H.; Mansouri, K.; Khazaei, M. Active Targeting Carrier for Breast Cancer Treatment: Monoclonal Antibody Conjugated Epirubicin Loaded Nanoparticle. *J. Drug Deliv. Sci. Technol.* **2019**, *53*, 101136. [CrossRef]
20. Wang, Y.; Qian, J.; Yang, M.; Xu, W.; Wang, J.; Hou, G.; Ji, L.; Suo, A. Doxorubicin/Cisplatin Co-Loaded Hyaluronic Acid/Chitosan-Based Nanoparticles for in Vitro Synergistic Combination Chemotherapy of Breast Cancer. *Carbohydr. Polym.* **2019**, *225*, 115206. [CrossRef] [PubMed]
21. Zhong, S.; Ling, Z.; Zhou, Z.; He, J.; Ran, H.; Wang, Z.; Zhang, Q.; Song, W.; Zhang, Y.; Luo, J. Herceptin-Decorated Paclitaxel-Loaded Poly(Lactide-Co-Glycolide) Nanobubbles: Ultrasound-Facilitated Release and Targeted Accumulation in Breast Cancers. *Pharm. Dev. Technol.* **2020**, *25*, 454–463. [CrossRef] [PubMed]
22. Pilkington, G.A.; Pedersen, J.S.; Briscoe, W.H. Dendrimer nanofluids in the concentrated regime: From polymer melts to soft spheres. *Langmuir* **2015**, *31*, 3333–3342. [CrossRef]
23. Choi, J.-H.; Gu, H.-J.; Park, K.-H.; Hwang, D.-S.; Kim, G.-C. Anti-Cancer Activity of the Combinational Treatment of Noozone Cold Plasma with p-FAK Antibody-Conjugated Gold Nanoparticles in OSCC Xenograft Mice. *Biomedicines* **2022**, *10*, 2259. [CrossRef] [PubMed]
24. Lodhi, M.S.; Khalid, F.; Khan, M.T.; Samra, Z.Q.; Muhammad, S.; Zhang, Y.-J.; Mou, K. A Novel Method of Magnetic Nanoparticles Functionalized with Anti-Folate Receptor Antibody and Methotrexate for Antibody Mediated Targeted Drug Delivery. *Molecules* **2022**, *27*, 261. [CrossRef]
25. Wilcock, P.; Webster, R.M. The breast cancer drug market. *Nat. Rev. Drug Discov.* **2021**, *20*, 339–340. [CrossRef]
26. Antibody Drug Conjugate Market, a $13.15 billion Industry by 2030 with a CAGR of 14.12%. Available online: https://www.globenewswire.com/en/news-release/2022/06/21/2465821/0/en/Antibody-Drug-Conjugate-Market-a-13-15-billion-Industry-by-2030-with-a-CAGR-of-14-12.html (accessed on 21 June 2022).
27. Pazo, C.D.; Nawaz, K.; Webster, R.M. The oncology market for antibody–drug conjugates. *Nat. Rev. Drug Discov.* **2021**, *20*, 583–584. [CrossRef] [PubMed]

28. Antibody-Drug Conjugates (ADCs) List Approved by FDA (2000–2023). Available online: https://axispharm.com/antibody-drug-conjugatesadcs-list-approved-by-fda2000-2022/ (accessed on 27 December 2022).
29. Mitchell, M.J.; Billingsley, M.M.; Haley, R.M.; Wechsler, M.E.; Peppas, N.A.; Langer, R. Engineering precision nanoparticles for drug delivery. *Nat. Rev. Drug Discov.* **2020**, *20*, 101–124. [CrossRef] [PubMed]
30. Afzal, O.; Altamimi, A.S.A.; Nadeem, M.S.; Alzarea, S.I.; Almalki, W.H.; Tariq, A.; Mubeen, B.; Murtaza, B.N.; Iftikhar, S.; Riaz, N.; et al. Nanoparticles in Drug Delivery: From History to Therapeutic Applications. *Nanomaterials* **2022**, *12*, 4494. [CrossRef] [PubMed]
31. Roma-Rodrigues, C.; Rivas-García, L.; Baptista, P.V.; Fernandes, A.R. Gene Therapy in Cancer Treatment: Why Go Nano? *Pharmaceutics* **2020**, *12*, 233. [CrossRef]
32. Mosleh-Shirazi, S.; Abbasi, M.; Moaddeli, M.R.; Vaez, A.; Shafiee, M.; Kasaee, S.R.; Amani, A.M.; Hatam, S. Nanotechnology Advances in the Detection and Treatment of Cancer: An Overview. *Nanotheranostics* **2022**, *6*, 400–423. [CrossRef]
33. Nakamura, T.; Yamada, Y.; Sato, Y.; Khalil, I.A.; Harashima, H. Innovative nanotechnologies for enhancing nucleic acids/gene therapy: Controlling intracellular trafficking to targeted biodistribution. *Biomaterials* **2019**, *218*, 119329. [CrossRef] [PubMed]
34. Chizenga, E.P.; Abrahamse, H. Nanotechnology in Modern Photodynamic Therapy of Cancer: A Review of Cellular Resistance Patterns Affecting the Therapeutic Response. *Pharmaceutics* **2020**, *12*, 632. [CrossRef] [PubMed]
35. Hu, D.; Pan, M.; Yu, Y.; Sun, A.; Shi, K.; Qu, Y.; Qian, Z. Application of nanotechnology for enhancing photodynamic therapy via ameliorating, neglecting, or exploiting tumor hypoxia. *View* **2020**, *1*, e6. [CrossRef]
36. Goldberg, M.S. Improving cancer immunotherapy through nanotechnology. *Nat. Rev. Cancer* **2019**, *19*, 587–602. [CrossRef] [PubMed]
37. Akkın, S.; Varan, G.; Bilensoy, E. A Review on Cancer Immunotherapy and Applications of Nanotechnology to Chemoimmunotherapy of Different Cancers. *Molecules* **2021**, *26*, 3382. [CrossRef]
38. Deng, C.; Xu, C.; Zhou, Q.; Cheng, Y. Advances of nanotechnology in osteochondral regeneration. *WIREs Nanomed. Nanobiotechnol.* **2019**, *11*, e1576. [CrossRef]
39. Kumar, R.; Aadil, K.R.; Ranjan, S.; Kumar, V.B. Advances in nanotechnology and nanomaterials based strategies for neural tissue engineering. *J. Drug Deliv. Sci. Technol.* **2020**, *57*, 101617. [CrossRef]
40. Körhegyi, Z.; Rózsa, D.; Hajdu, I.; Bodnár, M.; Kertész, I.; Kerekes, K.; Kun, S.; Kollár, J.; Varga, J.; Garai, I.; et al. Synthesis of ^{68}Ga-Labeled Biopolymer-based Nanoparticle Imaging Agents for Positron-emission Tomography. *Anticancer Res.* **2019**, *39*, 2415–2427. [CrossRef]
41. Szabó, J.P.; Csige, K.; Kálmán-Szabó, I.; Arató, V.; Opposits, G.; Jószai, I.; Kertész, I.; Képes, Z.; Méhes, G.; Fenyvesi, F.; et al. In vivo assessment of tumor targeting potential of 68Ga-labelled randomly methylated beta-cyclodextrin (RAMEB) and 2-hydroxypropyl-β-cyclodextrin (HPβCD) using positron emission tomography. *Int. J. Pharm.* **2023**, *630*, 122462. [CrossRef]
42. Hajdu, I.; Trencsényi, G.; Bodnár, M.; Emri, M.; Bánfalvi, G.; Sikula, J.; Márián, T.; Kollár, J.; Vámosi, G.; Borbély, J. Tumor-specific Localization of Self-assembled Nanoparticle PET/MR Modalities. *Anticancer Res.* **2014**, *34*, 49–59. [PubMed]
43. Pallares, R.M.; Mottaghy, F.M.; Schulz, V.; Kiessling, F.; Lammers, T. Nanoparticle diagnostics and theranostics in the clinic. *J. Nucl. Med.* **2022**, *63*, 1802–1808. [CrossRef]
44. Luo, D.; Wang, X.; Burda, C.; Basilion, J.P. Recent development of gold nanoparticles as contrast agents for cancer diagnosis. *Cancers* **2021**, *13*, 1825. [CrossRef] [PubMed]
45. Grand View Research. Nanomedicine Market Size, Share & Trends Analysis Report by Product (Therapeutics, Diagnostics), by Application (Oncology, Infectious Diseases), by Region, and Segment Forecasts, 2021–2028. Report ID: 978-1-68038-942-5. 2021. Available online: https://www.grandviewresearch.com/industry-analysis/nanomedicine-market (accessed on 13 March 2023).
46. Marques, M.R.; Choo, Q.; Ashtikar, M.; Rocha, T.C.; Bremer-Hoffmann, S.; Wacker, M.G. Nanomedicines—Tiny particles and big challenges. *Adv. Drug Deliv. Rev.* **2019**, *151–152*, 23–43. [CrossRef]
47. Whaby, M.; Khan, I.; O'Bryan, J.P. Chapter Eight—Targeting The "Undruggable" RAS with Biologics. In *Advances in Cancer Research*; Academic Press: Cambridge, MA, USA, 2022; Volume 153, pp. 237–266, ISBN 9780128244852, ISSN 0065-230X. [CrossRef]
48. Singh, K.; Canakci, M.; Kanjilal, P.; Williams, N.; Shanthalingam, S.; Osborne, B.A.; Thayumanavan, S. Evaluation of Cellular Targeting by Fab′ vs Full-Length Antibodies in Antibody–Nanoparticle Conjugates (ANCs) Using CD4 T-cells. *Bioconjugate Chem.* **2022**, *33*, 486–495. [CrossRef]
49. Johnston, M.C.; Scott, C.J. Antibody conjugated nanoparticles as a novel form of antibody drug conjugate chemotherapy. *Drug Discov. Today Technol.* **2018**, *30*, 63–69. [CrossRef]
50. Abdollahpour-Alitappeh, M.; Lotfinia, M.; Gharibi, T.; Mardaneh, J.; Farhadihosseinabadi, B.; Larki, P.; Faghfourian, B.; Sepehr, K.S.; Abbaszadeh-Goudarzi, K.; Johari, B.; et al. Antibody–drug conjugates (ADCs) for cancer therapy: Strategies, challenges, and successes. *J. Cell. Physiol.* **2018**, *234*, 5628–5642. [CrossRef] [PubMed]
51. Kang, M.S.; Kong, T.W.S.; Khoo, J.Y.X.; Loh, T.-P. Recent developments in chemical conjugation strategies targeting native amino acids in proteins and their applications in antibody–drug conjugates. *Chem. Sci.* **2021**, *12*, 13613–13647. [CrossRef]
52. Oltolina, F.; Colangelo, D.; Miletto, I.; Clemente, N.; Miola, M.; Verné, E.; Prat, M.; Follenzi, A. Tumor Targeting by Monoclonal Antibody Functionalized Magnetic Nanoparticles. *Nanomaterials* **2019**, *9*, 1575. [CrossRef]
53. Żelechowska-Matysiak, K.; Wawrowicz, K.; Wierzbicki, M.; Budlewski, T.; Bilewicz, A.; Majkowska-Pilip, A. Doxorubicin- and Trastuzumab-Modified Gold Nanoparticles as Potential Multimodal Agents for Targeted Therapy of HER2+ Cancers. *Molecules* **2023**, *28*, 2451. [CrossRef]

54. Zumaya, A.L.V.; Rimpelová, S.; Štějdířová, M.; Ulbrich, P.; Vilčáková, J.; Hassouna, F. Antibody Conjugated PLGA Nanocarriers and Superparmagnetic Nanoparticles for Targeted Delivery of Oxaliplatin to Cells from Colorectal Carcinoma. *Int. J. Mol. Sci.* **2022**, *23*, 1200. [CrossRef] [PubMed]
55. Ioele, G.; Chieffallo, M.; Occhiuzzi, M.A.; De Luca, M.; Garofalo, A.; Ragno, G.; Grande, F. Anticancer Drugs: Recent Strategies to Improve Stability Profile, Pharmacokinetic and Pharmacodynamic Properties. *Molecules* **2022**, *27*, 5436. [CrossRef] [PubMed]
56. Khongorzul, P.; Ling, C.J.; Khan, F.U.; Ihsan, A.U.; Zhang, J. Antibody–Drug Conjugates: A Comprehensive Review. *Mol. Cancer Res.* **2020**, *18*, 3–19. [CrossRef]
57. Hasan, M.; Laws, M.; Jin, P.; Rahman, K.M. Factors influencing the choice of monoclonal antibodies for antibody–drug conjugates. *Drug Discov. Today* **2021**, *27*, 354–361. [CrossRef]
58. Nessler, I.; Khera, E.; Vance, S.; Kopp, A.; Qiu, Q.; Keating, T.A.; Abu-Yousif, A.O.; Sandal, T.; Legg, J.; Thompson, L.; et al. Increased Tumor Penetration of Single-Domain Antibody–Drug Conjugates Improves In Vivo Efficacy in Prostate Cancer Models. *Cancer Res.* **2020**, *80*, 1268–1278. [CrossRef]
59. Salvador, J.-P.; Vilaplana, L.; Marco, M.-P. Nanobody: Outstanding features for diagnostic and therapeutic applications. *Anal. Bioanal. Chem.* **2019**, *411*, 1703–1713. [CrossRef]
60. Ståhl, S.; Gräslund, T.; Eriksson Karlström, A.; Frejd, F.Y.; Nygren, P.Å.; Löfblom, J. Affibody Molecules in Biotechno-logical and Medical Applications. *Trends Biotechnol.* **2017**, *35*, 691–712. [CrossRef]
61. Conilh, L.; Sadilkova, L.; Viricel, W.; Dumontet, C. Payload diversification: A key step in the development of antibody–drug conjugates. *J. Hematol. Oncol.* **2023**, *16*, 3. [CrossRef]
62. Sletten, E.M.; Bertozzi, C.R. Bioorthogonal Chemistry: Fishing for Selectivity in a Sea of Functionality. *Angew. Chem. Int. Ed.* **2009**, *48*, 6974–6998. [CrossRef] [PubMed]
63. Bargh, J.D.; Isidro-Llobet, A.; Parker, J.S.; Spring, D.R. Cleavable linkers in antibody–drug conjugates. *Chem. Soc. Rev.* **2019**, *48*, 4361–4374. [CrossRef] [PubMed]
64. Huang, B.C.B.; Kim, Y.C.; Bañas, S.; Barfield, R.M.; Drake, P.M.; Rupniewski, I.; Haskins, W.E.; Rabuka, D. Antibody-drug conjugate library prepared by scanning insertion of the aldehyde tag into IgG1 constant regions. *mAbs* **2018**, *10*, 1182–1189. [CrossRef]
65. Marei, H.E.; Cenciarelli, C.; Hasan, A. Potential of antibody–drug conjugates (ADCs) for cancer therapy. *Cancer Cell Int.* **2022**, *22*, 255. [CrossRef]
66. Tsuchikama, K.; An, Z. Antibody-drug conjugates: Recent advances in conjugation and linker chemistries. *Protein Cell* **2016**, *9*, 33–46. [CrossRef]
67. Csikós, Z.; Fazekas, E.; Rózsa, D.; Borbély, J.; Kerekes, K. Crosslinked poly-γ-glutamic acid based nanosystem for drug delivery. *J. Drug Deliv. Sci. Technol.* **2018**, *48*, 478–489. [CrossRef]
68. Csikós, Z.; Kerekes, K.; Fazekas, E.; Kun, S.; Borbély, J. Biopolymer based nanosystem for doxorubicin targeted delivery. *Am. J. Cancer Res.* **2017**, *7*, 715–726.
69. Marques, A.C.; Costa, P.C.; Velho, S.; Amaral, M.H. Lipid Nanoparticles Functionalized with Antibodies for Anticancer Drug Therapy. *Pharmaceutics* **2023**, *15*, 216. [CrossRef]
70. Aptamer Group and PinotBio Extend Collaboration for the Development of Optimer-Drug Conjugates. Available online: https://aptamergroup.com/aptamer-group-and-pinotbio-extend-collaboration-for-the-development-of-optimer-drug-conjugates/ (accessed on 23 May 2022).
71. Aptamer Group and PinotBio Extend Collaboration. Available online: https://markets.ft.com/data/announce/detail?dockey=1323-15463214-6M6DQKFL242JNP55NP7RKRB7S6 (accessed on 23 May 2022).
72. ADC Therapeutics Announces Results from Pivotal Phase 2 Clinical Trial of Camidanlumab Tesirine (Cami) in Relapsed or Refractory Hodgkin Lymphoma. Available online: https://ir.adctherapeutics.com/press-releases/press-release-details/2022/ADC-Therapeutics-Announces-Results-from-Pivotal-Phase-2-Clinical-Trial-of-Camidanlumab-Tesirine-Cami-in-Relapsed-or-Refractory-Hodgkin-Lymphoma/default.aspx (accessed on 10 June 2022).
73. Spirea Raises £2.4M in Funding. Available online: https://www.finsmes.com/2022/06/spirea-raises-2-4m-in-funding.html (accessed on 14 June 2022).
74. FDA Grants Regular Approval to Fam-Trastuzumab Deruxtecan-Nxki for Breast Cancer. Available online: https://www.fda.gov/drugs/resources-information-approved-drugs/fda-grants-regular-approval-fam-trastuzumab-deruxtecan-nxki-breast-cancer (accessed on 5 November 2022).
75. Burden, N.; Aschberger, K.; Chaudhry, Q.; Clift, M.J.D.; Fowler, P.; Johnston, H.; Landsiedel, R.; Rowland, J.; Stone, V.; Doak, S.H. Aligning nanotoxicology with the 3Rs: What is needed to realise the short, medium and long-term opportunities? *Regul. Toxicol. Pharmacol.* **2017**, *91*, 257–266. [CrossRef]
76. Hogle, L.F. Concepts of Risk in Nanomedicine Research. *J. Law Med. Ethics* **2012**, *40*, 809–822. [CrossRef]
77. Accomasso, L.; Cristallini, C.; Giachino, C. Risk assessment and risk minimization in nanomedicine: A need for predictive, alternative, and 3Rs strategies. *Front. Pharmacol.* **2018**, *9*, 228. [CrossRef]
78. Schuhmacher, A.; Gassmann, O.; Hinder, M. Changing R & D models in research-based pharmaceutical companies. *J. Transl. Med.* **2016**, *14*, 105. [CrossRef]
79. Domonkos, D.; Hronszki, I. *Risk Assessment of Innovations in The Biopharmaceutical Industry/Book 2*; IntechOpen: Rijeka, Croatia, 2012; pp. 85–104.

80. Baah, S.; Laws, M.; Rahman, K.M. Antibody–Drug Conjugates—A Tutorial Review. *Molecules* **2021**, *26*, 2943. [CrossRef]
81. Fu, Z.; Li, S.; Han, S.; Shi, C.; Zhang, Y. Antibody drug conjugate: The "biological missile" for targeted cancer therapy. *Signal Transduct. Target. Ther.* **2022**, *7*, 93. [CrossRef] [PubMed]
82. Mathur, R.; Weiner, G.J. Picking the Optimal Target for Antibody-Drug Conjugates. *Am. Soc. Clin. Oncol. Educ. Book* **2013**, *33*, e103–e107. [CrossRef] [PubMed]
83. Haubenreisser, S. EMA perspective on the development of Nanomedicines. In Proceedings of the PQRI Nanotechnology Workshop, Washington, DC, USA, 14–15 January 2014.
84. Pepić, I.; Hafner, A.; Lovrić, J.; Lakos, G.P. Nanotherapeutics in the EU: An overview on current state and future directions. *Int. J. Nanomed.* **2014**, *9*, 1005–1023. [CrossRef]
85. Hardison, S. Oncology Dealmaking in 2020. *Biopharm. Deal.* **2021**, B5. Available online: https://www.nature.com/articles/d43747-021-00024-y (accessed on 27 December 2022).
86. Sambandan, P.; Raja, H.B. Open Innovation in Pharmaceutical Industry, a Case Study of Eli Lilly: How Do Big Pharmas Implement OI?—A Critical Analysis of the Current OI Practices through a Case Study. Master's Thesis, Department of Industrial Economics and Management (INDEK), KTH School of Industrial Engineering and Management (ITM), Stockholm, Sweden, 2015.
87. Manning, S. The rise of project network organizations: Building core teams and flexible partner pools for interorganizational projects. *Res. Policy* **2017**, *46*, 1399–1415. [CrossRef]
88. Tong, J.T.W.; Harris, P.W.R.; Brimble, M.A.; Kavianinia, I. An Insight into FDA Approved Antibody-Drug Conjugates for Cancer Therapy. *Molecules* **2021**, *26*, 5847. [CrossRef] [PubMed]
89. Yu, S.; Lim, A.; Tremblay, M.S. Next Horizons: ADCs Beyond Oncology. In *Innovations for Next-Generation Antibody-Drug Conjugates*; Damelin, M., Ed.; Humana Press: Cham, Switzerland, 2018; pp. 321–347. [CrossRef]

pharmaceutics

MDPI

Review

Intranasal Drug Administration in Alzheimer-Type Dementia: Towards Clinical Applications

Raquel Taléns-Visconti [1,*], Jesus Vicente de Julián-Ortiz [2,*], Ofelia Vila-Busó [3], Octavio Diez-Sales [1,4] and Amparo Nácher [1,4]

1 Department of Pharmacy and Pharmaceutical Technology and Parasitology, Faculty of Pharmacy, University of Valencia, Av. Vicent Andrés Estellés s/n, 46100 Valencia, Spain; octavio.diez@uv.es (O.D.-S.); amparo.nacher@uv.es (A.N.)

2 Molecular Topology and Drug Design Research Unit, Department of Physical Chemistry, Faculty of Pharmacy, University of Valencia, Av. Vicent Andrés Estellés s/n, 46100 Valencia, Spain

3 Colloids Research Unit, Department of Physical Chemistry, Faculty of Pharmacy, University of Valencia, Av. Vicent Andrés Estellés s/n, 46100 Valencia, Spain; ofelia.vila@uv.es

4 Instituto Interuniversitario de Investigación de Reconocimiento Molecular y Desarrollo Tecnológico (IDM), Universitat Politècnica de València, Av. Vicent Andrés Estellés s/n, Burjassot, 46100 Valencia, Spain

* Correspondence: raquel.talens@uv.es (R.T.-V.); jesus.julian@uv.es (J.V.d.J.-O.)

Abstract: Alzheimer-type dementia (ATD) treatments face limitations in crossing the blood–brain barrier and systemic adverse effects. Intranasal administration offers a direct route to the brain via the nasal cavity's olfactory and trigeminal pathways. However, nasal physiology can hinder drug absorption and limit bioavailability. Therefore, the physicochemical characteristics of formulations must be optimized by means of technological strategies. Among the strategies that have been explored, lipid-based nanosystems, particularly nanostructured lipid carriers, are promising in preclinical investigations with minimal toxicity and therapeutic efficacy due to their ability to overcome challenges associated with other nanocarriers. We review the studies of nanostructured lipid carriers for intranasal administration in the treatment of ATD. Currently, no drugs for intranasal administration in ATD have marketing approval, with only three candidates, insulin, rivastigmine and APH-1105, being clinically investigated. Further studies with different candidates will eventually confirm the potential of the intranasal route of administration in the treatment of ATD.

Keywords: intranasal delivery; Alzheimer; blood–brain barrier; nanoparticles; intranasal devices; nanostructured lipid carriers

1. Introduction

The general objective of this work is to carry out a review on the intranasal administration of drugs in the treatment of Alzheimer's disease and related dementias. This review aims to offer a critical analysis of the most promising trends in nose–brain administered lipid nanoparticles that are likely to yield therapeutically useful outcomes, highlighting the most relevant and significant findings based on current knowledge in the field. To this end, technological strategies will be identified to overcome the limitations associated with this route of administration, as well as the most promising type of formulation for drug access to the brain from the nasal cavity. Finally, the current situation of the intranasal route of administration in the treatment of Alzheimer's disease and related dementias is evaluated.

1.1. The Pathophysiology of Alzheimer-Type Dementias

Alzheimer-type dementia (ATD) is a condition of the central nervous system (CNS) characterized by cognitive and behavioral deterioration that is usually associated with age, with a long and progressive course that ultimately leads to the death of the patient [1]. It is estimated that 60–80 percent of dementia cases are caused by ATD [2,3]. There were over 50 million ATD cases globally as of 2020 [4].

Dementia is more common in some parts of the world than others, with a prevalence of 4.7% in Central Europe and an 8.7% prevalence in North Africa and the Middle East [5,6]. For example, in Spain, the prevalence of dementia in the population over 65 years of age is between 4% and 9%, and is higher in the female sex [2], and the seventh most common cause of mortality in the USA is currently ATD [7]. Although up to 10% of cases have an early beginning and affect people in their 30s to mid-60s, adults over 65 years old account for the majority of cases [7,8]. Around 6% of adults 65 and older are affected, with women being more frequently affected than males [9]. Between 1990 and 2019, the incidence of Alzheimer's disease and other dementias increased by 147.95% worldwide [4]. This disease imposes a significant financial burden on society, with a global cost estimated to be 1 trillion euros per year [4].

The disease is named Alzheimer's disease in honor of Alois Alzheimer, a German psychiatrist and pathologist, who in 1906, first described the case of a 50-year-old woman with presenile dementia [10]. This disease is characterized by the progressive and irreversible damage and destruction of nerve cells, which leads to brain atrophy and death of neurons. The exact pathophysiological cause is not yet confirmed, although it is believed to be a multifactorial disorder. It occurs due to a series of physiological changes, including the formation of ß-amyloid plaques, tau hyperphosphorylation (a protein that is part of the cytoskeleton of neurons), mitochondrial dysfunction, neuronal excitotoxicity, cholinergic imbalance, neuroinflammation, oxidative stress and other factors. Various hypotheses have been put forward to explain the pathogenesis of ATD, including the cholinergic hypothesis, the tau hypothesis and the amyloid cascade hypothesis. In addition, among the main biomarkers of ATD are the deposition of the ß-amyloid peptide, which leads to the formation of ß-amyloid plaques, and the hyperphosphorylation of tau, which leads to the formation of neurofibrillary tangles [11]. Figure 1 shows the main differences between a healthy brain and the brain of a patient with ATD.

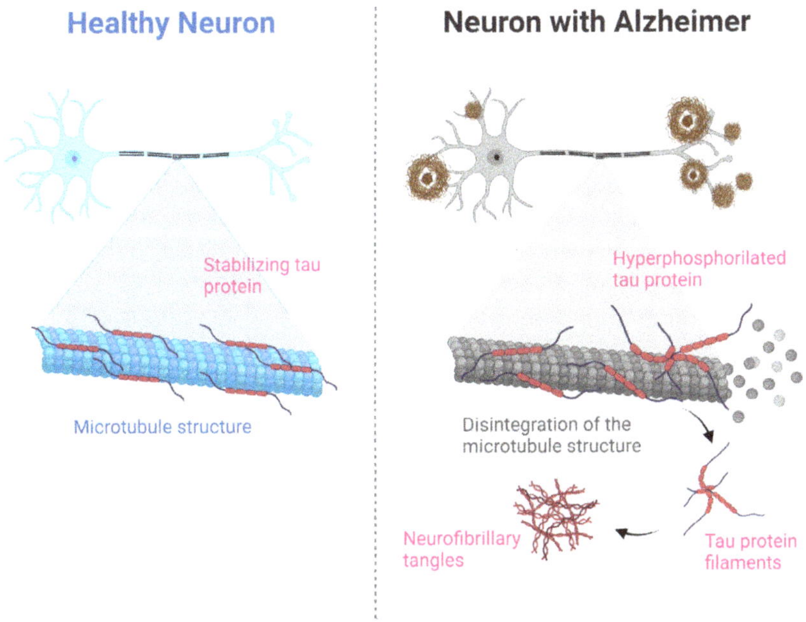

Figure 1. Main differences between a healthy brain and a brain with Alzheimer-type dementias.

1.2. Current Treatment of Alzheimer-Type Dementias

This complex aetiology poses a great challenge to finding a potential therapy that allows treating or correcting abnormal physiological conditions. Currently, the drugs approved by the FDA to treat the cognitive symptoms of ATD are based on the modulation of neurotransmitters or enzymes, among which are the following: acetylcholinesterase (AChE) inhibitors: donepezil [12,13], galantamine [14] and rivastigmine [15,16]; and the receptor antagonist N-methyl-d-aspartate also known as memantine [17]. However, none of these drugs can prevent the progression of the disease, but instead act by temporarily increasing cognitive functions through a partial improvement of cholinergic and glutamic neurotransmission [18].

In June 2021, the FDA approved the monoclonal antibody Aducanumab for the treatment of ATD [19]. However, the EMA denied marketing authorization for this drug because the results of the main studies were conflicting, and the drug was not shown to be sufficiently safe and effective [20].

As seen in Table 1, the available treatments for ATD use generally the peripheral route for the administration of drugs. The main drawback of peripheral administration is the limited accessibility of the drug to the brain from the blood due to the presence of a safety barrier that restricts the entry of molecules into the brain: the blood–brain barrier (BBB). At the same time, hepatic first-pass metabolism and enzymatic degradation (in the case of oral administration) and systemic clearance (in the oral and parenteral routes) also significantly reduce the bioavailability of the drug. In addition, systemic administration may present other limitations such as binding to plasma proteins, low volume of distribution, delay in drug delivery to the brain through the blood, and systemic adverse effects, especially at the gastrointestinal level (such as nausea and vomiting) [21].

Table 1. FDA-approved drugs for the treatment of Alzheimer-type dementias.

Active Ingredient	Route of Administration	Pharmaceutical Form	FDA Reference
Rivastigmine	Oral	Capsules Oral solution	[15,16]
	Transdermal	Transdermal patches	
Galantamine	Oral	Tablets Extended release capsules	[14]
Memantine	Oral	Tablets Oral solution Extended release capsules	[17]
Donepezil	Oral	Tablets Orodispersible tablets	[12,13,22]
	Transdermal	Transdermal patches	
Combination of memantine and donepezil	Oral	Controlled release capsules	[23]
Aducanumab	Parenteral	IV bolus	[19]

Therefore, there is great interest in identifying new alternative routes of administration within order to overcome the limitations of those currently available, increase the bioavailability of drugs at the site of action and obtain a better benefit/risk ratio in treatments.

In this context, nasal administration offers an alternative and novel approach for direct drug access to the brain.

1.3. Barriers That Prevent Drug Access to the Brain

There are two physiological barriers that the drug must overcome to reach the brain: the blood–brain barrier (BBB), which is an important limiting factor, and the blood–cerebrospinal fluid barrier, which separates the cerebrospinal fluid (CSF) systemic circulation and protects it from exposure to any exogenous substance that may be toxic.

The BBB is a complex structure made up of endothelial cells sealed by tight junctions, pericytes, basement membrane and astrocytes that create a continuous barrier [24]. The BBB separates brain tissue from the peripheral circulation allowing selective access of necessary nutrients and hormones to the CNS and, in turn, restricts the entry of foreign substances or neurotoxins, including drugs. In addition, it is responsible for maintaining an optimal ion composition for neuronal signaling [25].

Small lipophilic molecules and certain essential nutrients can pass through the blood–brain barrier (BBB) via passive diffusion (both paracellular and transcellular pathways) or through active transport mechanisms [26].

Molecules that can pass through the BBB via passive diffusion usually need to meet specific criteria, such as being lipid-soluble, having no electrical charge at physiological pH, having a molecular weight lower than 500 Da and presenting a partition coefficient between 0.5 and 6.0 [27]. In Figure 2, the structure of the BBB is shown.

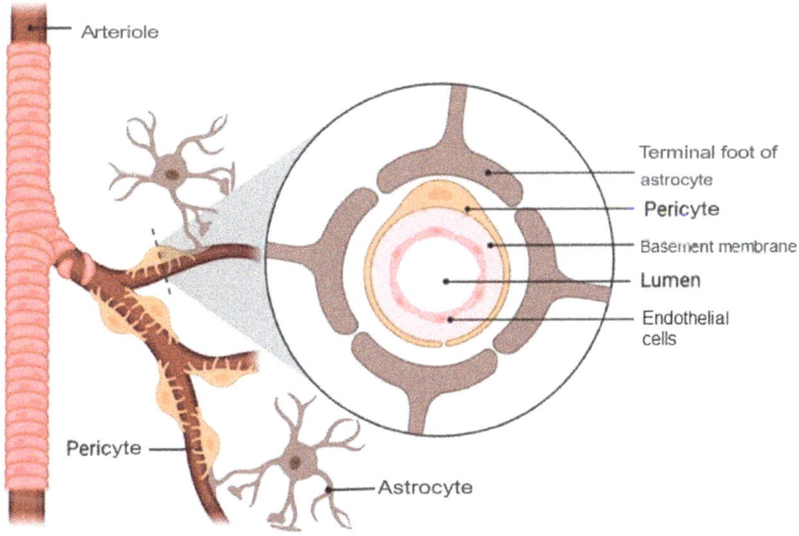

Figure 2. Structure of the blood–brain barrier.

1.4. Intranasal Route for Drug Administration

1.4.1. Nasal Cavity Anatomy

The main functions of the nasal cavity are breathing and smell, and, in addition, it exerts a protective action of filtering, warming and humidifying the inhaled air before it reaches the lower respiratory tract [28]. Likewise, it is composed of two symmetrical cavities that are divided by the nasal septum and lined with a layer of mucosa. These cavities in turn are divided into three regions:

- Vestibular region: This is the most anterior area, and contains nasal hairs to filter inhaled particles and the main cell type is squamous epithelial cells. In this region, drug absorption is very limited [29].
- Respiratory region: This is the area with the largest surface (about 130 cm^2) and the most vascularity, which makes it an ideal place for the systemic absorption of drugs. It encompasses the lateral walls of the nostrils, including the protruding nasal turbinates. There are four main cell types: goblet, ciliated, non-ciliated columnar and basal. In addition, the respiratory region is innervated by the maxillary and ophthalmic branches of the trigeminal nerve (V1, V2) [30].
- Olfactory region: This is located on the roof of the nasal cavity and covers only 10% of the total nasal area. There are different types of cells: basal cells, olfactory nerve cells,

support cells, cilia and trigeminal neurons. In addition, the olfactory region facilitates the transport of drugs to the brain having a direct connection with the brain through the olfactory and trigeminal neurons [31].

Figure 3 illustrates the anatomy of the nasal cavity and provides a schematic representation of the olfactory region.

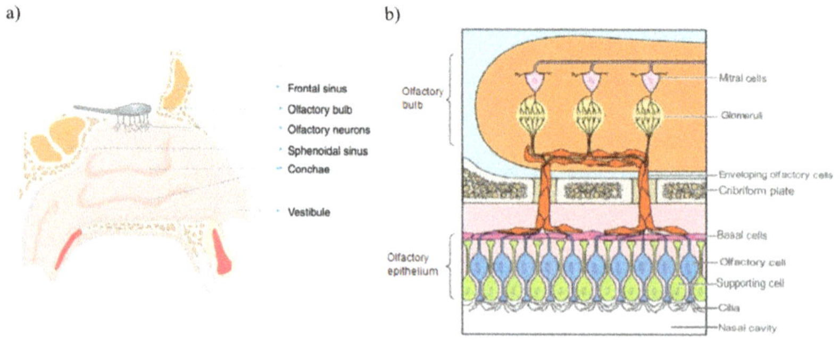

Figure 3. (**a**) Anatomy of the nasal cavity, adapted from [32]. (**b**) Schematic representation of the olfactory region, adapted from [33].

1.4.2. Intranasal Absorption

The first step in drug absorption from the nasal cavity is to cross the mucus layer. Mucin, which is the major mucus protein capable of binding solutes, hinders drug diffusion. Non-ionized, lipophilic and small particles easily pass through this layer of mucus. Thus, the molecular weight and lipophilicity of drugs can have a large impact on the rate and degree of nasal absorption of drugs. However, other physicochemical properties of the drug and the formulation must also be considered, as well as some physiological factors of the nose, which may also influence the bioavailability and transport of the drug from the nose to the brain [34].

Figure 4 summarizes the main factors that influence the nasal absorption of drugs administered by this route.

Once through the mucus, different mechanisms of absorption can be followed through the nasal mucosa: transcellular diffusion, paracellular transport, endocytic vesicle-mediated transport and carrier-mediated transport. Transcellular diffusion is the main mechanism of absorption and is the most suitable for lipophilic and low molecular weight molecules [26,35]. Figure 5 schematically represents the different mechanisms of absorption through the nasal mucosa.

1.4.3. Nose–Brain Absorption Pathways

(a) Direct Absorption Mechanism

- Olfactory Sensory Neurons

The olfactory nerve pathway is a direct and primary route of drug transport from the nasal cavity to the brain, preventing its passage through the BBB. When the drug reaches the olfactory receptors after passing through the nasal mucosa, there are two transport mechanisms to the brain: the intracellular route, in which the drug is taken up by the olfactory neurons through a process of endocytosis, and the extracellular route, in which the drug crosses the nasal epithelium through the tight junctions between the olfactory nerves and the supporting cells [36].

In both cases, the molecules travel through the cribriform plate, which separates the brain from the nasal cavity, to enter the olfactory bulb and the CSF. Subsequently, the drug can be distributed from the CSF to the brain by mixing with the interstitial fluid of the brain. The intracellular route requires from hours to days for the active fraction to reach the different regions of the brain and is adequate for the passage of lipophilic molecules, while the extracellular route only requires a few minutes for the active fraction to reach the brain and is suitable for the passage of hydrophilic molecules [37]. Figure 6 schematically represents the intranasal transport of drugs through the olfactory route to the CNS by the intracellular and extracellular route.

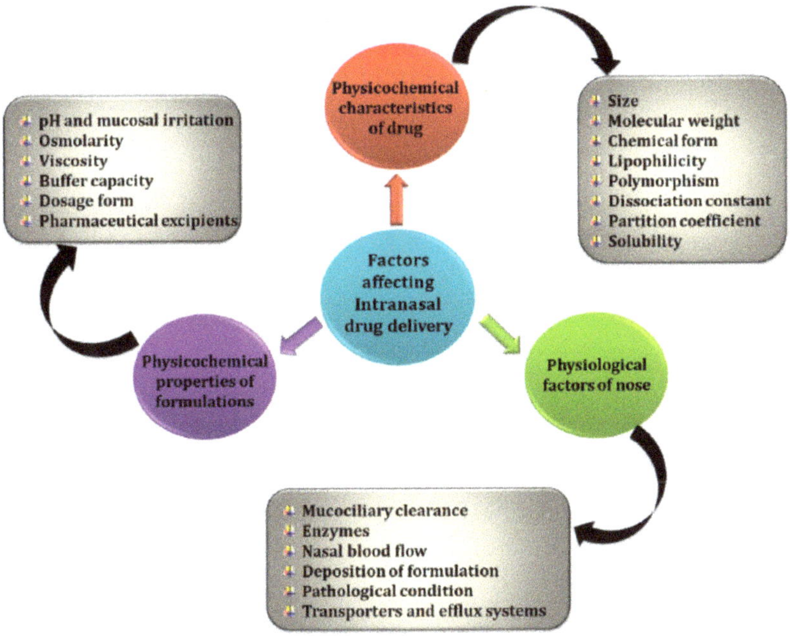

Figure 4. Factors affecting absorption through the nasal route [34].

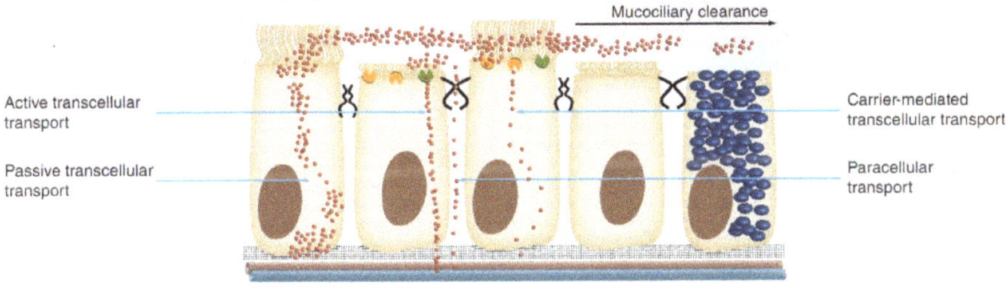

Figure 5. Mechanisms of drug absorption across the nasal epithelium [32].

- Trigeminal Neurons

Both the respiratory and olfactory epithelium are supplied with trigeminal nerves (Figure 7). Branches from the ophthalmic division (V1) of the trigeminal nerve innervate the dorsal and anterior part of the nasal mucosa, and those from the maxillary division (V2) innervate the turbinates.

Once the drug has diffused through the mucosa, it allows its transport directly to the brainstem at the level of the pons, avoiding its passage through the BBB. Transport through the trigeminal route can also occur intracellularly or extracellularly, as occurs with olfactory neurons [38].

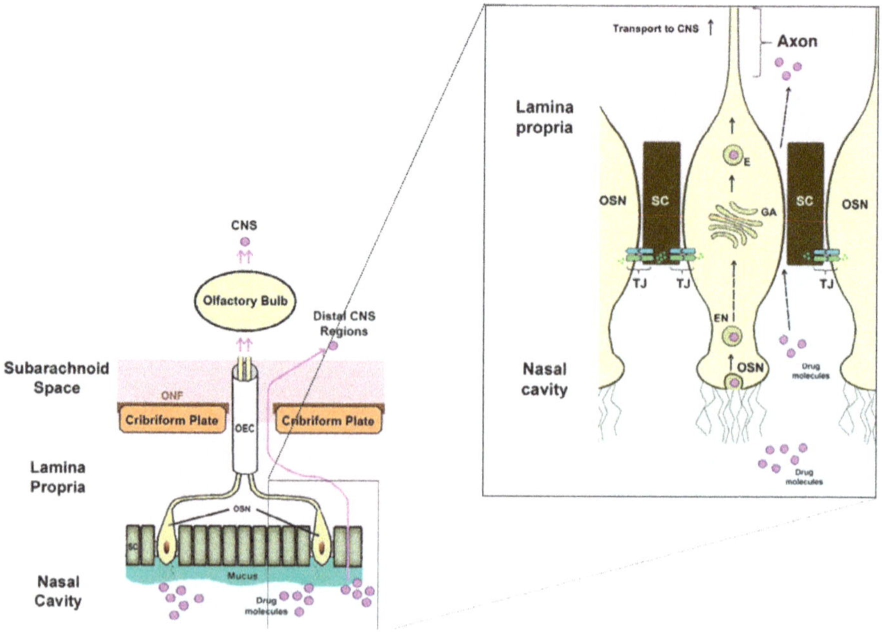

Figure 6. Intranasal transport of drugs through the olfactory route to the CNS by intracellular and extracellular routes [36].

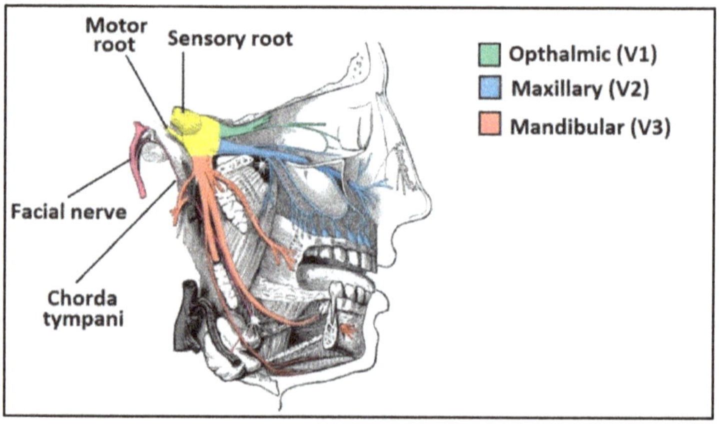

Figure 7. Innervation of the trigeminal nerve in the nasal cavity [39].

(b) Indirect Absorption Mechanism: The Systemic Route

When a drug is deposited in the nasal cavity and depending on its residence time, a part of it can be absorbed and enter the systemic circulation due to the rich vasculature of the respiratory epithelium and reach the brain through the BBB. This is a secondary route of drug transport to the brain after nasal administration [40].

Small lipophilic molecules gain access to the bloodstream and penetrate the BBB much more easily than high molecular weight and hydrophilic molecules. In addition, the active substances that use this route undergo renal and hepatic metabolism, so that the amount of drug that finally reaches the brain is very low and depends on the partition coefficient

(0.5–6.0) and the molecular weight of the drug (<500 Da). It also leads to a higher incidence of unwanted side effects in other peripheral organs [41].

Figure 8 schematically represents the structure of the nasal cavity and the different mechanisms of drug transport through the nasal mucosa to the brain.

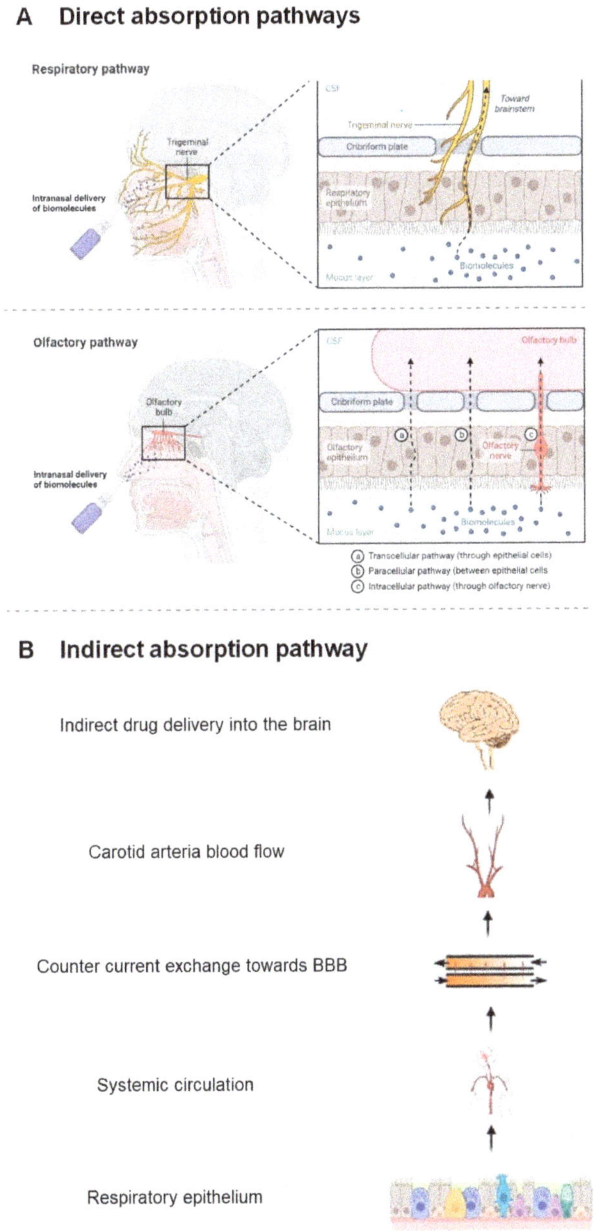

Figure 8. Schematic representation of the structure of the nasal cavity and the mechanisms of drug absorption through the nasal mucosa to the brain. (**A**) Direct absorption mechanism through olfactory nerve and trigeminal route. (**B**) Indirect absorption mechanism by systemic route [42].

1.4.4. Advantages and Disadvantages of the Nasal Route

To date, numerous techniques have been developed to circumvent the BBB and improve drug delivery to the brain, including invasive techniques, such as surgical exposure or intrathecal and intraventricular routes of administration.

The intranasal route offers a non-invasive strategy to bypass the BBB and provide direct drug access to the brain in a short period of time, as it has a direct connection to the brain through neural channels. In addition, the nasal cavity has a large number of blood vessels with rapid blood flow and has a large available surface area for absorption. For this reason, it offers other advantages such as rapid drug absorption, avoidance of hepatic metabolism and easy administration, which increases patient compliance [43].

However, there are several obstacles that can limit absorption of drug through this pathway, such as poor drug permeability from the nasal mucosa, mucociliary clearance, enzymatic degradation of the drug, low drug residence time and possible toxicity that the formulation or drug can produce on the nasal mucosa. In addition, due to the limited volume of application through this route, the dose of the drug may be limited depending on its solubility in the vehicle, so it will only be applicable to potent drugs that allow obtaining sufficient pharmacological effect at low doses or by using dry powders. Likewise, it would be convenient to reserve this route for those active principles with a wide therapeutic margin due to the possible inaccuracy of the dosage after administration. All these characteristics make it difficult to obtain bioavailability into the brain via the intranasal route [29]. For this reason, numerous technological strategies have been investigated to overcome these drawbacks and improve the bioavailability of drugs administered intranasally, which will be detailed later.

Table 2 indicates the main advantages and disadvantages associated with the administration of drugs through the nose.

Table 2. Advantages and drawbacks of nasal drug administration.

Advantages	Disadvantages
• Non-invasive, safe and painless • Lower risk of infections • Easy administration and better patient compliance • Alternative to parenteral route of administration, especially for protein and peptide drugs • Large absorption area and extensive network of blood and lymphatic vessels • Rapid absorption and excellent bioavailability for low molecular weight drugs • Prevents first-pass hepatic metabolism and drug degradation in the gastrointestinal tract • Possible direct delivery to the brain bypassing the BBB via the olfactory and trigeminal nerves	• Only applicable to potent drugs with a wide therapeutic margin • Small volumes of administration (25–200 μL) • Mucociliary clearance and low residence time • Enzymatic degradation • Low permeability of hydrophilic drugs • Need for absorption promoters that can cause toxicity in the nasal mucosa • Epithelial mucous pH 5–6.5 • Interindividual variability • Runny nose influences absorption • Possible inaccuracy in dosage due to improper administration technique • Some drugs and excipients can cause irritation to the nasal mucosa

2. Materials and Methods

To carry out this work, a bibliographical review of scientific articles on the access of drugs to the brain through intranasal administration in the treatment of ATD was carried out, as well as strategies to improve absorption and available nanotherapeutic strategies. In addition, the website of the Spanish Agency for Medicines and Health Products (AEMPS) and the website of the European Medicines Agency (EMA) and the Food and Drug Administration (FDA) were consulted for those drugs that have been approved for the treatment of Alzheimer-type dementias. Likewise, the Clinical trials.gov website was consulted to search for existing clinical trials of intranasal administration in ATD.

Search keywords used were: "Alzheimer's disease" AND ("intranasal drug delivery" OR "nose-to-brain delivery" OR "intranasal devices") AND ("blood–brain barrier" OR "brain targeting" OR "nanoparticles" OR "lipid nanoparticles" OR "NLC" OR "intranasal devices" OR "nano-lipidic formulation").

Finally, the Biorender® website was used to create some of the figures used in the work (https://biorender.com, accessed from July 2022 to April 2023).

3. Results and Discussion

In this section, possible strategies for overcoming the limitation of nasal administration are shown. Moreover, the importance of new drug nanocarriers are discussed, indicating some of the more relevant actives that are likely to be clinically used for the intranasal treatment of Alzheimer-type-dementia. To do this, the aforementioned search keywords were used.

3.1. *Possible Strategies for Overcoming the Limitations of Nasal Administration*

The following are possible strategies that can be applied to overcome the obstacles that the nasal route presents to the delivery of drugs to the brain.

3.1.1. Systems That Reduce Mucociliary Clearance

Mucociliary clearance acts as a defense mechanism of the respiratory tract that allows the effective and rapid elimination of inhaled hazardous substances through the combined effect produced by the mucus and the cilia of the columnar cells, which trap and drag the particles to the gastrointestinal system to be swallowed. When a formulation is administered via the nose, it has to go through mucociliary clearance, which will eliminate a part of the drug and which, together with the short residence time of the drug in the nasal cavity, will limit its absorption through the mucosa [44].

To avoid rapid clearance and improve drug residence time in the nasal cavity, bioadhesive systems formulated with mucoadhesive polymers, viscosifiers, nasal gels that are generated in situ in response to certain stimuli (e.g., temperature, ionic strength, pH), hydrogels, etc., which favor nasal absorption of drugs as a result of increased residence time in the area of administration, can be used [45]. Another strategy is to provide the particles with a positive surface charge, for example, with stearylamine, to increase mucoadhesiveness in the nasal cavity by interacting with the negatively charged sialic acid of the mucous layer. Table 3 lists several examples of mucoadhesion-enhancing excipients that have been explored for the development of intranasal drug delivery systems. Among them, the most widely used have been chitosan and cyclodextrins, due to their low toxicity and absorption-promoting effect.

In addition, the use of adequate administration devices that allow deep access of the drug to the olfactory region, associated with head tilt, can minimize mucociliary clearance, favoring greater circulation of the drug to the brain. Several delivery devices have been developed (listed in Table 3), of which only the Optinose®, Precision Olfactory Delivery® and ViaNase® devices have been used in studies of intranasal administration to the brain in humans [46]. The ViaNase® device was used in the clinical trial of intranasal insulin delivery for the treatment of ATD [47]. Likewise, although some of these devices, such as the Optinose®, have been approved by the FDA [48] for the treatment of migraine, none of them have been authorized for nasal administration in ATD.

3.1.2. Systems That Reduce Enzymatic Degradation

The epithelial barrier and the lumen of the nasal cavity contain some metabolic enzymes such as cytochrome P450 (CYP450) isoforms, transferases, carboxyesterases and peptidases, responsible for the degradation of various drugs administered through the nasal cavity, especially proteins and peptides [49].

Table 3. Strategies to improve nasal absorption.

Strategies to Improve Nasal Absorption					
Promoters of Absorption	**Promoters of Absorption**	**Promoters of Absorption**	**Promoters of Absorption**	**Promoters of Absorption**	**Promoters of Absorption**
Sodium lauryl sulfate Poloxamer Tween® Span® Sodium glucodeoxycholate Sodium taurodeoxycholate Taurodihydrofusidate Oleic acids Lauric acid Caprylic acid Phosphatidylcholine EDTA Citric acid Sodium salicytate Peppermint oil Cyclodextrins Chitosan Carbopol Starch Aminated gelatin	Comostat amylase Bestatin Aprotinin Amastatin Boroleucine Bacitracin Puromycin EDTA	Carboxymethylcellulose Microcrystalline cellulose Hydroxypropylcellulose Carbopol 971P Carbopol 934P Carbopol 981P Eudragrit Chitosan Cyclodextrins Stearylamine	MEMOGAIN® (Gln-106	Optinase nasal device ViaNase-electronic atomizer Precision-Olfactory device Naltos device	**Polymeric nanoparticles** Polymeric nanogels Core/shell nanoparticles Polymeric micelles Dendrimers **Inorganic nanoparticles** Silica nanoparticles Carbon nanotubes Magnetic nanoparticles Gold nanoparticles **Lipid nanoparticles** Nanoemulsions Microemulsions Liposomes Solid lipid nanoparticles Nanostructured lipid carriers

These enzymes affect the bioavailability and integrity of the drug when metabolizing it, which could be overcome through the use of enzyme inhibitors, prodrugs and new drug carriers, which will be specified later. In Table 3, numerous examples of each of these strategies are presented.

Among the prodrugs that have been studied, Gln-1062 (MEMOGAIN®), a lipophilic prodrug of galantamine, stands out. A randomized double-blind study was conducted in subjects aged ≥65 years in which some cognitive improvement was observed, minimizing complications associated with oral administration of galantamine, such as syncope and bradycardia. However, some adverse effects related to the gastrointestinal tract and nasal mucosal irritation, both dose-dependent, were reported, which were the cause of withdrawal in some subjects [50]. Moreover, it has not been approved by the FDA, nor are there any current studies related to this prodrug.

3.1.3. Systems That Reduce Irritation of the Nasal Mucosa

The irritation of the nasal mucosa is a critical parameter that must be taken into account during the development of a formulation. When deposited in the nasal cavity, the formulation comes into contact with the intact mucosa. Most drugs and excipients can exert some toxic effect, irritation or damage to the mucosa, which can range from mild sneezing to skin rash, itching and even severe or systemic toxicity. Toxicity is often exacerbated by the use of polymers, mucoadhesive agents or any other novel approach that can dramatically increase the contact time of the mucosal layer with the drug to increase its absorption across the nasal epithelium [51].

To minimize the incidence of nasal toxicity, the use of drugs, polymers or surfactants physiologically compatible and non-irritating to the nasal mucosa is recommended. In addition, the pH, isotonicity and viscosity of the formulation must be taken into account during drug development to prevent irritation of the mucosa.

Apart from this, the mode of administration of the drug, the size of the droplets or the particle size if using dry powder also has a significant impact on toxicity. Administration by nasal drops achieves maximum exposure to the respiratory region, directing the drug to the systemic circulation and to the pulmonary organs, while the administration of the formulation with the head inclined upwards with suitable devices ensures maximum focus on the olfactory region, facilitating the absorption of the drug directly by the brain [52]. However, maintaining the position of the head inclined upwards is complicated and uncomfortable for the patient, which makes it necessary to develop administration devices that facilitate drug deposition in the olfactory region, such as those mentioned in Section 3.1.1.

3.1.4. Systems That Promote Nasal Permeability

The nasal cavity is surrounded by a thin layer of mucus of a lipophilic nature. The presence of tight junctions in the olfactory and respiratory epithelium, together with the protective mucosal lining, act as a selective filter that delays the penetration and diffusion of drugs through the nasal mucosa. In addition, transcellular diffusion is the main transport mechanism from the nose to the brain, which mainly allows the passage of lipophilic and low molecular weight substances [40]. For this reason, the nasal route is more suitable for smaller lipophilic molecules, whereas large, polar and hydrophilic drugs, such as proteins and peptides, experience poor nasal permeability. The nasal absorption of polar drugs can be improved though the use of suitable absorption promoters and also by new drug carriers, which will be specified later. In Table 3, several examples of these strategies are shown.

3.2. New Drug Carriers as a Potential Strategy

Among all the strategies to improve nasal absorption that have been discussed, new nanotechnology-based drug carriers have shown the ability to overcome the resistance generated by mucociliary clearance, secretion systems and epithelial tight junctions. In addition, they allow drugs to be protected from chemical or metabolic degradation and improve solubility or transport through biological membranes, leading to better orientation and fewer systemic adverse effects [53]. For this reason, in recent years they have gained the greatest interest in research. Likewise, it is necessary to consider the appropriate selection of excipients, the physicochemical properties (particle size, size distribution, zeta potential and drug charge), biodegradability, biocompatibility, the release profile and also the safety profile [54].

Various nanotherapeutic strategies have been explored for intranasal treatment of ATD, including polymeric and inorganic nanoparticles, and lipid-based nanocarriers. Lipid-based nanocarriers are promising candidates for drug delivery from the nose to the brain in the treatment of ATD. They offer greater brain-targeting potential, rapid brain uptake, biodegradability and bioaccessibility, and overcome challenges associated with polymeric nanoparticles, such as cytotoxicity. Lipid-based nanocarriers can incorporate both hydrophilic and lipophilic drugs, provide protection and offer controlled drug release by immobilizing the drug in the solid lipid matrix, which biodegrades easily [55].

There are several types of lipid-based nanocarriers, such as liposomes [56], niosomes [57,58], microemulsions [59], nanoemulsions [60] and solid-matrix lipid nanoparticles [61] as shown in Figure 9. The latter have generated greater interest in recent years due to the advantages they present and will be detailed in the following section.

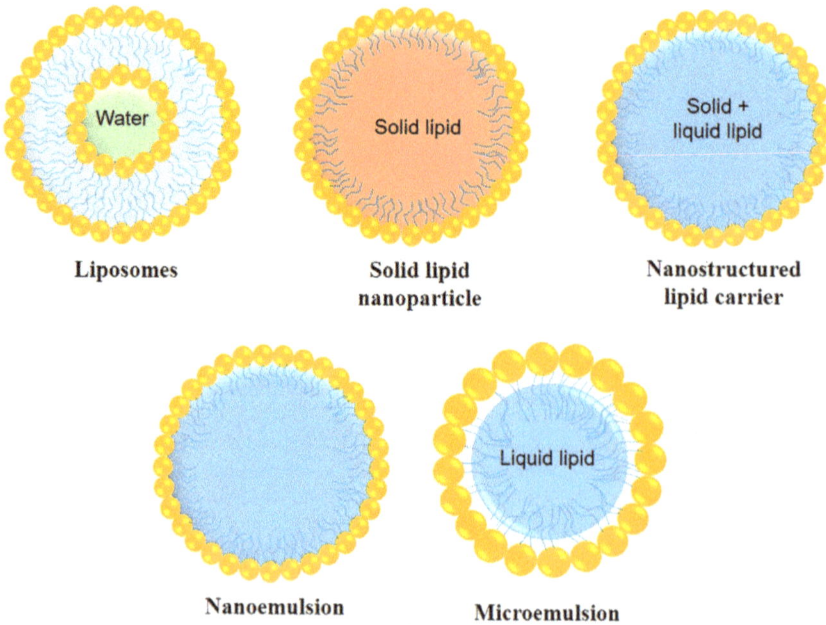

Figure 9. Lipid-based nanocarriers that have been explored for drug access to the brain from the nasal route [62].

3.2.1. Nanostructured Lipid Carriers for Intranasal Delivery

Solid matrix lipid nanoparticles are aqueous dispersions of solid particles that may be composed of physiological lipids and stabilized by one or several emulsifiers, with an average particle size that can range between 100 and 300 nm. There are two types, solid lipid nanoparticles (SLNs) and nanostructured lipid carriers (NLCs), as seen in Figure 9.

SLNs present a lipid matrix composed of a single solid lipid that has a highly organized structure and are considered the first generation of lipid nanoparticles, while NLCs present a disorganized internal lipid matrix, formed by a mixture of solid and liquid lipids and are known as second-generation lipid nanoparticles [63].

In the case of NLCs, the disruption of the lipid matrix caused by the liquid lipid allows for higher encapsulation efficiency (EE), since most hydrophobic molecules show better solubility in liquid than in solid lipids, and a low expulsion of the encapsulated drug during storage, i.e., a higher stability. For this reason, current research focuses on NLCs [64,65]. Figure 10 shows the main differences between SLNs and NLCs.

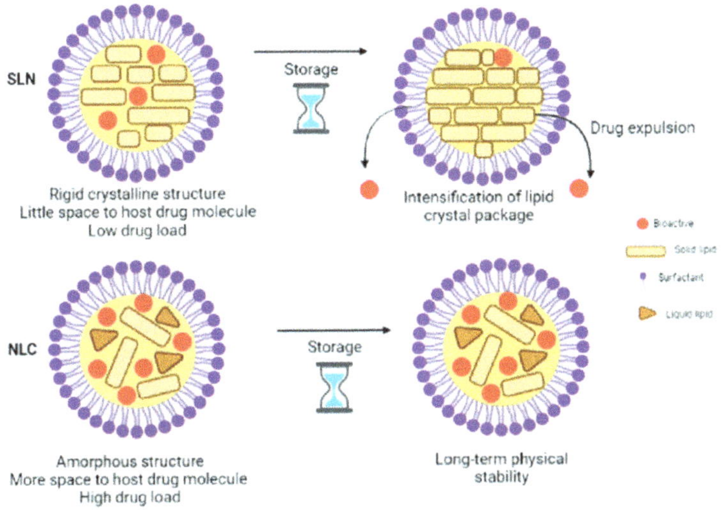

Figure 10. Main differences between solid lipid nanoparticles (SLN) and nanostructured lipid carriers (NLCs) [66].

Thus, NLCs have important advantages over the first generation of lipid nanoparticles. Likewise, NLCs offer advantages over other nanosystems for nasal administration, since they can be manufactured from biocompatible and biodegradable components, such as physiological lipids and other GRAS (Generally Recognized As Safe) excipients [67].

In addition, they protect drugs against enzymatic degradation, increasing the residence time in the nasal cavity and improving bioavailability. Similarly, it is possible to produce NLCs with the desired features for nasal administration to the brain, namely a mean particle size $\leq$ 200 nm, a polydispersity index (PDI) of 0.3 and a zeta potential (ZP) of 30 mV [68].

However, the low viscosity of NLCs may decrease the bioavailability of the drug in the brain as the residence time reduced. To overcome this drawback, the inclusion of NLC formulations in mucoadhesive materials has been described as an appropriate strategy [69].

Moreover, during storage, NLCs can aggregate, reducing their potential as controlled-release drug carriers, which can be avoided by modifying the stability of formulations through storage temperature and pH. When it comes to safety, it is important to consider that the type and amount of surfactant used in formulations can sometimes lead to sensitivity, irritation or cytotoxicity. For this reason, it is essential to develop detailed nanotoxicological studies to identify the specific elements that guarantee its safety, such as the biological compatibility of the formulation components, the effect of particle size and surface charge [42].

Different experiments have been used to evaluate pharmacokinetic parameters after the intranasal administration of NLCs in Alzheimer-type dementia compared with free drug solution or other administration routes [70]. In this sense, preclinical studies have been carried out with this type of formulation using different bioactive compounds, which have obtained promising results. Some of these studies are reflected in Table 4 and are detailed below.

Rivastigmine

Rivastigmine (RV) is an inhibitor of the AChE enzyme that has been shown to be useful in the treatment of ATD, which is marketed as oral capsules, oral solution and transdermal patches, as shown in Table 1. However, due to its hydrophilic nature, it exhibits poor permeability through the BBB and has a short half-life (90 min) and extensive first-pass metabolism. It has, therefore, a low bioavailability and requires frequent administration, causing the side effects of accumulation at a systemic level [71]. For this reason, researchers have tried to explore alternative administration routes that would make it possible to overcome these limitations of the formulations marketed with this drug.

In 2017, Wavikar et al. developed an in situ gelling system of rivastigmine NLCs for nasal administration using the ethanol injection method [72]. The resulting formulation showed a double improvement in drug permeability due to the mucoadhesive behavior of the gel in the upper nasal cavity. Subacute and nasal toxicity studies showed that the formulation was safe for nasal administration to the brain. Additionally, administration of the gel with RV-loaded NLCs improved pharmacokinetic characteristics and biodistribution in the mouse brain, improving cognition and memory compared to intravenous formulations and simple intranasal solutions.

In 2019, Anand et al. assessed the efficacy of rivastigmine hydrogen tartrate (RHT)-loaded lipid-based nanocarriers (NLCs) via the nasal route for treating Alzheimer's disease-associated dementia [73]. The formulation showed favorable physicochemical properties and ex vivo diffusion studies revealed sustained release of RHT from the NLCs. In vivo studies in male and female Wistar rats demonstrated improved neuronal function, accompanied by significant enhancements in memory, learning and cognitive response, indicating the therapeutic potential for ATD treatment.

Pioglitazone

Various studies have shown that impaired insulin signaling in the brain is one of the main factors behind the development of ATD. For this reason, some of the antidiabetic molecules have been studied for the treatment of ATD, including pioglitazone (PIO) [74].

In preclinical studies, pioglitazone was found to be effective in significantly reducing oxidative stress, as well as improving cognition and glucose use in the brain. However, subsequent clinical trials conducted with regular doses of oral PIO showed limited success due to an inadequate drug concentration reaching the brain, as it barely crosses the BBB. To achieve a therapeutic concentration at the site of action, a higher dose of the drug would be required; however, higher doses in human volunteers are restricted due to the serious dose-dependent peripheral side effects of PIO, such as bladder cancer and peripheral edema. Therefore, to improve the bioavailability of the drug in the brain, Jojo et al. developed a formulation with NLCs loaded with PIO for nasal administration [75].

The formulation obtained had good physicochemical and mucoadhesive characteristics, which significantly increased the permeability of the drug compared to its administration in an intranasal solution. Likewise, little or no toxicity was observed in the nasal epithelium.

On the other hand, the biodistribution in the brain as well as the brain/plasma ratio obtained after intranasal administration of NLC dispersion was higher compared to intravenous or intranasal PIO solution in male Wistar rats. This would contribute to achieving therapeutic concentration at the site of action, without the need for a higher dose of drug or increased systemic adverse effects, thus becoming a potential therapeutic alternative for the treatment of ATD.

Resveratrol

Resveratrol is a natural compound that has antiproliferative, anti-inflammatory and antioxidant properties and exerts neuroprotective effects due to the elimination of the ß-amyloid peptide and the breakdown of the amyloid precursor protein, so it could be used for prevention and treatment of ATD, alone or combined with other active ingredients. However, it exhibits poor aqueous solubility and chemical stability, along with high first-pass metabolism, which contributes to poor bioavailability after oral administration [76].

For this reason, Rajput et al. formulated an in situ gel to incorporate resveratrol-loaded NLCs for administration from the nose to the brain [77]. The formulation had good characteristics and nasal ciliotoxicity studies demonstrated the safety of the formulation.

Pharmacokinetic studies in male Sprague Dawley rats revealed faster absorption and greater bioavailability of the drug in the brain when administered intranasally in the form of a gel loaded with NLCs compared to an oral suspension of the same active ingredient. All these would demonstrate the potential brain selectivity of the intranasal formulation.

Berberine

Recent research has shown that the isoquinoline alkaloid berberine (BER) may be effective in the treatment of ATD as it inhibits the formation of ß-amyloid plaques, decreases tau hyperphosphorylation and reduces oxidative stress [78]. To precisely target BER to the brain in order to maximize therapeutic effects and minimize potential adverse effects, Abo El-Enin et al. developed a formulation of BER-loaded chitosan-coated NLCs for intranasal administration [79]. Chitosan (CTS) is a mucoadhesive polymer that also acts as an absorption promoter, thereby enhancing drug penetration through the nasal mucosa and reducing mucociliary clearance.

The BER-CTS-NLCs obtained presented good physicochemical characteristics (PS < 200 nm and ZP > +30; see Table 4), and, in addition, showed prolonged release behavior and enhanced drug permeability through the nasal mucosa of sheep. Furthermore, histopathological evaluation indicated that the BER-CTS-NLCs system was safe for nasal administration.

Pharmacokinetic and brain/plasma ratio studies in male Wistar rats showed that those treated intranasally with BER-CTS-NLCs had significantly higher drug levels in the brain compared to intranasal BER solution. In Table 4, the most important aspects of the aforementioned preclinical studies are summarized.

Astaxanthin

Astaxanthin (AST) is a second-generation antioxidant with anti-inflammatory and neuroprotective properties and could be a promising candidate for Alzheimer's disease (AD) therapy, but is shows poor oral bioavailability due to its high lipophilicity [80].

Shehata et al. (2023) [81] prepare and evaluate AST-loaded nanostructured lipid carriers (NLCs) for enhanced nose-to-brain drug delivery to improve the therapeutic efficacy in a rat model of AD. The Korsmeyer–Peppas model was the best fitted for AST release from NLCs and the release mechanism involves diffusion and matrix erosion.

Intranasal treatment of AD-like male albino rats with the optimized AST-NLCs improved the nasal permeability and nose-to-brain drug delivery. In fact, the rats treated with AST-NLC showed significant higher percentage of time spent in brain than AST solution. The neuroprotective role of AST against amyloid beta can be due to its potent antioxidant effects and reduction of ROS mitochondrial. In summary, AST-NLC provided significantly decreased oxidative stress, amyloidogenic pathway, neuroinflammation and apoptosis, and significantly improved the cholinergic neurotransmission compared to AST solution.

Donepezil

Donepezil is approved for treatment of dementia of the Alzheimer-type and is currently available in tablet form and transdermal patch in the United States [19,22,23].

Oral administration of this drug presents many drawbacks, resulting in treatment non-adherence among patients. Thus, the development of transdermal formulations for donepezil delivery is important.

Corplex™ donepezil transdermal delivery system (TDS), approved as Adlarity® for the treatment of mild, moderate and severe dementia of the Alzheimer-type, is designed to deliver donepezil through the skin over a 7-day wear period and is available in 5 and 10 mg/d doses.

Butani et al. (2018) [82] developed donepezil-loaded NLCs and incorporated the optimized NLCs into the ionic-triggered gellan gum matrix for IN delivery. Pharmacokinetic study of donepezil in brain and plasma were studied (dose 1 mg/kg) in male Sprague–Dawley rats. The $AUC_{0–8h,brain}$ of the NLC gel (IN) was 1.26-fold higher than that of the tablet (oral). This slight increase indicated that the NLC gel was not highly potential for brain targeting. However, this ratio is 1.5-fold lower in plasma, thus reducing toxicity to other organs. No PK data were provided for IV or IN administrations of the free drug. In a rat model of scopolamine-induced amnesia, the NLC gel (IN) improved cognitive function as compared with the marketed tablet (oral).

Table 4. Preclinical studies with nanostructured lipid carriers for intranasal administration in the treatment of Alzheimer-type dementias.

Preclinical Studies with Nanostructured Lipid Carriers					
Active Ingredient	**Preparation Method**	**Physicochemical Properties**	**In Vitro Results**	**In Vivo Results**	**Reference Year**
Rivastigmine	Ethanol injection method	**PS**: 123.2 nm **ZP**: 32 mV **EE**: 68.3%	Non-hemolytic, safe for intranasal administration	AUC: 506,731.3 ng.min/mL Cmax: 1984.23 ng/mL Tmax: 30 min $t_{1/2}$: 347.65 min Increase up to 4.6–5.3 brain concentration Increased bioavailability Improves memory and cognition No or very little nasal toxicity	[72] 2017
Rivastigmine hydrogen tartrate	Emulsification/ solvent evaporation	**PS**: 266 nm **PDI**: 0.23 **ZP**: −16.58 mV **EE**: 61.8%	Controlled and sustained release of 80% of the drug for 12 h Controlled diffusion, initially through surface erosion	Significant reduction in AChE1 and AchE2 activity, which enhances neuronal function Significant improvement of memory, learning and cognitive response	[73] 2019
Pioglitazone	Microemulsification	**PS**: 211.4 nm **PDI**: 0.25 **ZP**: 14.9 mV **EE**: 70.18%	Sustained release for 24 h Fickian Diffusion Higuchi kinetics Increased permeability (1789 μg/cm^2) No cytotoxic effect	Cmax: 1.62 μg/mL Brain/plasma ratio: 1.6 Increased bioavailability Nose–brain direct delivery	[75] 2019
Resveratrol	Ultrasonic emulsification	**PS**: 132 nm **PDI**: 0.165 **ZP**: −23 mV **EE**: 74%	Drug loading of 10 ± 3% and a mucoadhesive force of the gel of 4087 ± 115 dynes/cm^2 Long term stability at 2–8 °C	Cmax 2.5-fold higher in brain than NLC-based in situ gel Higher concentration in the brain with the gel in situ No nasal toxicity Faster absorption	[77] 2019
Berberine	Hot homogenization and ultrasonics	**PS**: 180.9 nm **ZP**: 36.8 mV	Sustained release properties and increased ex vivo permeability through nasal mucosa	The Cbrain/Cplasma ratio at 30 min was much higher in the intranasal NLC dispersion formulation (4.56) than in the IN plain solution (2.14) or IV solution (0.26)	[79] 2022

Table 4. *Cont.*

Preclinical Studies with Nanostructured Lipid Carriers					
Active Ingredient	**Preparation Method**	**Physicochemical Properties**	**In Vitro Results**	**In Vivo Results**	**Reference Year**
Astaxanthin	Hot high-pressure homogenization	PS: 142.8 nm PDI: 0.247 ZP: −32.2 mV EE: 94.1%	Initial burst release followed by sustained release for 24 h AST-NLCs were stable at 4–8 °C for six months	Significantly improved the cholinergic neurotransmission compared to AST solution	[81] 2023
Donepezil	Melt emulsification-probe sonication method	PS: 112 nm PDI: 0.114 ZP: −35 mV EE: 79.25%	Drug loading of 7.9 ± 2.1% and a mucoadhesive force of the gel of 3189 ± 84.26 dynes/cm^2	Higher drug distribution in brain (20% increase NLC vs marketed tablets) and lower drug concentration in plasma (25% reduction NLC vs tablet) NLC administration increased residence time and penetration through olfactory lobe Improved cognitive function (IN vs. oral tablet)	[82] 2018

PS: particle size; ZP: zeta potential; EE: encapsulation efficiency; PDI: polydispersity index.

These preclinical studies show that formulations with nanostructured lipid carriers have great potential to improve the bioavailability in the brain of drugs used to treat ATD through intranasal administration. Thus, all the NLCs studied present a similar line of action, since:

- They promote greater cerebral uptake than simple drug solutions, both intravenously and intranasally, leading to increased drug concentrations in the brain.
- They are more efficiently absorbed after intranasal administration and their small diameter facilitates their passage through the neuronal pathway.
- They provide a controlled and sustained release profile of the drug, unlike conventional drug delivery systems.
- They can be coated with polymers, such as chitosan, or dispersed in hydrogels, giving the formulation excellent mucoadhesive properties, prolonging the residence time in the nasal cavity and improving permeability through the nasal epithelium.
- They offer better drug residence and formulation stability during storage and greater physiological compatibility than other formulations.

3.3. Current Situation of Intranasal Administration in the Treatment of ATD

Preclinical studies with NLCs have shown promising results for treating ATD with low toxicity. However, there are currently no ongoing clinical trials with intranasal NLCs. One reason for this could be limitations in manufacturing methods, which require controllable, reproducible and large-scale preparation technologies. Another factor is the differences between animal and human models, which may not ensure reproducibility in humans. Finally, the lack of widely accepted international standards for regulating nanomedicines also poses a challenge [83].

Therefore, the intranasal administration of nanostructured lipid carriers for the treatment of ATD can be considered, today, as a promising alternative to currently marketed drugs, although it still requires a lot of scientific effort to put it into clinical practice.

Figure 11 summarizes the current situation in the investigation of therapies for ATD by the nasal route. As previously discussed, to date no drug has obtained marketing approval for intranasal administration in the treatment of ATD.

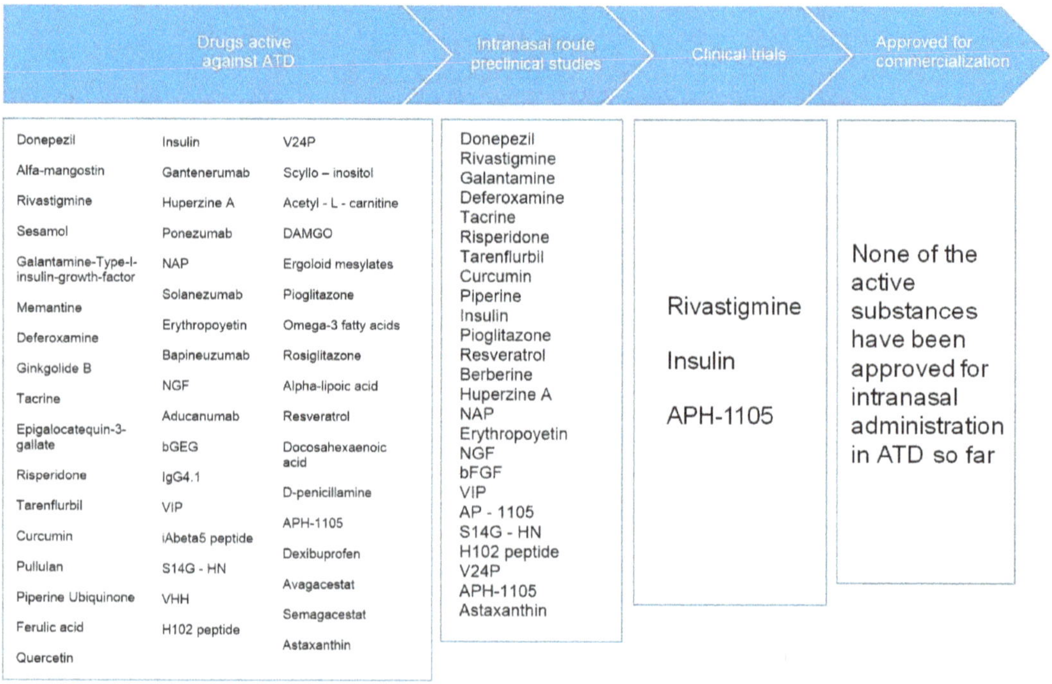

Figure 11. Current status of the investigation of nasal therapies for Alzheimer-type dementia.

As can be seen, most of the trials with bioactive compounds that can be administered through the nose to treat ATD stop at their preclinical phase, and only three candidates have managed to reach the clinical investigation phase, namely insulin [84], rivastigmine [85] and APH-1105 [86].

Although no formulation for intranasal administration in ATD has yet been marketed, there is great interest in the intranasal route as an alternative administration route for the treatment of Alzheimer-type dementias, due to the advantages it entails and the very promising results that have been obtained in preclinical investigations. In the year 2016, a bioavailability and safety study was conducted by administering a unit dose of rivastigmine intravenous solution (1 mg) and nasal spray (3126 mg) to eight healthy individuals aged 58–75 years. In this study, rivastigmine nasal spray provided an effective dose for the treatment of ATD-associated dementia [87], as shown in Table 5.

Likewise, several recent studies have directly associated insulin resistance with ATD pathology, since it has been correlated with poorer cognition and higher concentrations of hyperphosphorylated tau protein. This is due to the fact that insulin influences the production and elimination of the Aβ peptide and also prevents tau hyperphosphorylation, among other effects relevant to ATD [88]. For this reason, in recent decades, clinical trials have been carried out with intranasal insulin for the treatment of ATD (Table 5).

Only one clinical trial in phase 2 with intranasal insulin is currently being carried out "SNIFF–Combo INI+EMPA" trial that began in October 2021, with an estimated completion date of October 2026 [89]. The study consists of a randomized, double-blind trial with 60 participants, in which the effects of treatment will be compared for 4 weeks with intranasal insulin (40 IU, four times daily), empagliflozin (10 mg daily) and the combination of intranasal insulin (INI) and empagliflozin (EMPA), compared with placebo, on cerebrospinal fluid (CSF) biomarkers and cognition. On the other hand, there is another clinical trial in phase 2, "SNIFF–3-Week Aptar CPS Device" trial, whose start date is expected in May 2025 and its estimated completion date is May 2029 [90]. This study will examine whether the Aptar Pharma CPS intranasal delivery device can be used in adults with preclinical ATD (i.e., cognitively normal, but with abnormal brain Aβ-peptide levels) to reliably deliver insulin or placebo four times daily for a four-week period. In addition, the effects of treatment on cognition, CSF biomarkers and cerebral perfusion will be examined. If successful, the information obtained from the study will be useful in designing future phase 3 trials of intranasal insulin, but with abnormal brain levels of Aβ peptide, to reliably administer insulin or placebo four times a day for a period of four weeks.

In addition, a clinical trial is expected to start in 2023 to assess the safety, tolerability and efficacy of intranasal administration of APH-1105 for the treatment of mild to moderate ATD [74], as shown in Table 5. The APH-1105 is a modulator of the α-secretase, an enzyme that cleaves the amyloid precursor protein into a more soluble product that is more easily cleared from the brain and does not lead to the formation of insoluble amyloid plaques.

In Table 5, both the results and the identifiers of the clinical trials that have been carried out for intranasal administration in Alzheimer-type dementias are collected.

Table 5. Clinical trials evaluating the intranasal administration of drugs for the treatment of Alzheimer-type dementia.

Clinical Trials for Nasal Administration in Alzheimer-Type Dementias				
Active Ingredient	**Date**	**Phase**	**Results**	**Reference**
Insulin	2007–2012	Phase 2	An improvement in memory retention and attention was observed after administration of 20 IU daily of intranasal insulin compared to placebo.	NCT00438568 [91]
Insulin	2012	Phase 2	No results have been published.	NCT01547169 [92]
Insulin	2012–2018	Phase 2	Treatment with 20 IU of insulin improved delayed memory, and both insulin doses (20 and 40 IU) preserved general cognition as assessed via the ADAS-cog score and functional abilities as assessed via the ADCS-ADL scale. There were no treatment-related serious adverse events.	NCT01595646 [47,93]
Insulin	2013–2020	Phase 2	A total of 289 participants were randomized. Among the first 49 participants, the ViaNase device was used to deliver intranasal insulin, while the rest of the participants used a second device (device 2). After one year of treatment, participants using device 2 did not show cognitive or functional benefits, while patients using the ViaNase device did show a slowdown in cognitive decline according to the ADAS-cog-12 scale and in everyday activities. No clinically important adverse events were associated.	NCT01767909 [94]
Insulin	2015–2020	Phase 2	No results have been published.	NCT02462161 [95]
Insulin	2015–2020	Phase 2	Intranasal insulin glulisine had no significant effect on improving cognition or mood; however, the number of subjects successfully enrolled and the duration of the study were limited. Intranasal insulin glulisine was relatively safe and well tolerated.	NCT02503501 [96]

Table 5. *Cont.*

Clinical Trials for Nasal Administration in Alzheimer-Type Dementias				
Active Ingredient	**Date**	**Phase**	**Results**	**Reference**
Insulin	2021	Phase 2	In recruiting status.	NCT05081219 [89]
Insulin	2022	Phase 2	Not recruiting yet.	NCT05006599 [90]
Rivastigmine	2016	Phase 1	Rivastigmine nasal spray had superior absolute bioavailability compared to historical values for the oral capsule and transdermal patch determined by other investigators. It had a rapid onset of action, and a favorable safety and tolerability profile. No clinically important adverse events were associated.	ACTRN12614001313628 [87]
APH-1105	2023–2024	Phase 2	Not recruiting yet.	NCT03806478 [86]

It should be noted that among those active principles with which clinical trials have been carried out, the one that has generated the most interest would be intranasal insulin, which is the one with which the greatest number of clinical trials have been carried out and which has produced quite encouraging results.

4. Conclusions

The intranasal route has great potential as an alternative route of administration in the treatment of ATD, since it has numerous advantages over conventional routes of administration, among them, direct transport from the nasal cavity to the brain avoiding crossing the BBB. To obtain good bioavailability in the brain, the physicochemical characteristics of the formulations must be optimized through technological strategies (mucoadhesive polymers, prodrugs, absorption promoters, etc.) so that they are suitable for this route of transport and avoid the physiological mechanisms of elimination from the nasal cavity. Among the various technological strategies investigated to optimize intranasal formulations, lipid-based nanosystems, particularly nanostructured lipid carriers, have shown the most promise as effective systems for achieving drug access to the brain from the nasal cavity. They have generated considerable interest due to their stability, physiological compatibility and ability to overcome challenges associated with other nanocarriers. At the moment, no drugs have obtained marketing approval for intranasal administration in the treatment of ATD, and only three candidates have made it to the clinical investigation phase, namely insulin, rivastigmine and APH-1105. Further studies with different candidates will eventually confirm the potential of the intranasal route of administration in the treatment of ATD.

Author Contributions: R.T.-V. and A.N. contributed equally to this work. Conceptualization, R.T.-V. and A.N.; methodology, J.V.d.J.-O., O.D.-S. and O.V.-B.; investigation, J.V.d.J.-O., O.D.-S. and O.V.-B.; writing—original draft preparation, R.T.-V. and A.N.; writing—review and editing, J.V.d.J.-O., O.V.-B., R.T.-V. and A.N.; supervision, R.T.-V. and A.N. All authors have read and agreed to the published version of the manuscript.

Funding: This research received no external funding.

Institutional Review Board Statement: Not applicable.

Informed Consent Statement: Not applicable.

Data Availability Statement: Not applicable.

Conflicts of Interest: The authors declare no conflict of interest.

References

1. Scheltens, P.; De Strooper, B.; Kivipelto, M.; Holstege, H.; Chételat, G.; Teunissen, C.E.; Cummings, J.; van der Flier, W.M. Alzheimer's disease. *Lancet* **2021**, *397*, 1577–1590. [CrossRef] [PubMed]
2. Villarejo-Galende, A.; Eimil-Ortiz, M.; Llamas-Velasco, S.; Llanero-Luque, M.; López de Silanes de Miguel, C.; Prieto Jurczynska, C. Informe de la Fundación del Cerebro. Impacto social de la enfermedad de Alzheimer y otras demencias. *Neurología* **2021**, *36*, 39–49. [CrossRef] [PubMed]
3. Available online: https://www.alz.org/alzheimers-dementia/what-is-alzheimers (accessed on 8 January 2023).
4. Breijyeh, Z.; Karaman, R. Comprehensive Review on Alzheimer's Disease: Causes and Treatment. *Molecules* **2020**, *25*, 5789. [CrossRef]
5. Prince, M.J.; Wimo, A.; Guerchet, M.M.; Ali, G.C.; Wu, Y.-T.; Prina, M. *World Alzheimer Report 2015. The Global Impact of Dementia: An Analysis of Prevalence, Incidence, Cost and Trends*; Alzehimer Disease International: London, UK, 2015; Available online: http://www.alz.co.uk/research/world-report-2015 (accessed on 3 October 2022).
6. Li, X.; Feng, X.; Sun, X.; Hou, N.; Han, F.; Liu, Y. Global, regional, and national burden of Alzheimer's disease and other dementias, 1990–2019. *Front. Aging Neurosci. Lausanne* **2022**, *14*, 1–16. [CrossRef] [PubMed]
7. Alzheimer's Disease Fact Sheet. National Institute on Aging. 2022. Available online: https://www.nia.nih.gov/health/alzheimers-disease-fact-sheet (accessed on 3 October 2022).
8. Mendez, M.F. Early-onset Alzheimer's disease: Nonamnestic subtypes and type 2 AD. *Arch. Med. Res.* **2012**, *43*, 677–685. [CrossRef] [PubMed]
9. Burns, A.; Iliffe, S. Alzheimer's disease. *Brit. Med. J.* **2009**, *338*, b158. [CrossRef]
10. Hippius, H.; Neundörfer, G. The Discovery of Alzheimer disease. *Dialogues. Clin. Neurosci.* **2003**, *5*, 101–108. [CrossRef]
11. Agrawal, M.; Dethe, M.R.; Ahmed, H.; Yadav, A.; Gupta, U.; Alexander, A. Nose-to-brain drug delivery for the treatment of Alzheimer's disease: Current advancements and challenges. *Expert Opin. Drug Deliv.* **2022**, *19*, 87–102.
12. Food and Drug Administration. ARICEPT (Donepezil Hydrochloride) Prescribing Information [Internet]. 2021. Available online: http://www.aricept.com (accessed on 3 October 2022).
13. Food and Drug Administration. ADLARITY (Donepezil Transdermal System) Prescribing Information [Internet]. 2022. Available online: https://www.accessdata.fda.gov/drugsatfda_docs/label/2022/212304s000lbl.pdf (accessed on 24 April 2023).
14. Food and Drug Administration. RAZADYNE®(Galantamine Hydrobromide) Prescribing Information. 2021. Available online: https://www.accessdata.fda.gov/drugsatfda_docs/label/2021/021169s035,021615s026lbl.pdf (accessed on 3 October 2022).
15. Food and Drug Administration. Exelon®Patch (Rivastigmine Transdermal System) Prescribing Information. 2020. Available online: https://www.novartis.com/us-en/sites/novartis_us/files/exelonpatch.pdf (accessed on 15 July 2022).
16. Food and Drug Administration. RIVASTIGMINE TARTRATE Capsules Prescribing Information. 2019. Available online: https://www.apotex.com/products/us/downloads/pre/riva_imcp_ins.pdf (accessed on 15 July 2022).
17. Food and Drug Administration. NAMENDA (Memantine HCl) Prescribing Information [Internet]. 2019. Available online: https://www.accessdata.fda.gov/drugsatfda_docs/label/2019/022525s015lbl.pdf (accessed on 15 July 2022).
18. Binda, A.; Murano, C.; Rivolta, I. Innovative Therapies and Nanomedicine Applications for the Treatment of Alzheimer's Disease: A State-of-the-Art (2017–2020). *Int. J. Nanomed.* **2020**, *15*, 6113–6135. [CrossRef]
19. Food and Drug Administration. ADUHELM®(Aducanumab-Avwa) Injection Prescribing Information. 2022. Available online: https://www.biogencdn.com/us/aduhelm-pi.pdf (accessed on 15 July 2022).
20. European Medicines Agency. Refusal of the Marketing Authorization for Aduhelm (Aducanumab). 2022. Available online: https://www.ema.europa.eu/en/medicines/human/summaries-opinion/aduhelm#opinion-section (accessed on 3 October 2022).
21. Agrawal, M.; Saraf, S.; Saraf, S.; Antimisiaris, S.G.; Chougule, M.B.; Shoyele, S.A.; Alexander, A. Nose-to-brain drug delivery: An update on clinical challenges and progress toward approval of anti-Alzheimer drugs. *J. Control. Release* **2018**, *281*, 139–177. [CrossRef]
22. Tarriot, P.N.; Braeckman, R.; Oh, C. Comparison of steady-state pharmacokinetics of Donepezil transdermal delivery system with oral Donepezil. *J. Alzheimer's Dis.* **2022**, *90*, 161–172. [CrossRef] [PubMed]
23. Food and Drug Administration. NAMZARIC (Memantine and Donepezil Hydrochlorides) Prescribing Information. 2019. Available online: https://www.rxabbvie.com/pdf/namzaric_pi.pdf (accessed on 15 July 2022).
24. Terstappen, G.C.; Meyer, A.H.; Bell, R.D.; Zhang, W. Strategies for delivering therapeutics across the blood–brain barrier. *Nat. Rev. Drug Discov.* **2021**, *20*, 362–383. [CrossRef]
25. Tripathi, S.; Gupta, U.; Ujjwal, R.R.; Yadav, A.K. Nano-lipidic formulation and therapeutic strategies for Alzheimer's disease via intranasal route. *J. Microencapsul.* **2021**, *38*, 572–593. [CrossRef]
26. Riccardi, C.; Napolitano, F.; Montesarchio, D.; Sampaolo, S.; Melone, M.A.B. Nanoparticle-Guided Brain Drug Delivery: Expanding the Therapeutic Approach to Neurodegenerative Diseases. *Pharmaceutics* **2021**, *13*, 1897. [CrossRef] [PubMed]
27. Bell, R.D. Considerations When Developing Blood-Brain Barrier Crossing Drug Delivery Technology. *Handb. Exp. Pharmacol.* **2022**, *273*, 83–95.
28. Sobiesk, J.L.; Munakomi, S. *Anatomy, Head and Neck, Nasal Cavity*; StatPearls Publishing: Treasure Island, FL, USA, 2022.
29. Akel, H.; Ismail, R.; Csóka, I. Progress and perspectives of brain-targeting lipid-based nanosystems via the nasal route in Alzheimer's disease. *Eur. J. Pharm. Biopharm.* **2020**, *148*, 38–53. [CrossRef] [PubMed]

30. Lobaina Mato, Y. Nasal route for vaccine and drug delivery: Features and current opportunities. *Int. J. Pharm.* **2019**, *572*, 118813. [CrossRef]
31. Lee, D.; Minko, T. Nanotherapeutics for Nose-to-Brain Drug Delivery: An Approach to Bypass the Blood Brain Barrier. *Pharmaceutics* **2021**, *13*, 2049. [CrossRef] [PubMed]
32. Agu, R.U. Challenges in nasal drug absorption: How far have we come? *Ther. Deliv.* **2016**, *7*, 495–510. [CrossRef] [PubMed]
33. Dando, S.J.; Mackay-Sim, A.; Norton, R.; Currie, B.J.; St John, J.A.; Ekberg, J.A.; Batzloff, M.; Ulett, G.C.; Beacham, I.R. Pathogens penetrating the central nervous system: Infection pathways and the cellular and molecular mechanisms of invasion. *Clin. Microbiol. Rev.* **2014**, *27*, 691–726. [CrossRef]
34. Dhas, N.L.; Kudarha, R.R.; Mehta, T.A. Intranasal Delivery of Nanotherapeutics/Nanobiotherapeutics for the Treatment of Alzheimer's Disease: A Proficient Approach. *Crit. Rev. Ther. Drug Carrier Syst.* **2019**, *36*, 373–447. [CrossRef] [PubMed]
35. Poovaiah, N.; Davoudi, Z.; Peng, H.; Schlichtmann, B.; Mallapragada, S.; Narasimhan, B.; Wang, Q. Treatment of neurodegenerative disorders through the blood-brain barrier using nanocarriers. *Nanoscale* **2018**, *10*, 16962–16983. [CrossRef] [PubMed]
36. Erdő, F.; Bors, L.A.; Farkas, D.; Bajza, Á.; Gizurarson, S. Evaluation of intranasal delivery route of drug administration for brain targeting. *Brain Res. Bull.* **2018**, *143*, 155–170. [CrossRef] [PubMed]
37. Selvaraj, K.; Gowthamarajan, K.; Reddy-Karri, V.V.S. Nose to brain transport pathways an overview: Potential of nanostructured lipid carriers in nose to brain targeting. *Artif. CellsNanomed. Biotechnol.* **2018**, *46*, 2088–2095. [CrossRef]
38. Bahadur, S.; Pardhi, D.M.; Rautio, J.; Rosenholm, J.M.; Pathak, K. Intranasal Nanoemulsions for Direct Nose-to-Brain Delivery of Actives for CNS Disorders. *Pharmaceutics* **2020**, *12*, 1230. [CrossRef]
39. The Mandibular Division of the Trigeminal Nerve (CNV3)—TeachMeAnatomy. Available online: https://teachmeanatomy.info/head/nerves/mandibular-nerve/ (accessed on 28 March 2023).
40. Bourganis, V.; Kammona, O.; Alexopoulos, A.; Kiparissides, C. Recent advances in carrier mediated nose-to-brain delivery of pharmaceuticals. *Eur. J. Pharm. Biopharm.* **2018**, *128*, 337–362. [CrossRef]
41. Bicker, J.; Fortuna, A.; Alves, G.; Falcão, A. Nose-to-brain Delivery of Natural Compounds for the Treatment of Central Nervous System Disorders. *Curr. Pharm. Des.* **2020**, *26*, 594–619. [CrossRef]
42. Cunha, S.; Forbes, B.; Sousa Lobo, J.M.; Silva, A.C. Improving Drug Delivery for Alzheimer's Disease Through Nose-to-Brain Delivery Using Nanoemulsions, Nanostructured Lipid Carriers (NLC) and in situ Hydrogels. *Int. J. Nanomed.* **2021**, *16*, 4373–4390. [CrossRef]
43. Fonseca, L.C.; Lopes, J.A.; Vieira, J.; Viegas, C.; Oliveira, C.S.; Hartmann, R.P.; Fonte, P. Intranasal drug delivery for treatment of Alzheimer's disease. *Drug Deliv. Transl. Res.* **2021**, *11*, 411–425. [CrossRef]
44. Bustamante-Marin, X.M.; Ostrowski, L.E. Cilia and Mucociliary Clearance. *Cold Spring Harb. Perspect Biol.* **2017**, *9*, a028241. [CrossRef]
45. Brannigan, R.P.; Khutoryanskiy, V.V. Progress and current trends in the synthesis of novel polymers with enhanced mucoadhesive properties. *Macromol. Biosci.* **2019**, *19*, 1900194. [CrossRef] [PubMed]
46. Wang, Z.; Xiong, G.; Tsang, W.C.; Schätzlein, A.G.; Uchegbu, I.F. Nose-to-Brain Delivery. *J. Pharmacol. Exp. Ther.* **2019**, *370*, 593–601. [CrossRef] [PubMed]
47. Craft, S.; Claxton, A.; Baker, L.D.; Hanson, A.J.; Cholerton, B.; Trittschuh, E.H.; Dahl, D.; Caulder, E.; Neth, B.; Montine, T.J.; et al. Effects of regular and long-acting insulin on cognition and Alzheimer's disease biomarkers: A pilot clinical trial. *J. Alzheimers Dis.* **2017**, *57*, 1325–1334. [CrossRef] [PubMed]
48. Food and Drug Administration. ONZETRATM XsailTM (Sumatriptan Nasal Powder) Prescribing Information. 2016. Available online: https://www.accessdata.fda.gov/drugsatfda_docs/label/2016/206099s000lbl.pdf (accessed on 31 March 2022).
49. Ghadiri, M.; Young, P.; Traini, D. Strategies to Enhance Drug Absorption via Nasal and Pulmonary Routes. *Pharmaceutics* **2019**, *11*, 113. [CrossRef]
50. Bakker, C.; van der Aart, J.; Hart, E.P.; Klaassen, E.S.; Bergmann, K.R.; van Esdonk, M.J.; Kay, D.G.; Groeneveld, G.J. Safety, pharmacokinetics, and pharmacodynamics of Gln-1062, a prodrug of galantamine. *A&D Transl. Res. Clin. Interv.* **2020**, *6*, e12093. Available online: https://onlinelibrary.wiley.com/doi/10.1002/trc2.12093 (accessed on 28 March 2023).
51. Movia, D.; Bruni-Favier, S.; Prina-Mello, A. In vitro Alternatives to Acute Inhalation Toxicity Studies in Animal Models—A Perspective. *Front. Bioeng. Biotechnol.* **2020**, *8*, 549. [CrossRef]
52. Pires, P.C.; Santos, A.O. Nanosystems in nose-to-brain drug delivery: A review of non-clinical brain targeting studies. *J. Control. Release* **2018**, *270*, 89–100. [CrossRef]
53. Vitorino, C.; Silva, S.; Bicker, J.; Falcão, A.; Fortuna, A. Antidepressants and nose-to-brain delivery: Drivers, restraints, opportunities and challenges. *Drug Discov. Today* **2019**, *24*, 1911–1923. [CrossRef]
54. Hong, S.S.; Oh, K.T.; Choi, H.G.; Lim, S.J. Liposomal Formulations for Nose-to-Brain Delivery: Recent Advances and Future Perspectives. *Pharmaceutics* **2019**, *11*, 540. [CrossRef]
55. Pokharkar, V.; Patil-Gadhe, A.; Palla, P. Efavirenz loaded nanostructured lipid carrier engineered for brain targeting through intranasal route: In-vivo pharmacokinetic and toxicity study. *Biomed. Pharmacother.* **2017**, *94*, 150–164. [CrossRef]
56. Ross, C.; Taylor, M.; Fullwood, N.; Allsop, D. Liposome delivery systems for the treatment of Alzheimer's disease. *Int. J. Nanomed.* **2018**, *13*, 8507–8522. [CrossRef] [PubMed]

57. Kulkarni, P.; Deepak Rawtani, D.; Barot, T. Design, development and in-vitro/in-vivo evaluation of intranasally delivered Rivastigmine and N-Acetyl Cysteine loaded bifunctional niosomes for applications in combinative treatment of Alzheimer's disease. *Eur. J. Pharm. Biopharm.* **2021**, *163*, 1–15. [CrossRef] [PubMed]
58. Gharbavi, M.; Amani, J.; Kheiri-Manjili, H.; Danafar, H.; Sharafi, A.J. Niosome: A Promising Nanocarrier for Natural Drug Delivery through Blood-Brain Barrier. *Adv. Pharmacol. Pharm. Sci.* **2018**, *2018*, 6847971. [CrossRef] [PubMed]
59. Ferraz de Souza, I.F.; Queiroz dos Santos, T.; Rodrigo Vicentino, P.; Aparecida Mangerona, B.; Chiva Carvalho, F.; Bergamin Boralli, V.; Morais Ruela, A.L.; Ribeiro Pereira, G. The liquid crystalline phase behaviour of a nasal formulation modifies the brain disposition of donepezil in rats in the treatment of Alzheimer's disease. *Colloids Surf. B Biointerfaces* **2021**, *203*, 111721. [CrossRef] [PubMed]
60. Kaur, A.; Nigam, K.; Bhatnagar, I.; Sukhpal, H.; Awasthy, S.; Shankar, S.; Tyagi, A.; Dang, S. Treatment of Alzheimer's diseases using donepezil nanoemulsion: An intranasal approach. *Drug Deliv. Transl. Res.* **2020**, *10*, 1862–1875. [CrossRef] [PubMed]
61. Haider, M.; Abdin, S.M.; Kamal, L.; Orive, G. Nanostructured Lipid Carriers for Delivery of Chemotherapeutics: A Review. *Pharmaceutics* **2020**, *12*, 288. [CrossRef]
62. Nguyen, T.T.; Nguyen, T.T.D.; Tran, N.M.; Van Vo, G. Lipid-Based Nanocarriers via Nose-to-Brain Pathway for Central Nervous System Disorders. *Neurochem. Res.* **2022**, *47*, 552–573. [CrossRef]
63. Cunha, S.; Amaral, M.H.; Lobo, J.M.S.; Silva, A.C. Lipid Nanoparticles for Nasal/Intranasal Drug Delivery. *Crit. Rev. Ther. Drug Carr. Syst.* **2017**, *34*, 257–282. [CrossRef]
64. Agrawal, M.; Saraf, S.; Saraf, S.; Dubey, S.K.; Puri, A.; Patel, R.J.; Ravichandiran, V.; Murty, U.S.; Alexander, A. Recent strategies and advances in the fabrication of nano lipid carriers and their application towards brain targeting. *J. Control. Release* **2020**, *321*, 372–415. [CrossRef]
65. Agrawal, M.; Saraf, S.; Saraf, S.; Antimisiaris, S.G.; Hamano, N.; Li, S.-D.; Chougule, M.; Shoyele, S.A.; Gupta, U.; Ajazuddin; et al. Recent advancements in the field of nanotechnology for the delivery of anti-Alzheimer drug in the brain region. *Expert Opin. Drug Deliv.* **2018**, *15*, 598–617. [CrossRef]
66. de Souza Guedes, L.; Martinez, R.M.; Bou-Chacra, N.A.; Velasco, M.V.R.; Rosado, C.; Baby, A.R. An Overview on Topical Administration of Carotenoids and Coenzyme Q10 Loaded in Lipid Nanoparticles. *Antioxidants* **2021**, *10*, 1034. [CrossRef] [PubMed]
67. Costa, C.; Moreira, J.N.; Amaral, M.H.; Sousa Lobo, J.M.; Silva, A.C. Nose-to-brain delivery of lipid-based nanosystems for epileptic seizures and anxiety crises. *J. Control. Release* **2019**, *295*, 187–200. [CrossRef] [PubMed]
68. Gadhave, D.G.; Kokare, C.R. Nanostructured lipid carriers engineered for intranasal delivery of teriflunomide in multiple sclerosis: Optimization and in vivo studies. *Drug Dev. Ind. Pharm.* **2019**, *45*, 839–851. [CrossRef] [PubMed]
69. Desfrançois, C.; Auzély, R.; Texier, I. Lipid Nanoparticles and Their Hydrogel Composites for Drug Delivery: A Review. *Pharmaceuticals* **2018**, *11*, 118. [CrossRef] [PubMed]
70. Nguyen, T.-T.-L.; Maeng, H.-J. Pharmacokinetics and Pharmacodynamics of Intranasal Solid Lipid Nanoparticles and Nanostructured Lipid Carriers for Nose-to-Brain Delivery. *Pharmaceutics* **2022**, *14*, 572. [CrossRef]
71. Nguyen, K.; Hoffman, H.; Chakkamparambil, B.; Grossberg, G.T. Evaluation of rivastigmine in Alzheimer's disease. *Neurodegener. Dis. Manag.* **2021**, *11*, 35–48. [CrossRef]
72. Wavikar, P.; Pai, R.; Vavia, P. Nose to Brain Delivery of Rivastigmine by In Situ Gelling Cationic Nanostructured Lipid Carriers: Enhanced Brain Distribution and Pharmacodynamics. *J. Pharm. Sci.* **2017**, *106*, 3613–3622. [CrossRef]
73. Anand, A.; Arya, M.; Kaithwas, G.; Singh, G.; Saraf, S.A. Sucrose stearate as a biosurfactant for development of rivastigmine containing nanostructured lipid carriers and assessment of its activity against dementia in *C. elegans* model. *J. Drug Deliv. Sci. Technol.* **2019**, *49*, 219–226. [CrossRef]
74. Saunders, A.M.; Burns, D.K.; Gottschalk, W.K. Reassessment of Pioglitazone for Alzheimer's Disease. *Front. Neurosci.* **2021**, *15*, 666958. [CrossRef]
75. Jojo, G.M.; Kuppusamy, G.; De, A.; Reddy-Karri, V.V.S.N. Formulation and optimization of intranasal nanolipid carriers of pioglitazone for the repurposing in Alzheimer's disease using Box-Behnken design. *Drug Dev. Ind. Pharm.* **2019**, *45*, 1061–1072. [CrossRef]
76. Rao, Y.L.; Ganaraja, B.; Joy, T.; Pai, M.M.; Ullal, S.D.; Murlimanju, B.V. Neuroprotective effects of resveratrol in Alzheimer's disease. *Front. Biosci. (Elite Ed.)* **2020**, *12*, 139–149. [CrossRef] [PubMed]
77. Rajput, A.P.; Butani, S.B. Resveratrol anchored nanostructured lipid carrier loaded in situ gel via nasal route: Formulation, optimization and in vivo characterization. *J. Drug Deliv. Sci. Technol.* **2019**, *51*, 214–223. [CrossRef]
78. Akbar, M.; Shabbir, A.; Rehman, K.; Akash, M.S.H.; Shah, M.A. Neuroprotective potential of berberine in modulating Alzheimer's disease via multiple signaling pathways. *J. Food Biochem.* **2021**, *45*, e13936. [CrossRef] [PubMed]
79. Abo-El-Enin, H.A.; Elkomy, M.H.; Naguib, I.A.; Ahmed, M.F.; Alsaidan, O.A.; Alsalahat, I.; Ghoneim, M.M.; Eid, H.M. Lipid Nanocarriers Overlaid with Chitosan for Brain Delivery of Berberine via the Nasal Route. *Pharmaceuticals* **2022**, *15*, 281. [CrossRef] [PubMed]
80. Kales, H.C.; Lyketsos, C.G.; Miller, E.M.; Ballard, C. Management of behavioral and psychological symptoms in people with Alzheimer's disease: An international Delphi consensus. *Int. Psychogeriatr.* **2019**, *31*, 83–90. [CrossRef]
81. Shehata, M.K.; Ismail, A.A.; Kamel, M.A. Nose to Brain Delivery of Astaxanthin–Loaded Nanostructured Lipid Carriers in Rat Model of Alzheimer's Disease: Preparation, in vitro and in vivo Evaluation. *Int. J. Nanomed.* **2023**, *18*, 1631–1658. [CrossRef]

82. Rajput, A.P.; Butani, S.B. Fabrication of an ion-sensitive in situ gel loaded with nanostructured lipid carrier for nose to brain delivery of donepezil. *Asian J. Pharm.* **2018**, *12*, 293.
83. Younis, M.A.; Tawfeek, H.M.; Abdellatif, A.A.H.; Abdel-Aleem, J.A.; Harashima, H. Clinical translation of nanomedicines: Challenges, opportunities, and keys. *Adv. Drug Deliv. Rev.* **2022**, *181*, 114083. [CrossRef]
84. Dubey, S.K.; Lakshmi, K.K.; Krishna, K.V.; Agrawal, M.; Singhvi, G.; Saha, R.N.; Saraf, S.; Saraf, S.; Shukla, R.; Alexander, A. Insulin mediated novel therapies for the treatment of Alzheimer's disease. *Life Sci.* **2020**, *249*, 117540. [CrossRef]
85. Morgan, T.M.; Soh, B. Absolute bioavailability and safety of a novel rivastigmine nasal spray in healthy elderly individuals. *Br. J. Clin. Pharmacol.* **2017**, *83*, 510–516. [CrossRef]
86. Aphios. Safety, Tolerability and Efficacy Assessment of Intranasal Nanoparticles of APH-1105, A Novel Alpha Secretase Modulator For Mild to Moderate Cognitive Impairment Due to Alzheimer's Disease (AD). clinicaltrials.gov; Report No.: NCT03806478. 2021. Available online: https://clinicaltrials.gov/ct2/show/NCT03806478 (accessed on 28 March 2023).
87. ANZCTR. Evaluation of the Pharmacokinetics and Safety of Rivastigmine Intranasal Spray. Report No.: ACTRN12614001313628. 2016. Available online: https://www.anzctr.org.au/Trial/Registration/TrialReview.aspx?id=367562 (accessed on 28 March 2023).
88. Hallschmid, M. Intranasal insulin. *J. Neuroendocrinol.* **2021**, *33*, e12934. [CrossRef] [PubMed]
89. Wake Forest University Health Sciences. Study of Nasal Insulin to Fight Forgetfulness—Combination Intranasal Insulin and Empagliflozin Trial. clinicaltrials.gov;. Report No.: NCT05081219. 2021. Available online: https://clinicaltrials.gov/ct2/show/NCT05081219 (accessed on 28 March 2023).
90. Wake Forest University Health Sciences. Study of Nasal Insulin to Fight Forgetfulness (SNIFF)—3-Week Aptar CPS Device. clinicaltrials.gov; Report No.: NCT05006599. 2021. Available online: https://clinicaltrials.gov/ct2/show/NCT05006599 (accessed on 28 March 2023).
91. University of Washington. Therapeutic Effects of Intranasal Insulin Administration in AD. clinicaltrials.gov; Report No.: NCT00438568. 2012. Available online: https://clinicaltrials.gov/ct2/show/NCT00438568 (accessed on 28 March 2023).
92. Craft, S. Study of Nasal Insulin to Fight Forgetfulness—Long-Acting Insulin Detemir—21 Days. clinicaltrials.gov;. Report No.: NCT01547169. 2012. Available online: https://clinicaltrials.gov/ct2/show/NCT01547169 (accessed on 28 March 2023).
93. Wake Forest University Health Sciences. Study of Nasal Insulin to Fight Forgetfulness—Long-Acting Insulin Detemir—120 Days (SL120). clinicaltrials.gov; Report No.: NCT01595646. 2020. Available online: https://clinicaltrials.gov/ct2/show/NCT01595646 (accessed on 29 March 2023).
94. Aisen, P. Therapeutic Effects of Intranasally-Administered Insulin in Adults with Amnestic Mild Cognitive Impairment (aMCI) or Mild Alzheimer's Disease (AD). clinicaltrials.gov; Report No.: NCT01767909. Available online: https://clinicaltrials.gov/ct2/show/NCT01767909 (accessed on 28 March 2023).
95. Wake Forest University Health Sciences. Study of Nasal Insulin to Fight Forgetfulness—Short-Acting Insulin Aspart. clinicaltrials.gov; Report No.: NCT02462161. 2020. Available online: https://clinicaltrials.gov/ct2/show/NCT02462161 (accessed on 28 March 2023).
96. Health Partners Institute. A Phase II, Single Center, Randomized, Double-Blind, Placebo-Controlled Study of the Safety and the Therapeutic Effectiveness of Intranasal Glulisine in Amnestic Mild Cognitive Impairment and Probable Mild Alzheimer's Disease. clinicaltrials.gov; Report No.: NCT02503501. 2020. Available online: https://clinicaltrials.gov/ct2/show/NCT02503501 (accessed on 29 March 2023).

pharmaceutics

MDPI

Review

Advances in Xanthan Gum-Based Systems for the Delivery of Therapeutic Agents

Mahima Jadav [1], Deep Pooja [2], David J. Adams [3,*] and Hitesh Kulhari [1,*]

[1] School of Nano Sciences, Central University of Gujarat, Gandhinagar 382030, Gujarat, India
[2] School of Pharmacy, National Forensic Science University, Gandhinagar 382007, Gujarat, India
[3] Illawarra Health and Medical Research Institute (IHMRI), University of Wollongong, Wollongong, NSW 2522, Australia
* Correspondence: djadams@uow.edu.au (D.J.A.); hitesh.kulhari@cug.ac.in (H.K.)

Abstract: In the last three decades, polymers have contributed significantly to the improvement of drug delivery technologies by enabling the controlled and sustained release of therapeutic agents, versatility in designing different delivery systems, and feasibility of encapsulation of both hydrophobic and hydrophilic molecules. Both natural and synthetic polymers have been explored for the delivery of various therapeutic agents. However, due to the disadvantages of synthetic polymers, such as lack of intrinsic biocompatibility and bioactivity, hydrophobicity, and expensive and complex procedure of synthesis, there is a move toward the use of naturally occurring polymers. The biopolymers are generally derived from either plants or microorganisms and have shown a wide range of applications in drug administration due to their hydrophilic nature, biodegradability, biocompatibility, no or low toxicity, abundance, and readily available, ease of chemical modification, etc. This review describes the applications of a biopolymer, xanthan gum (XG), in the delivery of various therapeutic agents such as drugs, genetic materials, proteins, and peptides. XG is a high molecular weight, microbial heteropolysaccharide and is produced as a fermented product of Gram-negative bacteria, *Xanthomonas campestris*. Traditionally, it has been used as a thickener in liquid formulations and an emulsion stabiliser. XG has several favourable properties for designing various forms of drug delivery systems. Furthermore, the structure of XG can be easily modified using different temperature and pH conditions. Therefore, XG and its derivatives have been explored for various applications in the food, pharmaceutical, and cosmetic industries.

Keywords: biopolymers; polysaccharides; drug delivery; controlled release; nanoparticles; physiochemical properties

1. Introduction

Natural polymers and their derivatives possess properties such as biocompatibility, biodegradability, high availability, and capability for chemical modifications, which makes them ideal for new drug delivery systems. Natural polymers are scientifically chosen over synthetic polymers due to their predictable outcomes, such as biodegradability and biocompatibility in the human body. Natural gums are made up of several sugar units connected by the glycosidic bonds known as polysaccharides. Plants comprising vascular tissues produce gums as a defence mechanism against any injury. The gums produced are heteropolysaccharides, but upon hydrolysis, they are converted into monosaccharides such as mannose, xylose, arabinose, glucose, uronic acid, and galactose [1]. These natural gums are extensively researched because they are biodegradable, biocompatible, and sustainable and are often used in different industries [2]. Natural gums, xanthan gum, locust bean gum, guar gum, and tara gum, have a comprehensive application in the pharmaceutical industries [3,4].

Xanthan gum (XG) is a naturally occurring extracellular heteropolysaccharide and is a fermentation product of *Xanthomonas campestris*, which is an aerobic, Gram-negative pathogen. Generally, the pathogen invades and multiplies in the plant vascular tissues of the crucifers such as cabbage, broccoli, and radish and produces V-shaped necrotic lesions at the foliar areas and darkens the veins, causing typical "black rot" symptoms. The bacterium invades the crucifers through stomata, roots, wounds, and hydathodes [5]. It can live in the soil for a year and spread through irrigation and surface water. The bacterium depends on the water for its survival [6]. The world has faced periodic epidemics of black rot disease, especially in agriculturally developing areas of Asia and Africa. Such incidences cause significant losses in agricultural products [5]. The average molecular weight of XG chains ranges from 1×10^6 to 20×10^6 g/mol, which varies according to its biosynthesis conditions and interchain linkages. It is a water-soluble carbohydrate with a cellulosic backbone and a side chain of β-D-mannose-β-D glucuronic acid-α-D-mannose [7].

XG is one of the most important commercially available microbial polysaccharides. In the 1950s, Allene Rosalind Jeanes, from the United States Department of Agriculture, discovered XG. After being approved by the United States Food and Drug Administration (FDA) in 1969 as a food additive (Fed. Reg. 345376), XG was extensively used in food industries as a thickener and a stabiliser. In the 1960s, Kelco company (CP Kelco, Atlanta, GA, USA) started the production of XG in the USA under the name Kelzan1. Jungbunzlauer Austria AG and Solvay in Europe, with the trade name of Rhodopol®, started the production of XG at an industrial scale. Since 2005, one of the biggest producers of XG has been China [8].

Due to its pseudoplastic behaviour and thermal stability, it is added to the water-based drilling fluids in the petroleum industry [9]. Essentially, XG plays an important role as an efficient thickener and favours emulsion stability in the food and cosmetic industries [10]. In the pharmaceutical industry, XG is considered an active excipient for the sustained release of drugs. Furthermore, being highly stable in a broad range of temperatures, pH, and ionic strength and having biocompatibility with the human body, it is used as a drug delivery system. XG is also used as a reducing and capping agent for nanoparticles to make them biodegradable and non-toxic [11].

This review offers a systematic and comprehensive summary of current advancements in XG modification for its use as an excipient in pharmaceutical formulation development. Further, the applicability of XG in the delivery of various therapeutic agents such as drugs, genetic materials, proteins, and peptides is highlighted.

2. Source and Physicochemical Properties of XG

Xanthan gum is an exopolysaccharide, mainly produced by a plant pathogenic Gram-negative bacterial genus *Xanthomonas*. *Xanthomonas* cells are usually found in a single rod shape with a single polar flagellum. Various strains of *Xanthomonas* are capable of production of XG, such as *X. arboricola*, *X. axonopodis*, *X. citri*, *X. phaseoli*, *X. vasculorium*, *X. gummisudans*, *X. axonopodis*, *X. fragaria*, *X. campestris*. Amongst all the species, *X. campestris* is mainly used for XG production in industries [7,12].

The production of XG is affected by various factors such as the type of carbon source (glucose and/or saccharose), microorganism, and the operating conditions (continuous or batch). To ensure the high yield and better-quality production of XG, it is necessary to evaluate the settings of a bioreactor, concentration of nutrients in the growth medium, time of agitation, optimal pH and temperature, aeration, and fermentation time [7,12].

Xanthan gum is a dry, yellow-white coloured powder of high molecular weight (1×10^6 to 20×10^6 Da) exopolysaccharide. It is highly water soluble in both hot and cold conditions due to its polyelectrolyte behaviour [13], though constant mixing is needed to prevent agglomeration. Water-dispersed XG shows viscous solution even at very low concentrations [14]. XG shows high stability in acidic as well as alkaline mediums. It is a cellulose derivative and, thus, is a biodegradable gum in nature. It has cellobiose as a repeating unit in its backbone, and the side chain is composed of D-mannose, D-glucuronic acid (β-1,2), and D-mannose (β-1,4), a trisaccharide, linked to each other by α-1,3 linkages as shown in Figure 1. Approximately one-half of the terminal D-mannose holds a pyruvic acid residue linked by the keto group at the 4th and 6th positions, with an indefinite distribution. The non-terminal D-mannose is connected with the core backbone chain and contains an acetyl group at the O-6 position. The relative composition of pyruvyl and acetyl residues in XG chains mostly depends on the *Xanthomonas* strain used for its production [7].

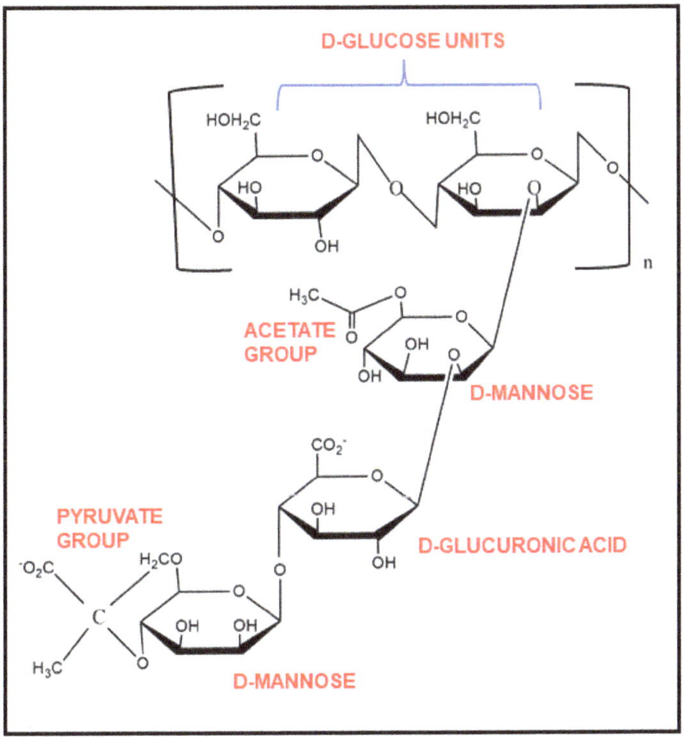

Figure 1. Chemical structure of xanthan gum and its functional groups.

XG behaves as a polyanion at a pH above 4.5 because of the deprotonation of O-acetyl and pyruvyl residues. After extraction from the broth, XG is present in a single helix form stabilised by already present Ca^{2+} ions. The helix form can be converted to coils irreversibly. The coiled conformation of XG can be converted into a double-helical structure reversibly, as illustrated in Figure 2. These conformational changes in the XG structure can be controlled by varying the ionic strength and temperature [7].

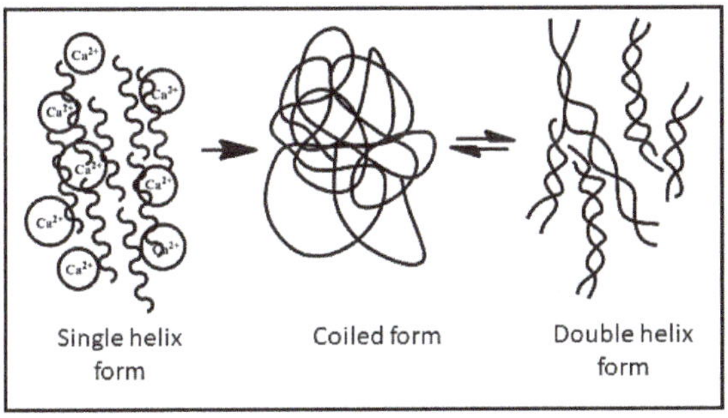

Figure 2. The conformational changes in xanthan gum.

XG exhibits pseudoplastic behaviour, which can be modified by varying temperature and shear rate, making XG useful in the food industry as a thickener, stabiliser, and suspending agent for food [14]. It shows a decrease in viscosity at high temperatures and remains stable when solubilised. XG also has high thermal stability, which prevents its degradation or hydrolysis, making it unique and more advantageous than other hydrophilic polysaccharides. The viscosity of the solutions containing XG is affected by various factors such as pH, the concentration of buffer solution, and the concentration of XG in the solution. Deacetylation of XG occurs in extreme alkaline pH (pH 9 or above) and forms a gel at higher pH (>pH 10) [15]. Table 1 shows the physicochemical properties of commercially available XG.

Table 1. Physicochemical properties of Xanthan gum.

Properties	Value
Physical state	Dry-solid powder
Colour	Yellow-white
Molecular weight	Ranges 1×10^6 to 2×10^6 g/mol
pH	6–7
Flash point	>100 °C
Solubility	Water, DMSO, DMF

3. Important Properties of XG for Designing Drug Delivery Systems

XG is essentially used as a carrier system for the delivery of drugs, proteins, and genes. The important characteristics of XG for this purpose are (i) high stability at low pH, which helps in the protection of the drug in gastric fluid from degradation, and (ii) the rate of drug release can be controlled by changing the pH of the release medium and ionic strength. XG has been used in the form of tablets, hydrogels, microspheres, and mucoadhesive gels for the sustained and controlled release of drugs [16]. For a targeted release of the drug, a controlled release is anticipated using different mechanisms. Such systems ensure an increase in the bioavailability of drugs at the targeted site and avoid drug loss in gastric degradation [17]. Xanthan gum acts as an acknowledged excipient for controlled and sustained-release formulations. Combining XG with other polysaccharides improves and achieves controlled release of the drugs. Even a small amount of XG improves the drug release and provides zero-order release kinetics. XG used as a mucoadhesive patch provided a controlled release of nicotine over a 10 h period with burst release of the drug initially [18]. Floating tablets consisting of a mixture of guar gum and XG encapsulating rosiglitazone maleate showed sustained release (98%) for up to 12 h [19]. For the controlled release of pentoxifylline (hydrophilic drug), XG-based tablets were prepared, which showed a decrease in drug release with an increase in polymer concentration [20]. pH-responsive interpenetrating network microspheres were prepared for the controlled delivery of diclofenac sodium. XG and polyvinyl alcohol (PVA) were used for the development of microspheres, and the ratio of polymers and extent of crosslinking greatly affected the release pattern of the drug [21].

3.1. Mucoadhesive Properties

To enhance the retention time of the drug and increase its therapeutic activity, mucoadhesion is a significant approach that holds the carrier to the biological site by interfacial forces. Mucoadhesive property is important for formulation such as ophthalmic, skin, gastro-retentive, and nasal formulations. Altering the physicochemical properties (concentration, molecular structure) of the polymer has a significant impact on the mucoadhesive property of the polymer [22]. Xanthan gum shows inherent mucoadhesive properties and is used for the synthesis of various drug delivery systems. Mucoadhesive buccal patches using XG were synthesised for the delivery of zolmitriptan. The drug permeability was increased by 3.3-fold [23]. The nanoemulgel system was prepared using XG as an anionic mucoadhesive polymer for the delivery of carbamazepine via olfactory mucosa to treat epilepsy. The results showed a decrease in peripheral actions of the drug [24]. Chitosan-XG-based mucoadhesive polyelectrolyte nasal insert was prepared for the delivery of promethazine hydrochloride (HCl). The surge in the concentration of XG delayed the release of the drug and showed 89% of mucoadhesion with nasal mucosa [25].

3.2. In Situ Gelling and Rheological Properties

Polymeric formulations are known as "in situ gelling systems" and show a transition from their sol state before entering the body to their gel state once inside. The pH shift, temperature modification, solvent exchange, UV radiation, and the presence of certain ions or molecules are only a few of the stimuli that can cause the sol-gel transition [26]. XG is present in the dry state; however, when it encounters water, it turns into the hydrated glassy state and, further, with an increase in swelling degree, changes to a more porous and rubbery state. The release of the drug can be dependent on the pH of the medium because there is a conformational change in XG with changes in ionic strength and pH [18].

At low ionic strength or high temperature, the XG chains are assumed to be present in coiled conformation, whereas, at high ionic strength or low temperature, the conformation of XG chains changes to the helical form. Hydrogen bonds stabilise the helical conformation of the gum, while electrostatic repulsion between carboxylate groups destabilises the structure [27]. XG already has calcium ions present, which destabilises the helical conformation. Further addition of Ca^{2+} ions in stoichiometric equivalence leads it to a gel-like state. However, adding Ca^{2+} ions more than its stoichiometric equivalence partially replaces the carboxyl groups and decreases the network formation and degree of complexation [28].

The low content of the pyruvyl group lowers the viscosity, and the high pyruvyl content increases the viscosity of the gel. Higher acetyl content decreases the gelling capacity of xanthan gum in an aqueous solution [29–31]. The viscosity of the XG solution greatly depends on various factors, including pH, the concentration of buffer solutions, and the concentration of XG [31].

4. Designing Various Forms of Delivery Systems Using XG

In the pharmaceutical and medical fields, natural polymers are preferred over semi-synthetic polymers as drug delivery systems because of properties such as low toxicity, low cost, high availability, and non-irritant nature. However, there are several drawbacks to using natural polysaccharides as an excipient, such as a wide variation, microbial contamination, decreasing viscosity with long-term storage, unsuitable mechanical properties, and uncontrollable rate of hydration. To overcome these disadvantages, chemical modification to the pure polysaccharides is carried out [32]. XG is used as a drug-loading excipient in many forms.

4.1. Matrix Systems

The regulated release of drugs is accomplished using biopolymeric matrix systems. Polymeric materials which swell in the biological fluids or water form matrix systems. The drug is evenly distributed across the matrix of polymeric material before being compressed to form a tablet. The drug that has been dissolved or dispersed follows a regulated drug release pattern. XG is used as a matrix tablet for the controlled release of the drug [16]. XG is used in tablets both as a binder and disintegrant. The polymer's ability to swell facilitates tablet breakdown and speeds up pharmacological absorption [33]. Graft copolymerisation of acrylamide onto XG was done for the delivery of antihypertensive drugs (atenolol and carvedilol). The release patterns were compared to that of the commercially available formulations, and a non-Fickian trend of transportation was followed [34].

4.2. Films

In addition to tablets, xanthan gum-based films have been thoroughly researched and used. Due to their adaptability, films can be used as a dosage form for several purposes (buccal, oral, sublingual, ophthalmic, transdermal, etc.). A drug delivery film is typically composed of a thin, flexible polymeric layer that may or may not contain a plasticiser. Because of these qualities, they are less obstructive for the patient, making them a more desirable mode of administration [35,36]. Mouth-dissolving films were prepared using XG for the rapid absorption of glibenclamide to treat diabetes. The solvent casting method was for the synthesis of films. The films showed optimum mechanical strength with instant drug release and rapid dissolution rate [37]. For the delivery of zolmitriptan, a bilayered mucoadhesive buccal film was prepared using XG, hydroxypropyl methylcellulose (HPMC)-E15, and polyvinyl alcohol (PVA). Instant drug release occurred in the initial 15 min, and further sustained release of the drug was observed. Upon the addition of dimethyl sulfoxide (DMSO), the permeability of the drug increased 3.2-fold without any damage to the buccal mucosa [23]. The gum's incorporation in the films reduced the moisture content and water vapour permeability in addition to the enhancement of mechanical and thermal qualities. In some cases, films are not an appropriate dosage form since they require a thick platform. The use of hydrogels in this situation might be an alternative.

4.3. Hydrogels

Hydrophilic polymer chains in an aqueous microenvironment crosslink to produce materials called hydrogels. Gelation is accomplished by a variety of methods, including covalent chemical cross-linking, electrostatic interactions, and physical cross-linking [38]. These polymeric networks' physicochemical characteristics and mechanical stiffness are determined by the procedure and cross-linking grade [39]. Hydrogels are useful for encapsulating pharmaceuticals because of their high hydrophilicity (ability to retain water from 70% to 99%), which preserves the structural integrity of the drugs and lowers the risk of enzymatic denaturation once inside the body [40]. Several studies on XG-based hydrogels for the delivery of drugs have been carried out. XG hydrogels provide a controlled release of drugs. Succinic anhydride-modified xanthan gum has been used to prepare ionic strength-sensitive hydrogel for the delivery of gentamicin. Initially, a rapid release of gentamicin was observed with increased ionic strength of the release medium. After that, a sustained release of the drug was observed because of the strong interaction between polymer molecules, forming a tight network [41]. To increase the bioavailability and contact time between the drug and the precorneal membrane, the in-situ hydrogel was synthesised and used to deliver the drug. XG and sodium alginate were used as a mucoadhesive polymer, and poloxamer 407 and poloxamer 188 were used to encapsulate moxifloxacin hydrochloride. The formulations prepared were transparent, uniform in consistency with good spreadability, and with optimum bioadhesion properties [42].

5. XG-Based Systems for the Delivery of Drugs

5.1. Delivery of Anti-Microbial Drugs

Infections brought on by various microorganisms pose a serious hazard to human health. The infections caused by microorganisms are treated using three different categories of drugs—antiviral, antibacterial, and antifungal [43]. These antimicrobial drugs have proven to be effective against a wide range of infectious diseases. However, patients' options for therapy are constrained since bacterial resistance develops faster than the development of antibacterial drugs. The number of multi-drug-resistant organisms is the reason behind mortality and morbidity worldwide. Nanoparticles have shown synergy when used with the best antibiotics, and they could help to control bacterial resistance development. XG is used as an anti-microbial drug carrier that exhibits a controlled and sustained release of drugs. XG also helps in the formation of different hydrogels, tablets, and mucoadhesive patches for the delivery of antibiotics. To improve the residence time and bioavailability of besifloxacin, in situ ocular gels were prepared using XG, sodium alginate, and ethyl cellulose in different ratios. Besifloxacin is a fluoroquinolone antibiotic that inhibits the DNA gyrase (topoisomerase II and IV) and is used against bacterial conjunctivitis. Different formulations were prepared and evaluated for pharmacopoeial specifications. It was observed that the optimised formulation was transparent, clear with high gelling capacity, non-irritant, and showed sustained drug release (for 8 h) with high drug content capacity (96.2–98.6%) [44].

Similarly, in situ ocular gel was synthesised using a combination of polymers: XG, hydroxypropyl methylcellulose (HPMC), and sodium alginate in different ratios. This in situ gel was employed for the delivery of flucytosine, a semisynthetic macrolide antibiotic for the treatment of trachoma (caused by *Chlamydia trachomatis*). The synthesised gel aimed to achieve prolonged pre-corneal retention time and improved bioavailability of the drug in comparison to the conventional dosage form. The drug-loaded formulation showed high drug content (95.2–98.6%), sustained drug release (up to 8 h), and stability at both room and higher temperatures [45]. In a study, a floating low-density tablet formulation of stavudine was prepared using different polymers: XG, HPMC K15, and HPMC K100M for HIV treatment. The formulation of 80 mg tablet with a flat face, 9 mm diameter, was prepared to increase the half-life of stavudine in gastric fluid. Sustained release of the antiviral drug stavudine was achieved with all the pre-compression and post-compression characteristics [46].

Modification in the XG structure is beneficial for the delivery of drugs. XG was modified and conjugated with magnetic nanoparticles. Implantable xanthan gum/Fe_3O_4—magnetic nanoparticle composite hydrogel was synthesised for a non-invasive imaging study as shown in Figure 3.

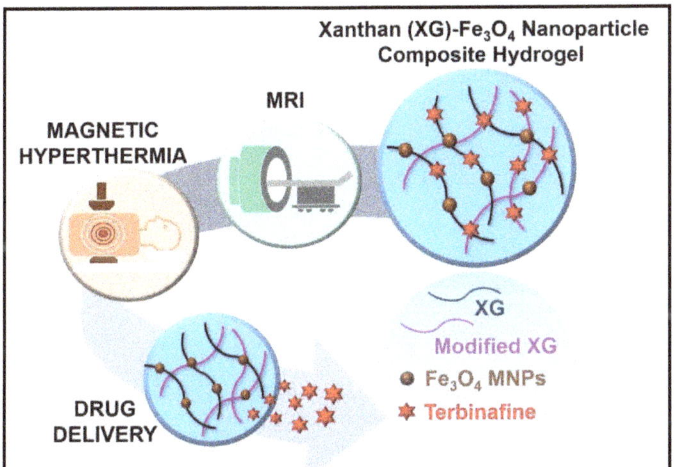

Figure 3. Xanthan gum-Fe_3O_4 nanoparticle composite hydrogel for the theranostics application [47]. Reproduced from ref. [47]. Copyright 2021 American Chemical Society.

The modified polymeric matrix encapsulated terbinafine (hydrophilic drug) and magnetic iron oxide. It showed rheological shear-thinning and enhanced swelling properties. The hydrogel also demonstrates thermally induced controlled drug delivery (3-fold) by magnetic hyperthermia and high efficacy compared to the pure drug for antifungal properties. The incorporation of Fe_3O_4 nanoparticles helped the hydrogel in functioning as non-invasive monitoring by magnetic resonance imaging (MRI). Magnetic hydrogels of different concentrations were compatible with normal human dermal fibroblasts and opened a new outlook in the biomedical application of XG [47].

In a study, two polymers, Chitosan and XG, were cross-linked with 2-acrylamide-2-methylpropane sulfonic acid (monomer). The polymeric network was synthesised using the free radical polymerisation method for the delivery of acyclovir, which is a highly selective antiviral drug, for the treatment of prophylaxis and *Herpes simplex* virus infections. The system showed pH-dependent drug release and swelling behaviour. It was observed that the hydrogel had a porous structure and was thermally stable [48].

5.2. *Delivery of Chemotherapeutic Agent*

XG is decomposed in the presence of colon enzymes rather than digested in the stomach or small intestine of humans. XG matrices are used to deliver the drug to the colon while protecting it from the stomach and small intestine environments. When they enter the colon, they are reduced into smaller monosaccharides by the anaerobic microflora present in the colon and either used as an energy source by the bacteria or broken down by an enzyme [49]. Microspheres and hydrogels were synthesised using XG to be targeted against colon cancer. α-Linolenic acid and docosahexaenoic acid were loaded in the formulations and were evaluated for their antioxidant and antineoplastic activities. The loaded formulations: microspheres encapsulated α-linolenic acid efficiently, whereas hydrogels were loaded with both the polyunsaturated fatty acids. The antioxidant activity and anti-neoplastic activity of α-linolenic acid were increased when loaded into the XG-based nanoformulations, whereas the activity of docosahexaenoic acid remained the same when tested against colon cancer cells [50]. 5-Fluorouracil is a chemotherapeutic drug used in the treatment of various cancers. The delivery of this drug causes serious side effects in the body, such as mucositis, alteration in the normal microflora environment of the colon, and diarrhoea [51]. For the target-specific treatment of colon cancer, 5-fluorouracil was encapsulated within XG-poly(-vinylpyrrolidone-co-poly) (PVP)—acrylic acid-based hydrogels. The pH-sensitive hydrogel was prepared using different concentrations of polymer, monomer, and crosslinker. The gel fraction was directly proportional to the concentration of polymer, monomer, and crosslinker, whereas the swelling capacity decreased with the increased concentration of polymer and crosslinker. It showed stability in a wide range of pH media and maximum swelling in a basic medium, confirming its use as a pH-responsive hydrogel. In rabbit models, acute oral toxicity studies were performed, demonstrating that the hydrogels were non-toxic. It was inferred that prepared pH-sensitive hydrogels were ideal for the delivery of 5-fluorouracil to the colon for sustained release [52]. Treatment of colorectal cancer by encapsulating 5-fluorouracil prebiotics and probiotics was an approach taken by Singh et al. [51]. The system with three components was taken: drug-loaded nanoparticles coated with prebiotics (XG, guar gum) and probiotics (*Bifidobacterium bifidum*). XG and guar gum were used to enhance the delivery of 5-fluorouracil in the colon and protect it from gastric juices. In vitro and in vivo studies showed intact microflora of the intestine, which fulfils the aim of the research of overcoming the side effects of 5-fluorouracil by co-administration of probiotics (*B. bifidum*) [51]. For the delivery of a well-known chemotherapeutic agent, curcumin, XG-chitosan nanofibers were prepared. The developed nanofibers were stable at pH 6.5 and pH 7.4. When tested against Caco-2 human colon cancer cell line, the loaded curcumin showed more efficacy than the pure drug. Further, the permeability of curcumin was increased 3.4-fold in the presence of XG-chitosan nanofibers [53].

pH-responsive nanoparticles [54] and nanogels [55] of XG were prepared for the delivery of doxorubicin, an anticancer drug used for the treatment of different cancers. The nanoparticles (110–180 nm) showed drug loading of 15.2% and pH-sensitive drug release. The nanoparticles were found to be biocompatible [54]. The nanogels were synthesised by amidation of xanthan gum with cysteamine tetra-hydrazide. The nanogels showed good pH sensitivity due to the presence of disulfide bonds. The disulfide bonds could be reduced by glutathione, causing drug release from the nanogels. Nanogels were also nontoxic and biocompatible, proving to be an ideal carrier for the delivery of chemotherapeutic drugs [55]. Gold nanoparticles were synthesised for the delivery of doxorubicin, and XG was used to stabilise the nanoparticles and as a reducing agent.

The XG stabilised gold nanoparticles were obtained with 5 mL of 1.5 mg/mL XG solution. The nanoformulation was stable under different pH conditions (pH 5–9). In vitro studies showed that nanoparticles coated with xanthan gum had a 3-fold increase in efficacy of the drug against the A549 human lung cancer cell line [56].

Similarly, thermos-responsive gold nanoparticles were synthesised and coated with carboxymethyl xanthan gum by microwave irradiation method. The release of the doxorubicin was extensive at acidic pH and negligible at physiological pH. The in vitro studies of doxorubicin-loaded nanoparticles showed a 4.6-fold increase in the anticancer activity of the drug compared to the free drug [57].

5.3. *Delivery of Anti-Diabetic Drugs*

Diabetes mellitus (DM) has emerged during the past few decades as one of the primary fatalities on a global scale [58]. Although many hypoglycaemic drugs are frequently used in diabetes control, a cure for diabetes is still a long way off because of the various inherent flaws and negative side effects of the drugs [59]. Due to proteolytic degradation and chemical instability in a severe pH environment, the necessary optimal drug concentration cannot reach focal regions. Additionally, substantial fluctuations in glucose that cause severe hypoglycaemia cannot be addressed with traditional dosing forms [60]. Other problems with pharmacological therapy, such as poor drug absorption, a short half-life, poor solubility, poor bioavailability, and negative effects on other organs, make treatment difficult [61]. XG owing to its various benefits is used for the delivery of anti-diabetic drugs to improve its bioavailability and solubility. Glimepiride is a hypoglycaemic drug with low solubility and bioavailability. It is extensively used in the treatment of diabetes (type 2). Transdermal delivery of the drug is desired with high permeation and increased solubility. Topical nanoemulgel was formulated using clove oil, PEG-400, and Tween-80 and further gelled with xanthan gum (3% *w/w*). The formulation significantly increased the permeation and bioavailability of a drug [62]. Glibenclamide is also a hypoglycaemic drug and is used to manage type 2 diabetes. It tends to pass through gastrointestinal mucosa. XG-based mouth-dissolving films were synthesised, and glibenclamide was loaded onto them. The films were formed for the rapid release of the drug and to increase its bioavailability in the body. The optimised formulation showed drug loading of 96% and decomposed within 40 s. Stability studies proved that the formulation was stable even in extreme conditions [37]. Repaglinide is an antidiabetic drug used to manage diabetes in humans. A mucoadhesive buccal tablet of the drug was prepared for the prolonged release of the drug in the system. XG and HPMC K100M polymers were used for the assessment of bioadhesion and release of the drug. The drug loading was 96.55%, and the release of the drug followed non-Fickian diffusion. The drug release was directly proportional to the concentration of both the polymers used. Stability studies showed that the tablets were stable for up to 2 months without any decrease in the drug content [63]. Tolbutamide has less bioavailability, and thus its efficacy in the body decreases. To enhance the bioavailability of the drug, xanthan gum-stabilized gold nanoparticles were synthesised. The loaded formulation showed high loading efficiency without changing the structure of the drug. In animal studies, insulin secretion was found to be higher in the mice treated with the developed formulation in comparison to the mice treated with pure drug solution [64].

5.4. *Delivery of Drugs for the Treatment of Cardiovascular Diseases*

For the treatment of cardiovascular diseases, XG-based carrier systems were developed. Verapamil hydrochloride is a calcium channel blocker and is used to lower blood pressure in the body. To achieve a controlled and sustained release of the drug, a matrix was synthesised using different polymers. XG, HPMC, and corn starch were used to delay drug release. Using carboxymethylation, different formulations were made of the drug with different polymers. XG, among all, significantly reduced the surge in the release of a drug.

The drug release was dependent upon the concentration of the polymer [65]. Furthermore, for the sustained release of propranolol hydrochloride, tablets were produced using XG and chitosan. The drug release pattern was observed in both of the polymers individually and in a combination of both (1:1) as a matrix, and with the addition of lactose (75% *w/w*), the mixture of both polymers prolonged the time of drug released from the matrix in pH 1.2 [66]. For the delivery of the drug, carvedilol, transdermal patches were synthesised to increase its bioavailability and to protect the drug loss in the first pass event. Different formulations were prepared using XG, HPMC K4M, and sodium alginate in different ratios. Formulations containing xanthan gum showed more than 85% of drug release within 12 h and followed Korsmeyer–Peppas' kinetics during drug release [67]. To increase the bioavailability and for sustained release of diltiazem HCl, different membranes with high and low molecular weight xanthan gum and poly (-vinyl alcohol) were synthesised. The mechanical strength of a high molecular weight polymer-based membrane was enhanced compared to a low molecular weight polymer-based membrane, whereas the physical properties were similar. Thinner membranes showed high encapsulation, whereas the thicker membrane showed more sustained and slow release of the drug [68]. In a study, a controlled-release pharmaceutical solid dosage form including metoprolol succinate, low molecular weight chitosan, and xanthan gum was developed and evaluated (Figure 4) [69].

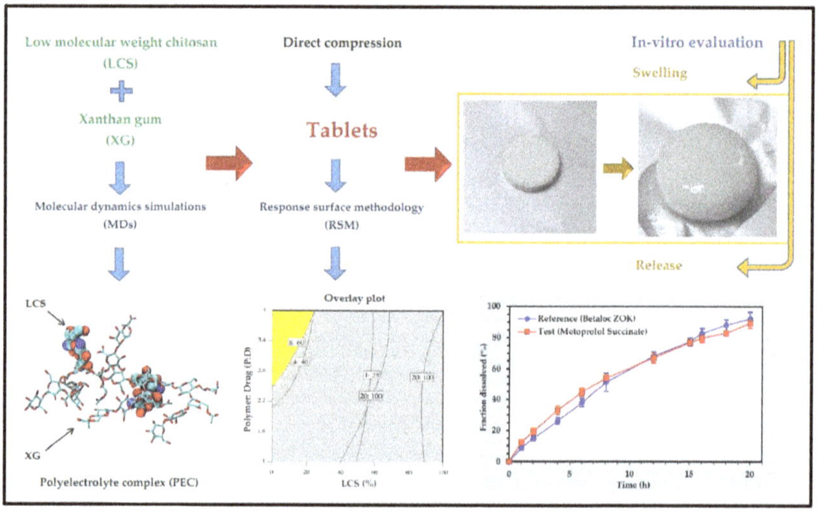

Figure 4. Development of xanthan gum-chitosan directly compressed tablets for the controlled release of metoprolol succinate [69]. Reproduced from ref. [69]. Copyright 2020 American Chemical Society.

Drug release and computational studies revealed low molecular weight chitosan performs better than high molecular wright chitosan in controlling the release of the drug when directly compressed with XG. Metoprolol succinate was found to be localized more in the core of the tablets in the beginning stages of dissolution because of film formation between XG and low molecular weight chitosan, which prevents the penetration of excessive water in the tablet matrix. The film begins to disintegrate or erode in the final stages of the breakdown process, enabling complete tablet hydration and a consistent drug distribution in the swelling tablet [69].

5.5. Anti-Spasmodic Drug Delivery

Irritable bowel syndrome (IBS) is a condition characterised by cramping, stomach pain, bloating, constipation, and diarrhoea. IBS affects women more than males, with a female-to-male ratio of 2–2.5:1 among those who seek medical care [70]. Anti-spasmodic drugs serve to prevent or treat painful cramping spasms in the intestines by relaxing the smooth muscles of the gut [71]. In a study, a floating tablet using XG and sodium alginate was prepared by an effervescent method [72]. Citric acid and sodium bicarbonate were used as gas-generating agents. Fenoverine is an antispasmodic medication used to treat muscle spasms, stomach cramps, and irritable bowel syndrome-related abdominal pain. Pre-compression studies were carried out, and for the directly compressed tablets, post-compression studies were done. The polymers enhanced the viscosity of the formulation in the dissolution medium and gave a sustained release of the drug efficiently for up to 12 h. Of the different formulations prepared, the formulation having 15% XG was determined to be the ideal formulation with 99.6% drug release in 12 h. It also showed a better swelling ratio with required drug release kinetics and floating behaviour. The floating tablets were ideal for the delivery of the drug and increased its bioavailability and sustained release in the system [72].

5.6. Delivery of Drugs for the Treatment of Inflammation, Rheumatoid Arthritis, and Gout

Rheumatoid arthritis and gout are common types of inflammatory arthropathy [73]. Non-steroidal anti-inflammatory drugs are used for the treatment of inflammatory disorders but have serious adverse effects, such as kidney damage [74]. To increase the efficacy of anti-inflammatory drugs and reduce their side effects in renal diseases, XG is used as a carrier system. It reduces the need for repetitive dosage and provides sustained release of the drug. Diclofenac potassium is a non-steroidal anti-inflammatory drug, and research was carried out to find the adsorption of graphene oxide nanoparticles on the drug [75]. To increase the adsorption of the graphene oxide, the co-polymeric hydrogel was prepared. XG and poly-acrylic acid were used to achieve sustained release of the drug by forming composite hydrogel with drug-loaded graphene oxide. The swelling ratio and drug release certainly increased with the increase in the pH. The drug-loaded composite hydrogel was pH-responsive. In vivo studies concluded that the bioavailability of the drug increased with an increase in the concentration of composite hydrogel [75]. Similarly, for the delivery of the same drug, double-walled carbon nanotubes enclosed with xanthan gum-based hydrogels were formed for the improvement of drug release. Carboxymethylation of the XG was carried out, and an increase in viscoelasticity was observed due to the presence of XG in the system. The encapsulation efficiency of the diclofenac was increased to 75%, and the sustained release of the drug was observed due to the presence of nanotubes. The composite was stable in extreme conditions similar to that which may be present on the skin [76]. For the delivery of lornoxicam-loaded thiolated XG as a topical gel formulation, it was synthesised by the microwave esterification reaction. The thiolation of the xanthan gum was carried out using thioglycolic acid [32]. The physicochemical properties of the gel formed were ideal, and the drug content in the thiolated gel was 98.4–99.7%. The in vitro drug release studies showed sustained release for up to 24 h. The thiolated gel was stable over a wide range of temperatures and was found to be ideal for topical drug delivery in the treatment of inflammation and pain [32].

Mucoadhesive buccal tablets were prepared using different concentrations of XG and *Vigna mungo* for the treatment of gout. The tablets were used to increase the bioavailability of febuxostat at the mucosal site for a prolonged time; the sustained release of the drug was 8 h. The tablets were stable at room temperature and showed good mucoadhesive properties [77].

To avoid an acidic environment and drug degradation in the stomach, XG-chitosan microparticles with pH-sensitive Eudragit®-L were prepared for colon-targeted delivery of the drug. Quercetin is a flavonoid with low bioavailability showing anti-inflammatory and antioxidant properties [78]. The system was prepared to increase oral drug bioavailability and enhance its release in the intestine for inflammatory-based disorders. The microparticles were loaded with quercetin, and the release was controlled by non-Fickian diffusion of the drug. The microparticles were compressed and converted into tablets with a coating of Eudragit [78]. To evaluate the effect of changing concentrations of hydrophilic and hydrophobic polymers on the sustained release of metoclopramide encapsulated in matrix tablets. The matrix tablets were prepared using XG and ethyl cellulose together and individually. Pre-compression and post-compression parameters of the tablets were found to be under the limit. The tablets prepared using XG and XG-ethyl cellulose both showed sustained drug release for up to 12 h. Drug release kinetics followed the Korsmeyer–Peppas model [79]. For the treatment of colitis, curcumin-loaded pH-responsive polyacrylamide-grafted-xanthan gum nanoparticles were prepared. The drug release was less (~8%) at acidic pH 1.2 in comparison to the drug released (~35%) at pH 7.2. These results indicated the pH-responsive nature of the formulation. Further, the drug release from the formulation was ~65% in rat caecal pH 6.8, suggesting microflora-dependent drug release. In in vivo studies, drug-loaded nanoparticles demonstrated a greater reduction in colonic inflammation, better prevention of weight loss, and more decrease in myeloperoxidase and nitrite levels than pure drug [80].

5.7. Delivery of Immunosuppressive Drugs

To assure therapeutic outcomes, immunosuppressant drugs are typically administered via conventional techniques that frequently demand high doses of the drugs, although this can lead to harmful side effects such as infections, cancers, and specific side effects in pregnancy [81,82]. Synthesis of formulations using XG carrying the optimum amount of drug to be effective at the target site was achieved by Sadeq et al. [83]. Ketotifen fumarate is an antihistamine (selectively blocks histamine H1 receptors) and is used in conjunctivitis and rhinitis. It has low solubility and belongs to Biopharmaceutical Classification System (BCS) class II drugs. To enhance its bioavailability in ocular tissues, it was loaded into in situ ocular gel for the treatment of eye infections. pH-sensitive ocular gels were prepared for better efficacy than the formulation present in the market. Different concentrations of polymers: xanthan gum, Carbopol, and gellan gum, were taken to form different formulations. Amongst the optimised formulations, formulation with 0.75% Carbopol and 0.3% XG was found to be optimum. It had a drug encapsulation efficiency of 99.74%, gel-forming strength in 46.6 s, and drug release of up to 8 h [83]. These findings demonstrate the impact of polymer combinations; it was observed that as XG or gellan gum concentrations rose, the percentage of the drug release reduced. This is due to increased gel strength and decreasing the size and quantity of gel structure channels, which limit the movement of the drug molecule and minimise dissolution media penetration, delaying the time at which the drug is released [83,84]. To enhance the bioavailability of the drug, Tacrolimus microbeads were synthesised. A highly effective immunosuppressive drug with demonstrated effectiveness in both in vivo and in vitro studies, Tacrolimus is also a calcineurin inhibitor. Varying concentrations of XG and sodium alginate were used to form microbeads. The sodium alginate was taken as the main polymer but to increase the drug-loading efficiency of microbeads, xanthan gum was added to the formulation. Using different concentrations of crosslinkers also affected drug release. Furthermore, the drug release is dependent on the concentration of the polymer and coating polymer. It was suggested that this system could be used as an alternative and cost-effective system for the delivery of drugs [85].

5.8. Delivery of Drugs for the Treatment of Skin Diseases

For topical delivery of a drug, the skin itself serves as a natural shield against particle penetration. It also provides a potential method for administering medications, particularly to damaged skin and through the apertures of hair follicles. Recent advancements in dermatological treatment may be made by improving the dermal localisation of bioactive components into the damaged skin region. These nanocarriers can concentrate at the target site after application by passing through the stratum corneum into viable skin [86]. Xanthan gum-based nanocarrier systems help in the efficient delivery of drugs for the treatment of various skin disorders. Pycnogenol is a French maritime pine bark extract that is a concentrate of phenolic compounds. It shows several biological activities such as anti-inflammatory, anti-oxidant, and anti-bacterial and is used in the treatment of open wounds.

For the enhanced release of the extract, a bioadhesive film using sodium alginate and XG was prepared in different ratios [87]. The film showed suitable mechanical properties such as high deformability, suggesting easy adaptability to any type of surface. The film shows high bioadhesion on the skin and absorption of exudates from the open wound. In vitro studies showed that drug-loaded films can inhibit the growth of bacterial strains, including *S. pyogenes*, *S. aureus*, and *E. faecalis*. Together with the help of sustained release of the drug, it was attributed that the formulation needs to be applied once a day, and it helps in the growth of keratinocytes, thus, healing the wound [87].

5.9. Delivery of Drugs for the Treatment of Central Nervous System-Related Disorders

The consequences of so-called modernised living include increased stress, mental disorders, and psychological misbehaviours due to humans' increasing desire and avarice. Every year, a wide range of mental illnesses claim millions of victims, and the need for treatment is only expanding. The blood-brain barrier and the blood-cerebrospinal fluid barrier restrict drug access to the central nervous system (CNS). These barriers can be overcome or bypassed by biomaterials, allowing for the controlled administration of medications into the CNS [88]. Quetiapine fumarate is a typical antipsychotic medication and is used in the management and treatment of psychosis, bipolar disorder, and other neuropsychological or affective disorders. To enhance the bioavailability of quetiapine fumarate and protect it from first-pass metabolism, mucoadhesive matrix tablets were prepared. The matrix tablets consisted of a mixture of polymers (xanthan gum, Carbopol 934P, HPMC K4M, polyvinyl pyrrolidone K30) and showed a sustained release of drug (96% in 10 h) along with ~77% permeation across the mucus membrane [89]. Furthermore, to increase the bioavailability of the drug and surface characteristics in the solid state, tablets—a blend of drugs with the XG and tragacanth gum were synthesised [90]. The results presented that the type of natural gum and its structure greatly affected drug release. XG exhibited a swelling mechanism for the controlled release of the drug. The binary tablets formed with XG showed pH-responsive drug release, fitting Higuchi and Korsmeyer–Peppas models [90].

Another drug, zolmitriptan, is a triptan that is used for the cure of migraine attacks with/without aura and clusters. To enhance the sustained release of this drug, oral tablets were synthesised for the treatment of migraine. Preparation of the tablets was carried out using xanthan gum, guar gum, and karaya gum by direct compression method. Formulations containing xanthan gum showed an increase in sustained release of the drug with the increase in the concentration of polymer [91]. For the treatment of Alzheimer's Disease, the administration of resveratrol-loaded lipid nano formulation via nasal route was prepared. The lipid carrier was incorporated in xanthan gum and gellan gum for in situ gel formation. The optimisation of in situ gel resulted in 5-fold higher permeation in the nasal mucosa when compared to pure drug suspension-based gel. The drug-loaded formulation was evaluated using the scopolamine-induced amnesia model in mice, and improvement in memory function was observed [92].

5.10. Delivery of Drugs for the Treatment of Obesity

Orlistat is a lipase inhibitor that blocks the absorption of fat from the body by blocking certain enzymes responsible for the breakdown of fats. It is used in the treatment of obesity and is taken with a low-calorie diet. The drug has several side effects, such as stool of fatty and oily consistency. To reduce side effects, an oral dosage formulation of orlistat was developed. Xanthan gum was used as an oil-trapping agent. Mini-tablets of orlistat (10 in no.) were co-administered with mini-tablets of XG (40 in no.). The drug release of the optimised mini-tablets was compared to commercial tablets available; both showed similar drug release profiles [93]. The formulation had a large amount of XG, but it has already been proven that XG is safe in food additives (FDA-approved). Comparative in vivo studies were performed between optimised mini-tablets and commercial forms of drug using Sprague-Dawley rats. The faeces of the treated rats were analysed, and the faeces were significantly less oily and fatty when treated with the optimised formulation of orlistat and xanthan gum. As a matter of patient compliance, optimised mini-tablets were easy to swallow. The overall conclusion was that the mini-tablets formed using XG as an oil-reducing agent might help in the development of many other delivery systems for anti-obesity drugs to reduce their side effects [93].

5.11. Delivery of Drugs for the Treatment of Glaucoma

Acetazolamide blocks carbonic anhydrase, which also reduces the build-up of certain fluids in the body. It is used to treat glaucoma, in which there is fluid build-up in the eye, creating pressure. To improve the efficacy of the drug and achieve its sustained release, the drug was encapsulated in a nanoemulsion-based in situ gel [94]. Different nanoemulsion formulations were prepared using different surfactants. For pH-responsive in situ gel formation, pure gellan gum and in combination with HPMC, xanthan gum, and Carbopol were used. The in situ gels showed improved sustained release than nanoemulsion. A combination of XG with gellan gum proved to be the ideal formulation, with increased mucoadhesive properties and a relative decrease in ocular pressure compared to commercial oral tablets and eye drops [94]. Dorzolamide hydrochloride is also used to treat glaucoma. A mucoadhesive in situ gelling formulation was prepared for the treatment via the ocular route. XG and poly acrylic acid were used in combination as well as individually for the synthesis of in situ gel. Formulation with 0.2% *w/v* XG showed 90.84% of drug release for 12 h. The drug release was also analysed using goat's cornea, and the formulation showed similar release kinetics to the in vitro release for up to 9 h [95].

5.12. Delivery of Drugs for the Treatment of Pulmonary Diseases

For the treatment of respiratory conditions associated with lung cancer, asthma, and chronic obstructive pulmonary disorders, pulmonary medication delivery allows targeted therapy. However, the rapid clearance mechanism, drug deposition, and instability mechanism present problems with this method. The effectiveness of pulmonary delivery systems is further influenced by additional elements such as the kind of inhaler device, patient compatibility, consistent device delivery, and inhaler technique [96]. In order to address these problems, nanotechnology has been used in pulmonary delivery systems, targeting the direct intake of aerosols into the respiratory epithelium and epithelial cells. The architecture of the aerosol formulation can be altered to offer either rapid absorption or prolonged retention depending on the desired drug release characteristics [97]. For the controlled release of theophylline, interpenetrating polymer network-based hydrogel beads were synthesised.

Sodium carboxymethyl xanthan and casein were crosslinked with aluminium chloride and glutaraldehyde to synthesise the hydrogel beads. Theophylline belongs to the class of drugs known as xanthines. It is a bronchodilator and is used to manage symptoms of Chronic Obstructive Pulmonary Disease (COPD), asthma, and other various lung conditions due to reversible airflow obstruction. Dissolution studies and drug release were carried out in both acidic and alkaline mediums, and delayed release of the drug was observed [98]. Furthermore, to control the dose dumping of theophylline, XG was used as a rate-controlling gum to synthesise alcohol-resistant mini-tablets and matrix tablets. It increased the solubility of the drug by 2.8 times in 40% alcohol. At lower concentrations of XG with large particles, the drug release was observed to be faster when compared to the higher concentration of polymer with finer particles (<75 μm) in dissolution studies [99]. A summary of the polymers, forms, and drugs targeting various diseases and their delivery is presented in Table 2.

Table 2. Summary of polymers, dosage forms, drugs, and their delivery.

Sr. No.	Polymers	Form	Drug	Remarks	Reference
			Delivery of antimicrobial drugs		
1.	XG, sodium alginate and ethyl cellulose	In situ ocular gels	Besifloxacin	The transparent, clear, non-irritant gel was synthesised with sustained drug release (for 8 h) and high drug content capacity.	[44]
2.	XG, HPMC, and sodium alginate	In situ ocular gels	Flucytosine	The drug-loaded formulation showed high drug content (95.2–98.6%), sustained drug release (up to 8 h) and stability at both room and higher temperatures.	[45]
3.	XG, HPMC, and sodium alginate	Floating low-density tablet	Stavudine	Sustained release of the antiviral drug stavudine was achieved with all the pre-compression and post-compression characteristics for the tablet.	[46]
4.	Xanthan gum, bovine serum albumin and magnetic nanoparticles	Magnetic field responsive antimicrobial patch	Amoxicillin	The adsorption of the drug was found to be more on XG and BSA films compared to others. Drug release followed a quasi-Fickian pattern. The formulation was tested for antimicrobial activity, and it showed more efficacy in the presence of an external magnetic field.	[100]
5.	XG		Combination of tobramycin and dexam-ethasone	The combination of both drugs loaded into xanthan gum was better than the commercially available treatment even when it contains half the amount of steroid.	[101]
6.	XG coated with Eudragit-L 100	Tablets	Metronidazole	Coated XG tablets showed delayed release. In the dissolution medium, decreased drug release was observed due to microbial degradation or polymer solubilisation. The release of the drug was greatly affected by the nature of the polymer used and the enteric coating of the tablet.	[102]
7.	XG and Chitosan	Polyelectrolyte hydrogel	Ciprofloxacin	Increased drug encapsulation with an increase in drug concentration was observed, and it followed zero-order drug release kinetics. The drug-loaded hydrogel inhibited the growth of bacteria and showed no effect against fungi.	[103]
8.	XG	Microspheres	Ciprofloxacin	The in-vitro drug release was carried out in both acidic as well as alkaline mediums; the release pattern followed a non-Fickian trend. Based on the results, it was concluded that the synthesised formulation was suitable for the sustained release of ciprofloxacin.	[104]
9.	XG	Hydrogel	Liranaftate	The microemulsion-based drug hydrogel showed increased antifungal activity in comparison to plain drug solution against *Candida albicans*. The microemulsion proved to increase the skin retention and permeability of the drug.	[105]
10.	XG, HPMC K15, and HPMC K100M	Floating low-density tablet	Stavudine	An increase in the half-life of stavudine in the gastric fluid was noted. Sustained release of the antiviral drug stavudine was achieved with all the pre-compression and post-compression characteristics.	[46]

Table 2. *Cont.*

Sr. No.	Polymers	Form	Drug	Remarks	Reference
11.	XG with magnetic nanoparticles (Fe_3O_4)	Hydrogel	Terbinafine	The hydrogel shows thermally induced controlled drug delivery (3-fold) by magnetic hyperthermia and high efficacy compared to the pure drug for antifungal properties. The incorporation of Fe_3O_4 nanoparticles helps the hydrogel function as non-invasive monitoring by magnetic resonance imaging (MRI).	[47]
12.	XG, poly(-vinyl alcohol), sodium alginate	Interpenetrating polymer network microbeads	Norfloxacin	Hydrogel showed high encapsulation efficiency. The microbeads had mucoadhesive properties at physiological pH, which was assessed with the use of a caprine intestine. The system showed pH-dependent drug release; it fitted the Higuchi and Korsemeyer–Peppas models.	[106]
13.	XG and Chitosan	Hydrogel	Acyclovir	The drug-loaded formulation showed pH-dependent drug release and swelling behaviour; it fitted the Higuchi and Korsemeyer–Peppas models. It was observed that the hydrogel had a porous structure and was thermally stable.	[49]
14.	XG and Chitosan	Hydrogel scaffold	Ampicillin, Rifampicin, and Minocycline.	Hydrogel showed good swelling and porosity. The in vitro drug release study of all the antibiotics showed fast drug release within 24 h in simulated gastric fluid. The gels prepared were stated to be non-toxic after the in vitro cell cytocompatibility test. Antibacterial activity of the loaded formulation was evaluated.	[107]
15.	XG	Solution	Linezolid	The drug-loaded formulation was analysed against *Staphylococcus aureus* infections in the ocular tissues. The in vitro antibacterial activity and in vivo evaluation of drug efficacy showed that the linezolid-loaded XG exhibits better topical instillation and ocular permeation with bioadhesion to the precorneal membrane.	[108]
			Delivery of chemotherapeutic drugs		
1.	XG	Microspheres and Hydrogels	α-linolenic acid and Docosahexaenoic acid	The antioxidant activity and anti-neoplastic activity of α-linolenic acid were increased when loaded into the XG-based nanoformulations whereas the activity of docosahexaenoic acid remained the same when treated against colon cancer.	[50]
2.	XG-chitosan	Nanofibers	Curcumin	The developed nanofibers were observed to be stable at pH 6.5 and pH 7.4. The drug-loaded formulation was tested on the colon cancer cell line; it was observed that the loaded polyphenol showed more efficacy than the pure drug. The permeation of the drug was increased by 3.4-fold.	[53]
3.	XG	pH-responsive nanoparticles	Doxorubicin	The nanoparticles showed high drug loading and pH-sensitive drug release. The drug-loaded formulation was biocompatible.	[54]

Table 2. *Cont.*

Sr. No.	Polymers	Form	Drug	Remarks	Reference
4.	XG	Nanogels	Doxorubicin	Nanogels were also nontoxic and biocompatible proving to be an ideal carrier for the delivery of chemotherapeutic drugs.	[55]
5.	XG coated Gold nanoparticles	-	Doxorubicin	XG is used as a stabilising and reducing agent. In vitro studies showed that nanoparticles coated with XG showed a 3-fold increase in the efficacy of the drug against the A549 cell line.	[56]
6.	Carboxymethyl XG coated Gold nanoparticles	-	-	Modified XG was used as a stabilising agent. The release of the drug was extensive at acidic pH and physiological pH. In-vitro studies of drug-loaded nanoparticles showed a 4.6-fold increase in the anticancer activity of the drug compared to pure drug.	[57]
7.	XG-poly(-vinylpyrrolidone-co-poly)-acrylic acid	pH-responsive hydrogel	5-fluorouracil	Controlled release of the drug was observed. It also showed stability in a wide range of pH media and maximum swelling in basic medium confirming its use as a pH-responsive hydrogel. In vivo studies showed acute oral toxicity studies were performed.	[52]
8.	XG and guar gum	Nanoparticles	5-fluorouracil	In vitro and in vivo studies showed intact microflora of the intestine, which fulfils the aim of the research of overcoming the side effects of 5-fluorouracil by co-administration of probiotics (*B. bifidum*)	[51]
9.	XG	Thermoreversible gel	Ondansetron hydrochloride	The addition of XG increased the viscosity of the gel, thus decreasing the release of the drug and achieving sustained drug release. The gel was stable at room temperature and freeze temperature for a period of a month.	[109]
			Delivery of anti-diabetic drugs		
1.	XG	Topical nanoemulgel	Glimepiride	XG (3% *w/w*) was used as a gelling agent. The formulation significantly increased the permeation and bioavailability of the drug.	[62]
2.	XG	Mouth dissolving films	Glibenclamide	The films were formed for the rapid release of the drug and to increase its bioavailability in the body. Stability studies proved the formulation was stable even in extreme conditions.	[37]
3.	XG and HPMC K100M	Mucoadhesive buccal tablet	Repaglinide	The drug release was directly proportional to the concentration of both the polymers used. The stability studies showed that the tablets were stable for up to 2 months without any decrease in the drug content.	[63]
4.	XG stabilised gold nanoparticles		Tolbutamide	The loaded formulation showed high loading efficiency without changing the structure of the drug. Drug-loaded formulation was investigated for its insulin secretion in mice models; it enhanced insulin secretion when compared to the plain drug solution.	[64]

Table 2. *Cont.*

Sr. No.	Polymers	Form	Drug	Remarks	Reference
			Delivery of drugs for the treatment of cardiovascular disease		
1.	XG, HPMC, and corn starch	Matrix	Verapamil hydrochloride	XG-based formulation significantly provided sustained release of the drug. The drug release was dependent upon the concentration of the polymer.	[65]
2.	XG and Chitosan	Tablets	Propranolol hydrochloride	The mixture of both polymers prolonged the time of drug release from the matrix at pH 1.2. Three layered polymer tablet pH-responsive was formed, with an increase in the barrier/middle layer of the matrix, and the lag in the release also increased.	[66]
3.	XG, HPMC K4M, and sodium alginate	Transdermal patches	Carvedilol	The XG-based formulation showed drug release within 12 h. It was stable in various physicochemical parameters and ideal for the delivery of carvedilol.	[67]
4.	XG and poly(-vinyl alcohol)	Membranes	Diltiazem HCl	Mechanical strength, physical properties, and drug release were observed to be dependent on the molecular weight of the polymers used for the preparation of membranes.	[68]
			Anti-spasmodic drug delivery		
1.	XG and sodium alginate	Floating tablets	Fenoverine	The formulation with 15% XG was ideal, with 99.6% drug release in 12 h. It also showed an improved swelling ratio with required drug release kinetics and floating behaviour.	[72]
2.	XG	Matrix Tablet	Mebeverine HCl	XG can hydrate more rapidly than the other three gums used. The resulting drug diffusional path length for XG was the longest. The in vitro drug release studies revealed that the level of the polymer in the matrix tablets played an important role in the modulation of drug release.	[110]
			Delivery of drugs for the treatment of inflammation, rheumatoid arthritis, and gout		
1.	XG and graphene oxide	Hydrogel	Diclofenac potassium	The swelling ratio and drug release certainly increased with the increase in pH. The drug-loaded composite hydrogel was pH-responsive. In vivo studies concluded the bioavailability of the drug increased with an increase in the concentration of the composite hydrogel.	[75]
2.	Carboxylated XG	Hydrogels	Diclofenac	An increase in viscoelasticity was observed due to the presence of XG in the system. The encapsulation efficiency of the drug was increased and the sustained release of the drug was due to the presence of nanotubes. The composite was stable in extreme conditions similar to that may be present on the skin.	[76]
3.	Thiolated XG	Topical gel	Lornoxicam	The physicochemical properties of the gel formed were ideal. The high drug content in the thiolated gel was observed. In-vitro drug release studies showed sustained release up to 24 h. The thiolated gel was ideal for topical drug delivery for the treatment of inflammation and pain.	[33]

Table 2. *Cont.*

Sr. No.	Polymers	Form	Drug	Remarks	Reference
4.	XG and *Vigna mungo*	Mucoadhesive bilayer buccal tablets	Febuxostat	The tablets were used to increase the bioavailability of febuxostat at the mucosal site for a prolonged time. The sustained release of the drug occurred for 8 h. The tablets were stable at room temperature and showed good mucoadhesive properties.	[77]
5.	XG and chitosan	Microparticles	Quercetin	The microparticles were loaded with quercetin and the release was controlled by non-Fickian diffusion of the drug. The microparticles were compressed and converted into tablets with a coating of Eudragit.	[78]
6.	XG and ethyl cellulose	Matrix tablet	Metoclopramide	Pre-compression and post-compression characteristics of the tablets were found to be ideal. The tablets prepared using XG and XG-ethyl cellulose both showed sustained drug release for up to 12 h.	[79]
7.	Polyacrylamide-grafted-XG	pH-responsive nanoparticles	Curcumin	In an alkaline medium, the drug release was excellent, with the presence of rat caecal suggesting microflora-dependent drug release. In vivo studies suggested that the drug-loaded nanoparticles showed better anti-inflammatory activity when compared to the pure drug.	[80]
			Delivery of immunosuppressive drugs		
1.	XG, Carbopol and gellan gum	In-situ ocular gel	Ketotifen fumarate	pH-sensitive ocular gels were prepared for better efficacy than the formulation present in the market. Different concentrations of polymers: xanthan gum, Carbopol®, and gellan gum were taken to form different formulations. Among the optimised formulations, formulation with 0.75% Carbopol and 0.3% xanthan gum was found to be the best. It had a drug encapsulation efficiency of 99.74%, gel-forming strength in 46.6 s, and drug release of up to 8 h.	[83]
2.	XG and sodium alginate	Microbeads	Tacrolimus	The sodium alginate was taken as the main polymer but to increase the drug-loading efficiency of microbeads, xanthan gum was added to the formulation. Using different concentrations of crosslinkers also affects drug release. Also, the drug release is dependent on the concentration of the polymer and coating polymer. It was suggested that this system could be used as an alternative and cost-effective system for the delivery of drugs.	[85]

Table 2. *Cont.*

Sr. No.	Polymers	Form	Drug	Remarks	Reference
			Delivery of drugs for the treatment of skin diseases		
	Sodium alginate and XG	Bioadhesive film	Pycnogenol	The film showed suitable mechanical properties such as high deformability, suggesting easy adaptability to any type of surface. The film shows high bioadhesion on the skin and absorption of exudates from the open wound. The drug-loaded films can inhibit the growth of bacterial strains.	[87]
			Delivery of drugs for the treatment of central nervous system-related disorders		
1.	XG, Carbopol 934P, HPMC K4M, and polyvinyl pyrrolidone K30	Mucoadhesive matrix tablet	Quetiapine fumarate	Sustained release (96%) was obtained for a prolonged time (10 h). The permeation in the mucus membrane was 77%.	[89]
2.	XG and tragacanth	Tablets		XG exhibited a swelling mechanism for the controlled release of the drug. The binary tablets formed with XG showed pH-responsive drug release, fitting Higuchi and Korsmeyer–Peppas models.	[90]
3.	XG, guar gum, karaya gum	Oral tablets	Zolmitriptan	Formulations containing xanthan gum showed an increase in sustained release of the drug with the increase in the concentration of polymer.	[91]
4.	xanthan gum and gellan gum	In-situ gel	Resveratrol	The optimisation of in situ gel resulted in 5-fold higher permeation in the nasal mucosa when compared to pure drug suspension-based gel. The drug-loaded formulation was evaluated using the scopolamine-induced amnesia model in mice. Improvement in memory function was observed.	[92]
			Delivery of drugs for the treatment of obesity		
	Xanthan gum	Mini-tablets	Orlistat	The mini-tablets formed using XG as an oil-reducing agent might help in the development of many other delivery systems for anti-obesity drugs for reducing their side effects.	[93]
			Delivery of drugs for the treatment of glaucoma		
1.	Pure gellan gum and in combination with HPMC, XG, and Carbopol	pH-responsive in situ gel	Acetazolamide	A combination of XG with gellan gum proved to be the best formulation among all, increasing mucoadhesive properties and a relative decrease in ocular pressure when compared to commercial oral tablets and eye drops.	[94]
2.	XG and poly acrylic acid	Mucoadhesive in situ gel	Dorzolamide hydrochloride	Formulation with 0.2% *w/v* XG showed 90.84% of drug release for 12 h. The drug release was also analysed using goat's cornea; formulation showed similar release kinetics to the in vitro release for up to 9 h.	[96]

Table 2. *Cont.*

Sr. No.	Polymers	Form	Drug	Remarks	Reference
Delivery of drugs for the treatment of pulmonary diseases					
1.	Sodium carboxymethyl XG and casein crosslinked with aluminium chloride and glutaraldehyde	Interpenetrating polymer network-based hydrogel beads	Theophylline	Dissolution studies and drug release carried out in both acidic and alkaline mediums delayed release of the drug was observed.	[98]
2.	XG	Mini-tablets and matrix tablets.	Theophylline	It increased the solubility of the drug by 2.8 times in 40% alcohol. At lower concentrations of XG with large particles, the drug release was observed to be faster when compared to the higher concentration of polymer with finer particles (<75 μm) in dissolution studies.	[99]

6. XG-Based Systems for the Delivery of Genetic Materials

Gene therapy is a novel method that uses genes to treat or prevent various diseases. Instead of employing drugs or surgery to treat a disease, doctors may be able to use the gene therapy procedure to implant a gene into the patient's cell. Gene therapy involves a variety of techniques, such as (i) "knocking out" or inactivating a defective gene, (ii) gene replacing a disease-causing mutant gene with a healthy gene, and (iii) introducing new genes to guard against any disease [111]. The grafting concept enables three-dimensional interaction with anionic surfaces of double-strand DNA chains by attaching side chain oligomers to either linear or branched hydrophilic polysaccharides. The polysaccharide-based gene delivery technology suggests that the cationic polysaccharide's structure has a vital impact on the transfection ability for gene delivery. The delivery of complexes such as peptides, oligonucleotides, proteins, and plasmids via the mucosa has been explored using colloidal polysaccharide particles [112]. To ensure a negative charge and hydrophilic nature on the surface of the shell for stabilisation and inherent capability to target endothelial cells, xanthan gum was incorporated in sorbitan monooleate nanoparticles.

The goal of this work was to add XG into created sorbitan monooleate nanoparticles to target liver sinusoidal endothelial cells (LSECs) by utilising the polysaccharide composition [113]. It was hypothesised that the mannose residues in XG composition would help target the mannose receptors, which are known to be overexpressed in LSECs. As a scavenger receptor, the mannose receptor recognises and promotes the absorption of a variety of glycoconjugate ligands [86]. However, after systemic injection, it was observed that XG span nanoparticles are internalised not only by LSECs but also by vascular endothelial cells that lack the mannose receptor. Enhanced green fluorescent protein plasmid was encapsulated into the nanoparticles. The cytotoxicity and transfection of the loaded nanoparticles were tested against human umbilical vein endothelial cells. Biocompatibility studies were carried out to confirm the biocompatibility of XG-functionalised nanoparticles with vascular endothelium of the liver, lung, and kidney [113]. Figure 5 illustrates the composition of XG-based span nanoparticles for the targeted gene delivery to LSECs.

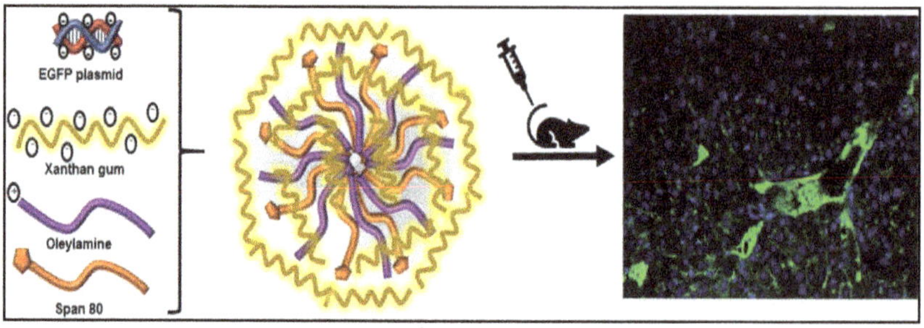

Figure 5. Xanthan gum-based span nanoparticles for gene delivery to endothelial cells [113]. Reproduced with permission from ref [113]. Copyright 2018 Elsevier.

7. XG-Based Systems for the Delivery of Proteins and Peptides

Since they are more effective and have lower toxicity than chemical medications, protein- and peptide-based therapeutics offer a wide range of possibilities as therapeutic agents. However, their use has been constrained due to delivery issues. Their oral bioavailability is very low, and transdermal administration is subject to absorption restrictions. Consequently, parenteral administration is used to deliver the majority of protein- and peptide-based medications. The discomfort of the patient is one issue with this method, particularly for paediatric use. Over the last few decades, intensive research has been conducted to create protein and peptide carriers that overcome the aforementioned issues. Nanoparticle carriers, absorption boosters, enzyme inhibitors, mucoadhesive polymers, and chemical alteration of protein or peptide structures are just a few of the methods that have been used [114]. Short half-life and physical-chemical changes in the protein structures in the gastrointestinal tract cause poor absorption. Crosslinked XG-poly(N-vinyl imidazole) based hydrogel was used to enhance the efficacy of bovine serum albumin (BSA). The carrier used was pH-responsive; further, drug release studies were performed in different pH mediums. The results showed that % release was faster at physiological pH (pH 7.4) than in an acidic medium (pH 1.2) following the non-Fickian release model. The intact structure of the released protein was confirmed by SDS-PAGE, and no disintegration of protein structure was observed. The loaded carrier was non-toxic against normal cells. Protein-loaded graft also showed antimicrobial activity against *Staphylococcus aureus* and *Aspergillus niger* and confirmed their utility as antimicrobial drug carriers [115]. Another carrier system prepared for the delivery of BSA was XG-PVI (poly-N-vinyl imidazole) hydrogel. The drug encapsulation and drug release were directly proportional to the gelling time and concentration of the drug, whereas they were inversely proportional to the concentration of the polymer. While in phosphate-buffered saline (PBS), the release of BSA was increased with the increased concentration of polymer. Cytotoxicity studies showed good biocompatibility of the hydrogel. Furthermore, structure intactness was confirmed by sodium dodecyl-sulfate polyacrylamide gel electrophoresis (SDS-PAGE). Given the results, it was concluded that XG-based drug delivery systems could be used for the delivery of proteins and peptides [116].

8. Conclusions and Future Perspective

Xanthan gum is a heteropolysaccharide produced by the fermentation of *X. campestris*. XG being inert and biocompatible in nature is used as an excipient for the sustained and controlled release of drugs. XG is also used as a reducing and capping agent in nanoparticle synthesis. Different forms (hydrogels, matrix tablets, gels, mucoadhesive patches, etc.) of XG are synthesised for the delivery of drugs for the treatment of various diseases. XG has shown applications in different industries, such as food, cosmetics, paints, ceramic glazes, and water-based drilling fluids. It is commercially used as a thickener and stabiliser. Inadequate thermal and mechanical qualities, microbiological contamination, unstable viscosity, low shear resistance, limited surface area, and an unpredictable degree of hydration are a few drawbacks of XG. Additionally, XG disintegrates slowly, particularly in cold water and at high concentrations. It disperses as fish eyeballs, which are water-based lumps, as a result of poor hydration. As a result, the surface of the particles quickly forms a gelatinous surface layer following dispersion. Water cannot pass through the gelatinous barrier, which delays the particle's complete disintegration. The improvement in physical and chemical characteristics (solubility, swelling, metal-induced gelling ability, mechanical and thermal stability) of XG is achieved by chemical alteration in the number of carboxyl and hydroxyl groups present in its structure. Etherification, esterification, acetylation, oxidation, peptide joining, ionic and covalent crosslinking, and mechanical modification are among the chemical modifications of XG, which might improve its overall physicochemical properties and broaden its applications.

The use of xanthan in the form of such luminous composites has immense promise for biomedical applications. Researchers may be challenged and inspired to develop new structures and functional systems based on xanthan gum because it is a fascinating biopolymer with a variety of intriguing features.

Author Contributions: Conceptualization, D.J.A., H.K.; Writing—original draft preparation, M.J., D.P.; writing—review and editing, M.J., D.P., D.J.A., H.K. All authors have read and agreed to the published version of the manuscript.

Funding: This research received no external funding.

Institutional Review Board Statement: Not applicable.

Informed Consent Statement: Not applicable.

Data Availability Statement: As no new data were created in this article, data sharing is not applicable to this article.

Acknowledgments: The authors thank the Central University of Gujarat, Gandhinagar, National Forensic Sciences University, Gandhinagar, and IHMRI, University of Wollongong, for providing the necessary facilities and support. M.J. acknowledges the University Grand Commission, New Delhi, for providing a PhD. Fellowship.

Conflicts of Interest: The authors declare no conflict of interest.

References

1. Jain, S.K.; Tiwari, A.; Jain, A.; Verma, A.; Saraf, S.; Panda, P.K.; Gour, G. Application potential of polymeric nanoconstructs for colon-specific drug delivery. In *Multifunctional Nanocarriers for Contemporary Healthcare Applications*; IGI Global: Hershey, PA, USA, 2018; pp. 22–49. [CrossRef]
2. Grenha, A.; Dionísio, M. Locust bean gum: Exploring its potential for biopharmaceutical applications. *J. Pharm. Bioallied Sci.* **2012**, *4*, 175–185. [CrossRef] [PubMed]
3. Cerqueira, M.A.; Pinheiro, A.C.; Souza, B.W.; Lima, M.; Ribeiro, C.; Miranda, C.; Teixeira, J.A.; Moreira, R.A.; Coimbra, M.A.; Gonçalves, M.P.; et al. Extraction, purification and characterization of galactomannans from non-traditional sources. *Carbohydr. Polym.* **2009**, *75*, 408–414. [CrossRef]
4. Verma, A.; Tiwari, A.; Panda, P.K.; Saraf, S.; Jain, A.; Jain, S.K. *Locust Bean Gum in Drug Delivery Application*; Elsevier Inc.: Amsterdam, The Netherlands, 2019; pp. 203–222. [CrossRef]
5. Patel, J.; Maji, B.; Moorthy, N.S.H.N.; Maiti, S. Xanthan gum derivatives: Review of synthesis, properties and diverse applications. *RSC Adv.* **2020**, *10*, 27103–27136. [CrossRef] [PubMed]
6. Koenraadt, H.; Van Betteray, B.; Germain, R.; Hiddink, G.; Jones, J.; Oosterhof, J. Development of specific primers for the molecular detection of bacterial spot of pepper and tomato. *Acta Hortic.* **2009**, *808*, 99–102. [CrossRef]
7. Petri, D.F.S. Xanthan gum: A versatile biopolymer for biomedical and technological applications. *J. Appl. Polym. Sci.* **2015**, *132*, 42035. [CrossRef]
8. Tao, F.; Wang, X.; Ma, C.; Yang, C.; Tang, H.; Gai, Z.; Xu, P. Genome Sequence of *Xanthomonas campestris* JX, an Industrially Productive Strain for Xanthan Gum. *J. Bacteriol.* **2012**, *194*, 4755–4756. [CrossRef]
9. Elkatatny, S.; Jafarov, T.; Al-Majed, A.; Mahmoud, M. Formation Damage Avoidance by Reducing Invasion with Sodium Silicate-Modified Water-Based Drilling Fluid. *Energies* **2019**, *12*, 1485. [CrossRef]
10. Hui, Y.H.; Corke, H.; de Leyn, I.; Nip, W.-K.; Cross, N. *Bakery Products: Science and Technology*; Wiley-Blackwell: Hoboken, NJ, USA, 2007. [CrossRef]

11. Deep, A.; Rani, N.; Kumar, A.; Nandal, R.; Sharma, P.C.; Sharma, A.K. Prospective of Natural Gum Nanoparticulate Against Cardiovascular Disorders. *Curr. Chem. Biol.* **2019**, *13*, 197–211. [CrossRef]
12. Lopes, B.M.; Lessa, V.L.; Silva, B.M.; Carvalho Filho, M.A.S.; Schnitzler, E.; Lacerda, L.G. Xanthan gum: Properties, production conditions, quality and economic perspective. *J. Food Nutr. Res.* **2015**, *54*, 185–194.
13. García-Ochoa, F.; Santos, V.E.; Casas, J.A.; Gomez, E. Xanthan gum: Production, recovery, and properties. *Biotechnol. Adv.* **2000**, *18*, 549–579. [CrossRef]
14. Habibi, H.; Khosravi-Darani, K. Effective variables on production and structure of xanthan gum and its food applications: A review. *Biocatal. Agric. Biotechnol.* **2017**, *10*, 130–140. [CrossRef]
15. Singhvi, G.; Hans, N.; Shiva, N.; Dubey, S.K. Chapt. 5: Xanthan gum in drug delivery applications. In *Natural Polysaccharides in Drug Delivery and Biomedical Applications*; Elsevier Inc.: Amsterdam, The Netherlands, 2019; pp. 121–124. [CrossRef]
16. Benny, I.S.; Gunasekar, V.; Ponnusami, V. Review on application of Xanthan gum in drug delivery. *Int. J. PharmTech Res.* **2014**, *6*, 1322–1326.
17. Philip, A.; Philip, B. Colon Targeted Drug Delivery Systems: A Review on Primary and Novel Approaches. *Oman Med. J.* **2010**, *25*, 70–78. [CrossRef]
18. Abu-Huwaij, R.; Obaidat, R.M.; Sweidan, K.; Al-Hiari, Y. Formulation and In Vitro Evaluation of Xanthan Gum or Carbopol 934-Based Mucoadhesive Patches, Loaded with Nicotine. *AAPS PharmSciTech* **2010**, *12*, 21–27. [CrossRef] [PubMed]
19. Kavitha, K.; Puneeth, K.P.; Tamizh, M.T. Development and evaluation of rosiglitazone maleate floating tablets using natural gums. *Int. J. PharmTech Res.* **2010**, *2*, 1662–1669.
20. El-Gazayerly, O.N. Release of Pentoxifylline from Xanthan Gum Matrix Tablets. *Drug Dev. Ind. Pharm.* **2003**, *29*, 241–246. [CrossRef]
21. Ray, S.; Banerjee, S.; Maiti, S.; Laha, B.; Barik, S.; Sa, B.; Bhattacharyya, U.K. Novel interpenetrating network microspheres of xanthan gum–poly(vinyl alcohol) for the delivery of diclofenac sodium to the intestine—In vitro and in vivo evaluation. *Drug Deliv.* **2010**, *17*, 508–519. [CrossRef] [PubMed]
22. Mann, A.; Campbell, D.; Tighe, B. Chapt. 2: The ageing ocular surface: Challenges for biomaterials design and function. In *Biomaterials and Regenerative Medicine in Ophthalmology*, 2nd ed.; Elsevier Ltd.: Amsterdam, The Netherlands, 2016; pp. 17–43. [CrossRef]
23. Shiledar, R.R.; Tagalpallewar, A.A.; Kokare, C.R. Formulation and in vitro evaluation of xanthan gum-based bilayered mucoadhesive buccal patches of zolmitriptan. *Carbohydr. Polym.* **2014**, *101*, 1234–1242. [CrossRef]
24. Samia, O.; Hanan, R.; Kamal, E.T. Carbamazepine Mucoadhesive Nanoemulgel (MNEG) as brain targeting delivery system via the olfactory mucosa. *Drug Deliv.* **2011**, *19*, 58–67. [CrossRef]
25. Dehghan, M.G.; Kazi, M. Lyophilized chitosan/xanthan polyelectrolyte complex based mucoadhesive inserts for nasal delivery of promethazine hydrochloride. *Iran. J. Pharm. Res.* **2014**, *13*, 769–784. [CrossRef]
26. Madan, M.; Bajaj, A.; Lewis, S.; Udupa, N.; Baig, J.A. In situ forming polymeric drug delivery systems. *Indian J. Pharm. Sci.* **2009**, *71*, 242–251. [CrossRef]
27. Mazeau, K.; Rinaudo, M. The prediction of the characteristics of some polysaccharides from molecular modeling. Comparison with effective behavior. *Food Hydrocoll.* **2004**, *18*, 885–898. [CrossRef]
28. Mohammed, Z.H.; Haque, A.; Richardson, R.K.; Morris, E.R. Promotion and inhibition of xanthan 'weak-gel' rheology by calcium ions. *Carbohydr. Polym.* **2007**, *70*, 38–45. [CrossRef]
29. Bergmann, D.; Furth, G.; Mayer, C. Binding of bivalent cations by xanthan in aqueous solution. *Int. J. Biol. Macromol.* **2008**, *43*, 245–251. [CrossRef]
30. Dário, A.F.; Hortêncio, L.M.A.; Sierakowski, M.R.; Neto, J.C.Q.; Petri, D.F.S. The effect of calcium salts on the viscosity and adsorption behavior of xanthan. *Carbohydr. Polym.* **2011**, *84*, 669–676. [CrossRef]
31. Fitzpatrick, P.; Meadows, J.; Ratcliffe, I.; Williams, P.A. Control of the properties of xanthan/glucomannan mixed gels by varying xanthan fine structure. *Carbohydr. Polym.* **2013**, *92*, 1018–1025. [CrossRef] [PubMed]
32. Goudanavar, P.; Sreeharsha, N.; Murtale, S.; Naveen, N.; Alsanie, W.; Alhomrani, M.; Alamri, A.; Asdaq, S.B.; Anwer, M.; Gharsan, M.; et al. Synthesis and evaluation of grafted xanthan gum as a drug carrier in developing lornoxicam gel formulations. *Pharmacogn. Mag.* **2022**, *18*, 271–278. [CrossRef]
33. Verhoeven, E.; Vervaet, C.; Remon, J. Xanthan gum to tailor drug release of sustained-release ethylcellulose mini-matrices prepared via hot-melt extrusion: In vitro and in vivo evaluation. *Eur. J. Pharm. Biopharm.* **2006**, *63*, 320–330. [CrossRef]
34. Mundargi, R.C.; Patil, S.A.; Aminabhavi, T.M. Evaluation of acrylamide-grafted-xanthan gum copolymer matrix tablets for oral controlled delivery of antihypertensive drugs. *Carbohydr. Polym.* **2007**, *69*, 130–141. [CrossRef]
35. Karki, S.; Kim, H.; Na, S.-J.; Shin, D.; Jo, K.; Lee, J. Thin films as an emerging platform for drug delivery. *Asian J. Pharm. Sci.* **2016**, *11*, 559–574. [CrossRef]
36. Tran, T.T.D.; Tran, P.H.L. Controlled Release Film Forming Systems in Drug Delivery: The Potential for Efficient Drug Delivery. *Pharmaceutics* **2019**, *11*, 290. [CrossRef] [PubMed]
37. Shukla, A.; Bhatt, N. Formulation and evaluation of glibenclamide loaded mouth dissolving films with xanthan gum as film forming polymer. *Eur. J. Mol. Clin. Med.* **2021**, *8*, 1016–1025.
38. Zhang, Y.S.; Khademhosseini, A. Advances in engineering hydrogels. *Science* **2017**, *356*, eaaf3627. [CrossRef]
39. Caliari, S.R.; Burdick, J.A. A practical guide to hydrogels for cell culture. *Nat. Methods* **2016**, *13*, 405–414. [CrossRef]

40. Li, J.; Mooney, D.J. Designing hydrogels for controlled drug delivery. *Nat. Rev. Mater.* **2016**, *1*, 16071. [CrossRef]
41. Wang, B.; Han, Y.; Lin, Q.; Liu, H.; Shen, C.; Nan, K.; Chen, H. In vitro and in vivo evaluation of xanthan gum–succinic anhydride hydrogels for the ionic strength-sensitive release of antibacterial agents. *J. Mater. Chem. B* **2016**, *4*, 1853–1861. [CrossRef]
42. Shastri, D.H.; Prajapati, S.; Patel, L. Design and Development of Thermoreversible Ophthalmic In Situ Hydrogel of Moxifloxacin HCl. *Curr. Drug Deliv.* **2010**, *7*, 238–243. [CrossRef]
43. Barzic, A.I.; Ioan, S. Chapt. 1: Antibacterial drugs—From basic concepts to complex therapeutic mechanisms of polymer systems. In *Concepts, Compounds and the Alternatives of Antibacterials*; Bobbarala, V., Ed.; IntechOpen: London, UK, 2015; pp. 3–28. [CrossRef]
44. Kala, S.; Gurudiwan, P.; Juyal, D. Formulation and Evaluation of Besifloxacin Loaded In Situ Gel For Ophthalmic Delivery. *Pharm. Biosci. J.* **2018**, *6*, 36–40. [CrossRef]
45. Garg, R.; Kumar, V.; Sharma, V. Design and Characterization of Flucytosine Loaded Bioadhesive In Situ Ophthalmic Gel for Improved Bioavailability. *Pharm. Biosci. J.* **2019**, *7*, 17–20. [CrossRef]
46. Gangane, P.S.; Pachpute, T.; Mahapatra, D.K.; Mahajan, N.M. HPMC Polymers and Xanthan Gum Assisted Development and Characterization of Stavudine Extended Release Floating Tablets. *Indian J. Pharm. Educ. Res.* **2021**, *55*, S681–S692. [CrossRef]
47. Ribeiro, M.; Boudoukhani, M.; Belmonte-Reche, E.; Genicio, N.; Sillankorva, S.; Gallo, J.; Rodríguez-Abreu, C.; Moulai-Mostefa, N.; Bañobre-López, M. Xanthan-Fe_3O_4 Nanoparticle Composite Hydrogels for Non-Invasive Magnetic Resonance Imaging and Magnetically Assisted Drug Delivery. *ACS Appl. Nano Mater.* **2021**, *4*, 7712–7729. [CrossRef]
48. Malik, N.S.; Ahmad, M.; Minhas, M.U.; Tulain, R.; Barkat, K.; Khalid, I.; Khalid, Q. Chitosan/Xanthan Gum Based Hydrogels as Potential Carrier for an Antiviral Drug: Fabrication, Characterization, and Safety Evaluation. *Front. Chem.* **2020**, *8*, 50. [CrossRef]
49. Sinha, V.R.; Kumria, R. Binders for colon specific drug delivery: An in vitro evaluation. *Int. J. Pharm.* **2002**, *249*, 23–31. [CrossRef] [PubMed]
50. Trombino, S.; Serini, S.; Cassano, R.; Calviello, G. Xanthan gum-based materials for omega-3 PUFA delivery: Preparation, characterization and antineoplastic activity evaluation. *Carbohydr. Polym.* **2019**, *208*, 431–440. [CrossRef] [PubMed]
51. Webster, T.J.; Singh, S.; Kotla, N.G.; Tomar, S.; Maddiboyina, B.; Sharma, D.; Sunnapu, O. A nanomedicine-promising approach to provide an appropriate colon-targeted drug delivery system for 5-fluorouracil. *Int. J. Nanomed.* **2015**, *10*, 7175–7182. [CrossRef]
52. Anwar, M.; Pervaiz, F.; Shoukat, H.; Noreen, S.; Shabbir, K.; Majeed, A.; Ijaz, S. Formulation and evaluation of interpenetrating network of xanthan gum and polyvinylpyrrolidone as a hydrophilic matrix for controlled drug delivery system. *Polym. Bull.* **2021**, *78*, 59–80. [CrossRef]
53. Faralli, A.; Shekarforoush, E.; Ajalloueian, F.; Mendes, A.C.; Chronakis, I.S. In vitro permeability enhancement of curcumin across Caco-2 cells monolayers using electrospun xanthan-chitosan nanofibers. *Carbohydr. Polym.* **2019**, *206*, 38–47. [CrossRef] [PubMed]
54. Feng, Z.; Xu, J.; Ni, C. Preparation of redox responsive modified xanthan gum nanoparticles and the drug controlled release. *Int. J. Polym. Mater. Polym. Biomater.* **2021**, *70*, 994–1001. [CrossRef]
55. Zhang, L.; Xu, J.; Wen, Q.; Ni, C. Preparation of xanthan gum nanogels and their pH/redox responsiveness in controlled release. *J. Appl. Polym. Sci.* **2019**, *136*, 6–11. [CrossRef]
56. Pooja, D.; Panyaram, S.; Kulhari, H.; Rachamalla, S.S.; Sistla, R. Xanthan gum stabilized gold nanoparticles: Characterization, biocompatibility, stability and cytotoxicity. *Carbohydr. Polym.* **2014**, *110*, 1–9. [CrossRef] [PubMed]
57. Alle, M.; Reddy, G.B.; Kim, T.H.; Park, S.H.; Lee, S.-H.; Kim, J.-C. Doxorubicin-carboxymethyl xanthan gum capped gold nanoparticles: Microwave synthesis, characterization, and anti-cancer activity. *Carbohydr. Polym.* **2020**, *229*, 115511. [CrossRef]
58. Saeedi, P.; Petersohn, I.; Salpea, P.; Malanda, B.; Karuranga, S.; Unwin, N.; Colagiuri, S.; Guariguata, L.; Motala, A.A.; Ogurtsova, K.; et al. Global and regional diabetes prevalence estimates for 2019 and projections for 2030 and 2045: Results from the International Diabetes Federation Diabetes Atlas, 9th edition. *Diabetes Res. Clin. Pract.* **2019**, *157*, 107843. [CrossRef] [PubMed]
59. Tao, Z.; Shi, A.; Zhao, J. Epidemiological Perspectives of Diabetes. *Cell Biochem. Biophys.* **2015**, *73*, 181–185. [CrossRef] [PubMed]
60. Kaul, K.; Tarr, J.M.; Ahmad, S.I.; Kohner, E.M.; Chibber, R. Introduction to Diabetes Mellitus. *Adv. Exp. Med. Biol.* **2012**, *771*, 1–11. [CrossRef] [PubMed]
61. Cooke, D.W.; Plotnick, L. Type 1 Diabetes Mellitus in Pediatrics. *Pediatr. Rev.* **2008**, *29*, 374–385. [CrossRef]
62. Razzaq, F.A.; Asif, M.; Asghar, S.; Iqbal, M.S.; Khan, I.U.; Khan, S.-U.; Irfan, M.; Syed, H.K.; Khames, A.; Mahmood, H.; et al. Glimepiride-Loaded Nanoemulgel; Development, In Vitro Characterization, Ex Vivo Permeation and In Vivo Antidiabetic Evaluation. *Cells* **2021**, *10*, 2404. [CrossRef]
63. Thakare, E.B.; Malpure, P.S.; Maru, A.D.; More, Y.M. Formulation and Evaluation of Mucoadhesive Buccal Tablet of Repaglinide. *J. Drug Deliv. Ther.* **2019**, *9*, 415–424. [CrossRef]
64. Imran, M.; Hameed, A.; Shafiullah Hafizur, R.M.; Ali, I.; Roome, T.; Shah, M.R. Fabrication of Xanthan stabilized green gold nanoparticles based tolbutamide delivery system for enhanced insulin secretion in mice pancreatic islets. *J. Macromol. Sci. Part A* **2018**, *55*, 729–735. [CrossRef]
65. Verma, A.K.; Kumar, A.; Ramasamy, S.; Yadav, A.K.; Bijauliya, R.K. Evaluation of Drug Release from Carboxymethyl Starch-Xanthan Gum-HPMC Matrix. *J. Drug Deliv. Ther.* **2021**, *11*, 27–32. [CrossRef]
66. Phaechamud, T.; Ritthidej, G.C. Sustained-release from Layered Matrix System Comprising Chitosan and Xanthan Gum. *Drug Dev. Ind. Pharm.* **2007**, *33*, 595–605. [CrossRef]
67. Ratnam, S.V.; Bhowmik, D.; Yadav, R.; Singh, D. Formulation and evaluation of carvedilol fast dissolving tablets. *J. Chem. Pharm. Sci.* **2014**, *7*, 85–88.

68. Bhunia, T.; Giri, A.; Nasim, T.; Chattopadhyay, D.; Bandyopadhyay, A. Uniquely different PVA-xanthan gum irradiated membranes as transdermal diltiazem delivery device. *Carbohydr. Polym.* **2013**, *95*, 252–261. [CrossRef]
69. Dadou, S.M.; El-Barghouthi, M.I.; Antonijevic, M.D.; Chowdhry, B.Z.; Badwan, A.A. Elucidation of the Controlled-Release Behavior of Metoprolol Succinate from Directly Compressed Xanthan Gum/Chitosan Polymers: Computational and Experimental Studies. *ACS Biomater. Sci. Eng.* **2019**, *6*, 21–37. [CrossRef]
70. Kim, Y.S.; Kim, N. Sex-Gender Differences in Irritable Bowel Syndrome. *J. Neurogastroenterol. Motil.* **2018**, *24*, 544–558. [CrossRef]
71. TarikaPriya, A.; Neeharika, M.S.; Sekhar, C.K. Formulation and Evaluation of Sustained Release Matrix Tablets of Mebeverine Hydrochloride Using Natural and Synthetic Polymers. *World J. Pharm. Pharm. Sci.* **2014**, *3*, 1424–1436.
72. Rashmitha, V.; Madhusudan, R.Y.; Pavani, S. Formulation and evaluation of fenoverine floating tablets. *Asian J. Pharm. Clin. Res.* **2021**, *14*, 175–180. [CrossRef]
73. Olaru, L.; Soong, L.; Dhillon, S.; Yacyshyn, E. Coexistent rheumatoid arthritis and gout: A case series and review of the literature. *Clin. Rheumatol.* **2017**, *36*, 2835–2838. [CrossRef]
74. Del Favero, A. Anti-inflammatory analgesics and drugs used in rheumatoid arthritis and gout. *Side Eff. Drugs Annu.* **1985**, *9*, 83–100. [CrossRef]
75. Li, L.; Zheng, X.; Pan, C.; Pan, H.; Guo, Z.; Liu, B.; Liu, Y. A pH-sensitive and sustained-release oral drug delivery system: The synthesis, characterization, adsorption and release of the xanthan gum-*graft*-poly(acrylic acid)/GO–DCFP composite hydrogel. *RSC Adv.* **2021**, *11*, 26229–26240. [CrossRef] [PubMed]
76. Stasaid, M.; Boutemak, K.; Ibtissem, L.; Flahaut, E.; Zafour, A.H.-Z. Synthesis of carboxymethyl Xanthan/ double-walled carbon nanotube hybrid hydrogel nanocomposite for transdermal release of drug. *Soft Mater.* **2021**, *20*, 168–182. [CrossRef]
77. Dhage, C.M.; Shinkar, D.M.; Pathan, V.T.; Gage, G.P.; Jadhav, A.G. Formulation and in Vitro Evaluation of Mucoadhesive Buccal Tablet of Febuxostat. LINO. Available online: https://www.sybespharmacy.com/asset/pdf/research-activities/publication/300156_10._dms.pdf (accessed on 3 August 2022).
78. Caddeo, C.; Nacher, A.; Díez-Sales, O.; Merino-Sanjuán, M.; Fadda, A.M.; Manconi, M. Chitosan–xanthan gum microparticle-based oral tablet for colon-targeted and sustained delivery of quercetin. *J. Microencapsul.* **2014**, *31*, 694–699. [CrossRef]
79. Jaya, S.; Divya, S. Formulation and in vitro evaluation of matrix tablets of metoclopramide hydrochloride. *Int. J. Appl. Pharm.* **2019**, *11*, 25–30. [CrossRef]
80. Mutalik, S.; Suthar, N.A.; Managuli, R.S.; Shetty, P.K.; Avadhani, K.; Kalthur, G.; Kulkarni, R.V.; Thomas, R. Development and performance evaluation of novel nanoparticles of a grafted copolymer loaded with curcumin. *Int. J. Biol. Macromol.* **2016**, *86*, 709–720. [CrossRef] [PubMed]
81. Hussain, Y.; Khan, H. Immunosuppressive Drugs. *Encycl. Infect. Immun.* **2022**, *4*, 726–740. [CrossRef]
82. Hwang, J.; Lee, E.; Kim, J.; Seo, Y.; Lee, K.H.; Hong, J.W.; Gilad, A.A.; Park, H.; Choi, J. Effective delivery of immunosuppressive drug molecules by silica coated iron oxide nanoparticles. *Colloids Surf. B Biointerfaces* **2016**, *142*, 290–296. [CrossRef]
83. Sadeq, Z.A.; Sabri, L.A.; Al-Kinani, K.K. Natural polymer Effect on gelation and rheology of ketotifen-loaded pH-sensitive in situ ocular gel (Carbapol). *J. Adv. Pharm. Educ. Res.* **2022**, *12*, 45–50. [CrossRef]
84. Azari, F.; Ghanbarzadeh, S.; Safdari, R.; Yaqoubi, S.; Adibkia, K.; Hamishehkar, H. Development of a Carrier Free Dry Powder Inhalation Formulation of Ketotifen for Pulmonary Drug Delivery. *Drug Res.* **2019**, *70*, 26–32. [CrossRef]
85. Bhupathyraaj, M.; Pole, S. Study on effect of combination of sodium alginate and xanthan gum on drug release from Tacrolimus microbeads. *Eur. J. Mol. Clin. Med.* **2020**, *7*, 4585.
86. Gupta, M.; Agrawal, U.; Vyas, S.P. Nanocarrier-based topical drug delivery for the treatment of skin diseases. *Expert Opin. Drug Deliv.* **2012**, *9*, 783–804. [CrossRef]
87. Pagano, C.; Puglia, D.; Luzi, F.; Michele, A.; Scuota, S.; Primavilla, S.; Ceccarini, M.; Beccari, T.; Iborra, C.; Ramella, D.; et al. Development and Characterization of Xanthan Gum and Alginate Based Bioadhesive Film for Pycnogenol Topical Use in Wound Treatment. *Pharmaceutics* **2021**, *13*, 324. [CrossRef]
88. Nance, E.; Pun, S.H.; Saigal, R.; Sellers, D.L. Drug delivery to the central nervous system. *Nat. Rev. Mater.* **2021**, *7*, 314–331. [CrossRef]
89. Mehetre, G.D.; Patki, S.S.; Thenge, R.R.; Shrikhande, V.N. Quetiapine Fumarate Buccoadhesive Tablet-Formulation and In Vitro Evaluation. *Res. J. Pharm. Technol.* **2020**, *13*, 5095–5102. [CrossRef]
90. Salamanca, C.H.; Yarce, C.J.; Moreno, R.A.; Prieto, V.; Recalde, J. Natural gum-type biopolymers as potential modified nonpolar drug release systems. *Carbohydr. Polym.* **2018**, *189*, 31–38. [CrossRef] [PubMed]
91. Pandala, S.; Bakshi, V.; Jadi, R.K. Formulation Development and In Vitro Characterization of Zolmitriptan Controlled Release Drug Delivery Systems. *INNOSC Theranostics Pharmacol. Sci.* **2019**, *2*, 550. [CrossRef]
92. Rajput, A.; Bariya, A.; Allam, A.; Othman, S.; Butani, S.B. In situ nanostructured hydrogel of resveratrol for brain targeting: In vitro-in vivo characterization. *Drug Deliv. Transl. Res.* **2018**, *8*, 1460–1470. [CrossRef] [PubMed]
93. Kim, S.; Kim, J.-H.; Seok, S.H.; Park, E.-S. Anti-obesity effect with reduced adverse effect of the co-administration of mini-tablets containing orlistat and mini-tablets containing xanthan gum: In vitro and in vivo evaluation. *Int. J. Pharm.* **2020**, *591*, 119998. [CrossRef]
94. Morsi, N.; Ibrahim, M.; Refai, H.; El Sorogy, H. Nanoemulsion-based electrolyte triggered in situ gel for ocular delivery of acetazolamide. *Eur. J. Pharm. Sci.* **2017**, *104*, 302–314. [CrossRef]

95. Deulker, A.L.; Sancoalcar, A.; Vaidya, S.; Gude, R. Formulation development and evaluation of long acting ophthalmic in-situ gelling system of Dorzolamide Hydrochloride. *Int. J. Drug Dev. Res.* **2013**, *5*, 156–163.
96. Mishra, B.; Singh, J. Chapt. 4: Novel drug delivery systems and significance in respiratory diseases. In *Targeting Chronic Inflammatory Lung Diseases Using Advanced Drug Delivery Systems*; Elsevier Inc.: Amsterdam, The Netherlands, 2020; pp. 57–95. [CrossRef]
97. Todoroff, J.; Vanbever, R. Fate of nanomedicines in the lungs. *Curr. Opin. Colloid Interface Sci.* **2011**, *16*, 246–254. [CrossRef]
98. Giri, T.K.; Thakur, A.; Tripathi, D.K. Biodegradable Hydrogel Bead of Casein and Modified Xanthan Gum for Controlled Delivery of Theophylline. *Curr. Drug Ther.* **2016**, *11*, 150–162. [CrossRef]
99. Lazzari, A.; Kleinebudde, P.; Knop, K. Xanthan gum as a rate-controlling polymer for the development of alcohol resistant matrix tablets and mini-tablets. *Int. J. Pharm.* **2018**, *536*, 440–449. [CrossRef]
100. Bueno, P.V.A.; Hilamatu, K.C.; Carmona-Ribeiro, A.M.; Petri, D.F. Magnetically triggered release of amoxicillin from xanthan/Fe3O4/albumin patches. *Int. J. Biol. Macromol.* **2018**, *115*, 792–800. [CrossRef]
101. Balzli, C.L.; McCormick, C.C.; Caballero, A.R.; Tang, A.; O'Callaghan, R.J. The effectiveness of an improved combination therapy for experimental *Staphylococcus aureus* keratitis. *Adv. Ther.* **2010**, *27*, 933–940. [CrossRef]
102. Mundargi, R.C.; Patil, S.A.; Agnihotri, S.A.; Aminabhavi, T.M. Development of Polysaccharide-Based Colon Targeted Drug Delivery Systems for the Treatment of Amoebiasis. *Drug Dev. Ind. Pharm.* **2007**, *33*, 255–264. [CrossRef]
103. Hanna, D.H.; Saad, G.R. Encapsulation of ciprofloxacin within modified xanthan gum- chitosan based hydrogel for drug delivery. *Bioorg. Chem.* **2019**, *84*, 115–124. [CrossRef] [PubMed]
104. Bhattacharya, S.S.; Mazahir, F.; Banerjee, S.; Verma, A.; Ghosh, A. Preparation and in vitro evaluation of xanthan gum facilitated superabsorbent polymeric microspheres. *Carbohydr. Polym.* **2013**, *98*, 64–72. [CrossRef]
105. Mishra, B.; Sahoo, S.K.; Sahoo, S. Liranaftate loaded Xanthan gum based hydrogel for topical delivery: Physical properties and ex-vivo permeability. *Int. J. Biol. Macromol.* **2018**, *107*, 1717–1723. [CrossRef]
106. Ghosal, K.; Adak, S.; Agatemor, C.; Kalarikkal, P.G.N.; Thomas, S. Novel interpenetrating polymeric network based microbeads for delivery of poorly water soluble drug. *J. Polym. Res.* **2020**, *27*, 1–11. [CrossRef]
107. Ngwabebhoh, F.A.; Zandraa, O.; Patwa, R.; Saha, N.; Capáková, Z.; Saha, P. Self-crosslinked chitosan/dialdehyde xanthan gum blended hypromellose hydrogel for the controlled delivery of ampicillin, minocycline and rifampicin. *Int. J. Biol. Macromol.* **2021**, *167*, 1468–1478. [CrossRef] [PubMed]
108. Zhang, F.; Jia, D.; Li, Q.; Zhang, M.; Liu, H.; Wu, X. Preparation and Evaluation of a Xanthan Gum–Containing Linezolid Ophthalmic Solution for Topical Treatment of Experimental Bacterial Keratitis. *Pharm. Res.* **2021**, *38*, 347–359. [CrossRef]
109. Parhi, R.; Reddy, S.S.; Swain, S. Transdermal Delivery of Ondansetron HCl from Thermoreversible Gel Containing Nanocomposite. *Curr. Nanomater.* **2019**, *4*, 137–147. [CrossRef]
110. Chauhan, D.; Patel, A.; Shah, S. Influence of selected natural polymers on in-vitro release of colon targeted mebeverine HCl matrix tablet. *Int. J. Drug Dev. Res.* **2012**, *4*, 247–255.
111. Sung, Y.K.; Kim, S.W. Recent advances in the development of gene delivery systems. *Biomater. Res.* **2019**, *23*, 8. [CrossRef]
112. Janes, K.; Calvo, P.; Alonso, M.J. Polysaccharide colloidal particles as delivery systems for macromolecules. *Adv. Drug Deliv. Rev.* **2001**, *47*, 83–97. [CrossRef] [PubMed]
113. Fernandez-Piñeiro, I.; Alvarez-Trabado, J.; Márquez, J.; Badiola, I.; Sanchez, A. Xanthan gum-functionalised span nanoparticles for gene targeting to endothelial cells. *Colloids Surf. B Biointerfaces* **2018**, *170*, 411–420. [CrossRef] [PubMed]
114. Giri, N.C. Chapt. 2: Protein and peptide drug delivery. In *Smart Drug Delivery*; Ahmad, U., Haider, M.F., Akhtar, J., Eds.; IntechOpen: London, UK, 2022; pp. 1–53. [CrossRef]
115. Abu Elella, M.H.; Sabaa, M.W.; Hanna, D.H.; Abdel-Aziz, M.M.; Mohamed, R.R. Antimicrobial pH-sensitive protein carrier based on modified xanthan gum. *J. Drug Deliv. Sci. Technol.* **2020**, *57*, 101673. [CrossRef]
116. Sabaa, M.W.; Hanna, D.H.; Abu Elella, M.H.; Mohamed, R.R. Encapsulation of bovine serum albumin within novel xanthan gum based hydrogel for protein delivery. *Mater. Sci. Eng. C* **2019**, *94*, 1044–1055. [CrossRef]

pharmaceutics

MDPI

Review

Delivery of Therapeutic Biopolymers Employing Silica-Based Nanosystems

Zoila Gándara [1,2,*], Noelia Rubio [1,2,*] and Rafael R. Castillo [1,2,*]

[1] Departamento de Química Orgánica y Química Inorgánica, Universidad de Alcalá, 28805 Alcalá de Henares, Spain
[2] Instituto de Investigación Química "Andrés M. del Río" (IQAR), Universidad de Alcalá, 28805 Alcalá de Henares, Spain
* Correspondence: zoila.gandara@uah.es (Z.G.); noelia.rubio@uah.es (N.R.); rafael.castillor@uah.es (R.R.C.)

Abstract: The use of nanoparticles is crucial for the development of a new generation of nanodevices for clinical applications. Silica-based nanoparticles can be tailored with a wide range of functional biopolymers with unique physicochemical properties thus providing several advantages: (1) limitation of interparticle interaction, (2) preservation of cargo and particle integrity, (3) reduction of immune response, (4) additional therapeutic effects and (5) cell targeting. Therefore, the engineering of advanced functional coatings is of utmost importance to enhance the biocompatibility of existing biomaterials. Herein we will focus on the most recent advances reported on the delivery and therapeutic use of silica-based nanoparticles containing biopolymers (proteins, nucleotides, and polysaccharides) with proven biological effects.

Keywords: mesoporous silica nanoparticles; biopolymers; nanomedicine; nanotechnology

1. Introduction

Biomacromolecules play a fundamental role in many biological aspects. For example, they are responsible for all signaling and metabolic processes, thus contributing to the structural integrity of eukaryotic and prokaryotic cells or encoding genetic information, among many other roles. Therefore, the use of biomacromolecules in advanced new-generation therapies is one of the most promising fields of research. For example, regarding nucleic acids, the transfection strategies that allowed to genetically modify unicellular organisms to produce human enzymes and hormones have been clearly outranged by current technology. Recently, the milestone of gene editing on multicellular organisms has also been achieved, opening the way to more complex therapies [1]. In addition to that, chemical modifications of nucleic acids have also arisen as a potent strategy to develop new generation biomedicines [2,3]. On the other hand, proteins and peptides have also contributed to the development of relevant immunotherapeutics [4,5], immunodetection methods, and to the treatment of chronic diseases such as diabetes, growth disorders, cancer [6] and new-generation antibacterial therapies [7,8]. In all these therapies, biomacromolecules with unique 3D structures play a crucial role in tuning biological processes in a way that typical small molecules and drugs are not able to exert. However, the use of those macromolecular biotherapeutics is extremely challenging, as they must penetrate across biological membranes or physiological barriers that they are not able to in order to perform their therapeutic action [9]. This is the case of the messenger RNA molecules of the anti-COVID Comirnaty® and Spikevax® vaccines, which require two protective strategies to carry out their function. In both cases, a liposomal coating is used to preserve the integrity of the mRNA during trafficking in the bloodstream. However, this modification alone is not enough to achieve a suitable effect and uracil must be replaced by pseudouridine within the sequence in order to intracellularly increase mRNA lifespan and thus produce enough amounts of viral protein to provoke immune responses [10,11].

Regarding cancer treatment, the discovery of the enhanced permeation and retention (EPR) effect [12] in solid tumors has made it possible to create a huge number of possible biomedical solutions. This is a consequence of how nanoparticles interact with tissues and living cells, which typically differ from common small molecule–cell interactions. However, since most of these nanoparticles are not of biogenic origin, there is a major limitation to their widespread use, especially when these systems are dosed intravenously or parenterally where immune responses are typically triggered [13]. Nevertheless, current nanotechnology also offers significant advances in immune stealthing materials [14,15], where polyethylene glycol outranges any other material. Nevertheless, some current developments in the field of biopolymers are also able to accomplish this task, but with additional functional features, which is of enormous interest for the development of therapeutic nanohybrids.

Regarding nanoparticles, there are an overwhelming number of them. They can be classified in many ways on the basis of their different chemical nature, shape, size, and topology [16,17]. However, among them, liposomes display the highest loading capacity and the most advanced clinical development [18], although they also present several limitations. For example, the poor thermal, mechanical, and chemical stability of liposomes complicates the development of multifunctional liposomal systems and the development of formulations for co-delivery, for which reason there is a need to improve the current drug delivery technology. Among all available nanosystems, porous silica is one of the most widely employed for the development of potential nanotherapeutics, even though mesoporous silica is still not approved for use in clinical practice [19,20]. Nevertheless, SiO_2 is generally considered a safe material by the FDA and is widely employed in food industry and even as a structural basis for certain living organisms [21]. Indeed, there are many examples in which biosafety and biocompatibility of silicas have been demonstrated [22–24], suggesting that silica may be used either alone or as a component for the development of new nanotherapeutics, being even employed by certain diatom algae as structural materials [21]. In this review, the most relevant advances based on porous silica particles able to deliver biomacromolecules with therapeutic potential are covered, as well as other targeting strategies [25–27] and those employed as diffusion barriers in the development of controlled release nanosystems [28,29] (Figure 1).

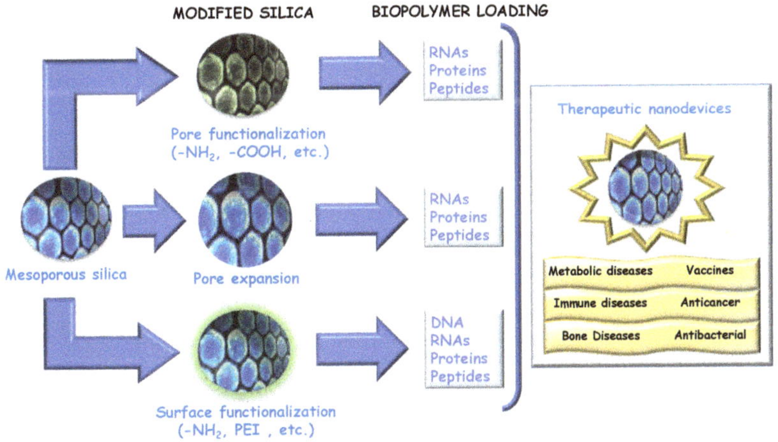

Figure 1. Workflow for building silica-based therapeutics for reviewed diseases. Different bioactive macromolecules could be incorporated onto the silica and efficiently dosed by properly engineering particles porous structure and surfaces.

2. Development of Silica-Based Nanomedicines Using Therapeutic Nucleotides

The huge potential of nucleotides for developing advanced therapies has resulted in a very fruitful field of research. Among the most relevant therapeutic applications, there could be highlighted the following: (1) gene transfection with plasmids [30,31], (2) development of vaccines and protein modulation therapies employing miRNAs [32,33], (3) gene silencing therapies [34,35], and (4) the most recent to arrive, CRISPR genome editing [36,37]. However, like with many other therapeutic biomolecules, the indiscriminate use of nucleotides is not easy and immediate, as they suffer from quick degradation in extracellular regions and intracellular environments and are not able to cross biological membranes. For such reasons, DNA and RNA-based biotechnology must be also accompanied by a development on delivery technologies. Historically, viral-based vectors allowed for development of transfection processes successfully, but the difficulty of handling and the triggering of immune responses, discouraged from their use. This has forced the scientific community to develop other non-immunogenic systems able to deliver these therapeutic biopolymers to both cytoplasm and cell nucleus. Among them, lipofectamine technology makes it possible to efficiently transfect plasmids in vitro but not in vivo; therefore, new systems need to be developed.

Chromosomes, based on DNA sequences, are responsible for keeping the genetic information. For such reason living cells have established biological mechanisms to preserve them unaltered within the nucleus but to destroy them in cytoplasmatic regions. On the other hand, RNA nucleotides responsible for encoding protein expression must be also removed from the cytoplasm once their message is delivered, as otherwise they may induce homeostasis imbalances due to excessive protein expression. The consequence is that both species are quickly metabolized by living organisms to maintain homeostasis. For this reason, it is important to create carriers able to protect these biopolymers when dosed in vivo. This problem has been addressed for more than a decade since the pioneering work of Lin [38] who opened the way to the use of mesoporous silica systems for nucleotide delivery. This topic was visited 5 years ago in a previous contribution by some of us; in that review [39], the known strategies for the preparation of DNA/RNA-based silica nanosystems were detailed [39]. Despite that review covering all relevant aspects of the development of nucleotide-based nanomedicines, we revisit them herein, focusing on the most recent advances (Table 1). Nevertheless, as this review is focused on the therapeutic profile of biopolymers, cell targeting [29,40–44], biosensors [45] and pore gating [46,47] strategies [29,40–47] will be omitted.

As highlighted elsewhere [39,48], effective transfection is mainly achieved in the presence of polycationic species, which upon electrostatic interaction with negatively charged DNA/RNA provide a complex structure. If this structure is still able to interact with the negatively charged membrane, it may favor cell uptake and deliver the nucleotide intracellularly. The main problem with this technology is the membrane-lytic effect caused by polycations; which could be partially avoided by using nanoparticles as nonviral vectors. In addition to the more convenient nanoparticulate formulation, the use of more benign biogenic polymers, i.e., polylysine (PLL) instead polyamidoamine (PAMAM) dendrimers or polyethyleneimine (PEI) [49], have significantly improved the applicability of these vectors.

In general, polycationic MSNs show lower toxicity levels than free polymers. This effect could be understood considering two different but complementary aspects: (1) the polycationic particle has a diffuse distribution of positive charges along the surface and hence lower destructive interactions with membranes and (2) when building multilayered assemblies, polycationic components are partially balanced with the negatively charged silica, thus reducing the overall toxicity by lowering the number of cationic groups. Both approaches have been successfully employed for the construction of transfection agents as previously reported [39]. Reported strategies for nucleotide delivery are summarized in Figure 2 and Table 1.

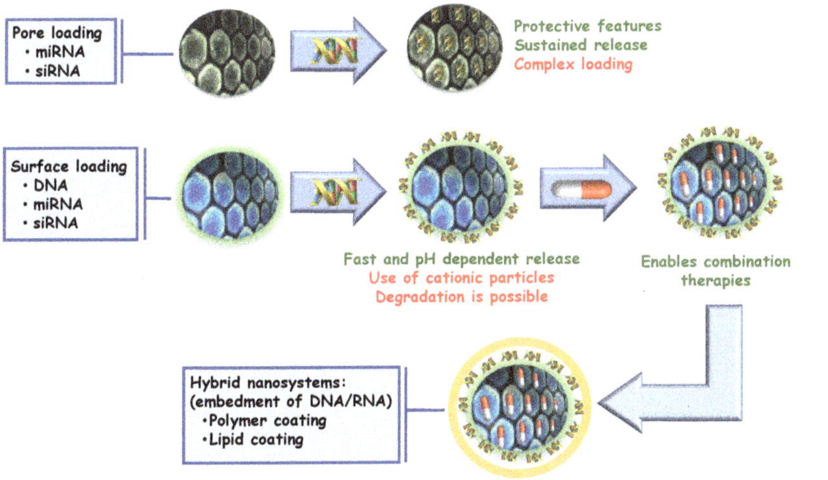

Figure 2. General strategies for nucleic acid drug delivery employing mesoporous silica nanoparticles.

Focusing on nanoparticle morphology, Wang et al. demonstrated that small (ca. 50 nm) dendritic mesoporous silica nanoparticles with large pores (>20 nm) were highly effective transfection agents when coated with 10 KDa PEI, as demonstrated with the green fluorescent protein plasmids [50]. In that work, the authors claimed that the dendritic morphology provided better transfection profiles than regular MSNs (up to ca. 40%). This effect, although not detailed in depth, could be a consequence of the better hosting and protection of plasmids at the larger mesopores. In another contribution, Su, Deng and coworkers, studied the performance of ultrasmall MSNs for siRNA delivery employing PEI too [51]. Their nanosystem, with an overall size of less than 12 nm, showed good biocompatibility on cancerous HeLa cultures (up to 200 $\mu g \cdot mL^{-1}$) and up to 80% knockdown onto glyceraldehyde-3-phosphate dehydrogenase (GADPH) onto HEK-293T cells, suggesting a great transfection efficiency similar to the reference, Lipofectamine 2000. This system employed a 1800 Da PEI chemically grafted onto the surface of modified aminopropyl triethoxysilane (APTES) together with a short, crosslinked polyethylene glycol (PEG) chain. This approach, based on the post-functionalization of MSNs with APTES, made it possible to significantly reduce the overall number of grafted PEI chains and thus to reduce the number of positive charges and hence, overall toxicity.

The modification of silica's textural properties aimed at maximizing the loading efficiency of DNA is also an interesting aspect. Along this line, an article by Yu, Song, and coworkers studied with the binding behavior of PEI and DNA onto phosphonate-modified silica nanoparticles with different surface roughness [52]. Herein, the authors prepared different asymmetric nanoparticles with both smooth and spiky hemispheres within the same nanoparticle. According to their investigations, the larger spiky surface the better the deposition of PEI-DNA. In addition to the better loading efficiency, larger spiky surfaces also provided lower hemolysis rates. As a result, the nanoparticle with ca. 93% spiky surface showed the best behavior with a loading rate of up to 100 $\mu g \cdot mL^{-1}$ and a red blood cell hemolysis rate of ca. 40% for 1200 μg of MSNs per mL concentration. These asymmetric particles were evaluated on the transfection of GFP in HEK-293T, reaching a 35% of effective transfection, which suggests that this kind of structured surfaces could help on the rational design of more biocompatible and efficient nonviral vectors. Perhaps, although not studied, the phosphonate groups present at the silica surface may be responsible for such good performance because of their ubiquity on cell membranes and biomolecules and its highly convenient interaction both in vitro and in vivo.

Despite the advances achieved by tuning the morphological properties of silica, it is also important to consider the chemical nature of PEI, which may also exert an additional toxic effect. As accurately reviewed by Taranejoo et al. [53], any increase in molecular weight or chemical stability, produce increments PEI toxicity; being only PEI chains below 1 KDa safe for in vivo applications. For this reason, other research groups have tried to develop silica coatings based on more biocompatible polymers. For instance, Nhavene et al. evaluated two biodegradable polymers as MSNs coatings [54]. The chosen highly biocompatible polymers polycaprolactone (PCL) and chitosan (CS) were grafted onto MCM-41 MSNs upon reaction of silica with an epoxide-containing (3-Glycidoxypropyl)trimethoxysilane. As a result, both nanosystems showed great biocompatibility on MCF-7 cell line (up to 100 $\mu g \cdot mL^{-1}$) and effective transfection of α-tubulin and laminB1 siRNAs. Unfortunately, no further analyses were given on the efficacy of such nanodevices, especially when employing the non-charged PCL polymer.

Despite the advances in multilayered nanosystems, this approach does not offer protection against nucleotide degradation; being pore hosting the most convenient strategy to preserve nucleic acid integrity. The development of RNA carriers based on large-pore MSNs has been successfully implemented by decorating the internal surface of mesopores with cationic groups are able to promote nucleotide threading into mesopores. However, despite electrostatically driven loading, the pH-driven release does not fully ensure biopolymer delivery on target. For this reason, Rosenholm and coworkers developed a promising release strategy based on intracellular glutathione-triggered (GSH) release. These authors modified the internal surface of mesopores with amino groups and cysteamine moieties which provided both the required positive charge and a disulfide bridge suitable for reductive cleavage [55]. The system was able to load antiGFP siRNA in the mesopores and produce effective gene silencing on genetically modified green-fluorescent MDA-MB-231 cells. This system was further coated with a PEI layer to favor carrier uptake, although without surpassing acceptable hemolytic levels.

On understanding the threading phenomenon, Rankin, Knutson and coworkers studied the behavior of amino-functionalized MSNs as carriers of double-stranded (ds) RNA oligonucleotides as a function of size (84 base pairs dsRNA with a 2.6 nm $\times$ 24 nm estimated size vs. 282 base pairs dsRNA with a 2.6 nm $\times$ 80 nm estimated size) and pore diameter (nonporous, 2.7, 4.3, and 8.1 nm pored silica particles) [56]. According to their results, both dsRNA were able to thread in all mesopores except in the case of 282 bp dsRNA which could not be loaded in the 2.7 nm wide mesopores, which set the basis for the further development of nucleotide nanocarriers able to circumvent the effect of nucleases. Nevertheless, in our opinion, this study must be completed with PEG modified MSNs more suitable for in vivo applications.

Another promising strategy for gene delivery is remote triggering. This topic was studied by Du et al., who designed an ultrasound-triggered device for plasmid DNA (pDNA) delivery [57]. Their system was built from pDNA encapsulated within mesopores (ca. 3.6 nm) of PEI-modified magnetic MSNs [58] which were further embedded within a solid lipidic microbubble able to be exploded in the presence of ultrasound activation. The resulting non-viral vector presented good biocompatibility due to the outer lipidic coating, good DNA binding stability due to the double protective layer and a remarkable magnetic targeting response. Nevertheless, despite the active magnetic targeting and the focused activation of microbubbles, this strategy must be still studied as no information is available on bioaccumulation and trafficking in vivo, especially when its size does not suggest intravenous applications.

Table 1. Reported devices based on the combination of silica nanoparticles and nucleotides.

Nanosystem	Assembly Strategy	Nucleotide	Secondary Therapeutic	Therapy	Release Stimulus	Action Mechanism	Biological Evaluation	Ref.
Strategies for carrying oligonucleotides and plasmids with silica nanosystems								
LP-MSNs	Electrostatic deposition (PEI)	Plasmid (eGFP)	None	None (GFP transfection)	pH-driven detachment	Transfection	In vitro: HEK-293	[50]
US-MSNs	PEI grafting (glutaraldehyde)	siRNA (siGADPH)	None	None (Knockdown quantification)	pH-driven detachment	Gene silencing	In vitro: HeLa, HEK-293	[51]
Asymmetric MSNs (Phosphonate coated)	Electrostatic deposition (PEI)	Plasmid (eGFP)	None	None (GFP transfection)	pH-driven detachment	Transfection	In vitro: HEK-293T	[52]
MSNs (MCM-41)	Chemical grafting (PCL or CS)	siRNAs (α-tubulin, laminB1)	None	None (cytoskeleton reduction)	pH-driven detachment	Gene silencing	In vitro: HeLa, MCF-7	[54]
LP-MSNs	Pore loading (Cystamine-PEI coated)	siRNA (anti-GFP)	None	None (Green florescence knockdown)	Redox (GSH) driven cleavage	Gene silencing	In vitro: MDA-MB-231	[55]
MSN@ lipid Microbubbles	Pore loading	Plasmid (eGFP)	None	None (GFP transfection)	Ultrasound	Transfection	In vitro: SKOV3, HEK-293T In vivo: Mice	[57]
Strategies for gene therapy employing silica nanoparticles as non-viral vectors.								
MSNs	Pore loading	siRNA (anti-miR-155)	None	AS1411-targeted oncogene silencing	pH-sensitive PDA coating.	Adjuvant gene silencing for 5-FU chemotherapy	In vitro: SW480, HT-29, SW620, Lovo, Caco-2, NCM460 In vivo: Mice	[59]
MSNs	Electrostatic deposition (PEI)	siRNA (anti-HER2)	None	Transtuzumab-targeted oncogene silencing	pH-driven siRNA detachment	Gene silencing	In vitro: BT474	[60]
MSNs	Pore loading	siRNA (anti-MDR1)	None	TAT-targeted anticancer siRNA delivery	Detachment of chitosan protecting layer	Gene silencing	In vitro: HeLa, EPG85.257	[61]
CC-MSNs	Electrostatic deposition (PEI)	miRNA (rno-miRNA-26a-5p)	None	Osteogenic therapy	None	Gene expression	In vitro: rBMSC	[62]
Dual MSNs (Amino modified)	Pore loading	Plasmids (GLP-1AR and FGF-21)	None	Antidiabetic therapy	None	Transfection	In vitro: Hepa1-6 In vivo: Mice	[63]
Rambutan-like MSNs	Electrostatic deposition (PEI)	Plasmid (OVA)	None	Vaccine	pH-driven detachment	Transfection of an immune stimulator	In vitro: HEK-293 In vivo: Mice	[64]
Porous silica microrods	Electrostatic deposition (PEI)	Plasmid (OVA)	None	Vaccine	pH-driven detachment	Transfection of an immune stimulator	In vivo: Mice.	[65]
Ca-dopped Si NPs (amino modified)	Not determined	Plasmid (PRPF31-GFP)	None	Blindness treatment	pH-driven detachment	Transfection	In vivo: Mice	[66]
Dendritic MSNs (amino modified)	Not determined	Antisense plasmid (ASvicR)	None	Antibiofilm (anticaries)	pH-driven detachment	Transfection	In vitro: S. mutans UA159	[67]
Strategies for combined gene and chemotherapy: siRNA delivery								
MSNs (amino modified)	Electrostatic deposition	siRNA (siVEGF)	Sorafenib	LA-targeted siRNA-drug combination	pH-driven detachment	Gene silencing plus kinase inhibition	In vitro: Huh7, HepG2, HeLa and A549	[68]

Table 1. *Cont.*

Nanosystem	Assembly Strategy	Nucleotide	Secondary Therapeutic	Therapy	Release Stimulus	Action Mechanism	Biological Evaluation	Ref.
MSNs	Electrostatic deposition (PEI)	siRNA (Survivin)	Docetaxel, etoposide and/or carfilzomib	siRNA-drug combination	pH-driven detachment	Gene silencing plus proteasome inhibition	In vitro: HEK-293, A549	[69]
Disulfide bridged MSNs	Electrostatic deposition (HA-PEI)	siRNA (Bcl-2)	Doxorubicin	siRNA-drug combination	pH-driven detachment, GSH-driven SiO2 degradation	Gene silencing plus chemotherapeutic	In vitro: MCF7	[70]
Ca2+ doped LP-MSNs	Pore loading (Ca2+)	siRNA (Bcl-2)	Chloroquine	siRNA-drug combination	Pore release	Gene silencing plus chemotherapeutic	In vitro: SKOV3	[71]
Strategies for combined gene and chemotherapy: plasmid delivery								
UCNP@MSNs (amino modified)	Electrostatic deposition (H2A)	Plasmid (p53)	Bortezomib	H2A-targeted, combined genic and chemotherapy	pH-driven detachment	Transfection plus chemotherapeutic	In vitro: NCI-H1299 and HeLa	[72]
MSNs	Electrostatic deposition onto pH-labile polymer coating	Plasmid (p53)	5-Fluoruracil	Combined genic and chemotherapy	pH-driven detachment	Transfection plus chemotherapeutic	In vitro: MCF-7	[73]
MSNs	Electrostatic deposition (PEI)	Plasmid (HNF4α)	Cisplatin	Combined genic and chemotherapy	pH-driven detachment	Transfection plus chemotherapeutic	In vitro: Huh7 In vivo: Mice	[74]
Rod-shaped MSNs	Not Specified	Plasmid (Survivin)	Camptothecin	AS1411-targeted, Combined genic and chemotherapy	Not Specified	Transfection plus chemotherapeutic	In vitro: C26, CHO In vivo: Mice	[75]
Strategies for combined gene and chemotherapy: miRNA delivery								
MSNs	Lipid coating (erythrocytes' membranes)	miRNA (miR137)	Indocyanine green (PDT)	RGD-targeted photothermal- miRNA	Detachment of protecting lipid layer	Photothermal and gene expression therapy.	In vitro: U87, RAW264.7 In vivo: Mice	[76]
MSNs	Pore loading	miRNA (hsa-miR-200c) siRNA (anti-Plk1)	Indocyanine green (PDT)	RGD-targeted photothermal-miRNA-siRNA	Detachment of protecting lipid layer	Photodynamic, gene silencing and expression	In vitro: MDA-MB-231 In vivo: Mice	[77]
MSNs	Pore loading	Dual miRNA (miR-34a and anti-miR-10b)	None	HA-targeted dual miRNA-combination	Detachment of protecting polymer layer (HA-PEG-PLGA)	Dual gene modulation	In vitro: MDA-MB-231, MDA-MB-468, HEK-293T, 4T1 In vivo: Chicken embryo In vivo: Mice	[78]
Strategies employing silica nanoparticles for gene therapy in bone diseases								
MSNs	Electrostatic deposition (PEI-PLL)	Plasmid (BMP-2)	Dexamethasone	Targeted and combined anti- osteoporotic treatment	pH-driven detachment	RGD-targeted osteoporosis gene silencing plus osteogenic stimulation	In vitro: HEK293T, BMSCs and RAW 264.7	[79]
MSNs	Pore loading	Anti-miRNA (miR26)	None	Targeted anti-osteoporotic treatment	pH-driven detachment	KALA-targeted osteogenic stimulation	In vitro: rBMSCs	[80]
LP-MSNs	Pore loading	Anti-miRNA (miR26)	None	Targeted anti-osteoporotic treatment	Pore release	Osteogenic stimulation	In vitro: rBMSCs	[62]
MSNs	Electrostatic deposition (PEI)	siRNA (SOST)	Osteostatin	Combined anti-osteoporotic treatment	pH-driven detachment	Osteoporosis gene silencing plus osteogenic stimulation	In vitro: MEF In vivo: Mice	[81]

Table 1. *Cont.*

Nanosystem	Assembly Strategy	Nucleotide	Secondary Therapeutic	Therapy	Release Stimulus	Action Mechanism	Biological Evaluation	Ref.
MSNs	Electrostatic deposition (PEI)	siRNA (SOST)	Osteostatin	Combined anti-osteoporotic treatment	pH-driven detachment	Alendronate-targeted Osteoporosis gene silencing plus osteogenic stimulation	In vitro: MC3T3-E1 In vivo: Mice	[82]
Other therapies								
MSNs	Electrostatic deposition	Plasmid (RhoG)	Curcumin	TAT-targeted neurite growth	pH-driven detachment	Neurite growth induction plus antioxidative effect	In vitro: N2a cells	[83]
MSNs	Electrostatic deposition	Anti-miRNA (miR33)	None	Metabolic lipid disorder treatment	pH-driven detachment	miRNA scavenging	In vitro: L02, LX02, RAW264.7 In vivo: Mice	[84]
LP-MSNs	Electrostatic driven loading	Plasmid (antisense vicR)	None	Biofilm disruption	pH-driven detachment	Reduction of extracellular polysaccharides	In vitro: *Streptococcus Mutans*	[67]

Abbreviations: BMDCs: Bone Marrow Dendritic Cells; CC-MSNs: Core-cone Mesoporous Silica Nanoparticles; CS: Chitosan; GFP: Green fluorescence protein; GSH: Glutathione; H2A: Histone 2A; HA: Hyaluronic acid; LA: lactobionic acid; LP-MSNs: large pore mesoporous silica nanoparticles; OVA: Chicken Ovalbumin; PCL: Poly-ε-caprolactone; PDA: polydopamine; PEI: polyethylene imine; PLL: polylysine; UCNPs: Upconversion nanoparticles; US-MSN: Ultrasmall mesoporous silica nanoparticles; rBMSC: rat bone marrow mesenchymal stem cells.

2.1. Gene Therapy and Gene Silencing

One of the most active areas of non-viral vectors research is gene therapy and silencing for cancer treatment, but not the only one as discussed below. The relevance of gene therapies for anticancer treatment relies on the genetic origin of tumors, which can be modulated with siRNAs and miRNAs to reduce disease's virulency and improve survival. In this section, we will focus on the most recent advances obtained for silica-based gene therapies employing oligonucleotides. However, as the porous structure of silica enables the loading of additional chemotherapeutic agents, it is logical to assume that most promising anticancer strategies result from drug-loaded and multifunctional nanosystems. Combination therapy, due to its importance and number of contributions during the last years, will be reviewed in a following section. In fact, pioneering works on MSNs as non-viral transfection and gene silencing were developed two decades ago [38] and have been visited in numerous contributions. For a review focused on the insights of such combinations, we recommend the reader a previous contribution by us in which many details were covered [39].

Nevertheless, during the last 5 years, significant advances also have been made on nanosilica-based gene therapies (Table 1). Regarding anticancer treatments, there have been reported interesting advances on targeted nanodevices as well as in combination therapies. Along these lines, Li, Duo and coworkers, reported an aptamer-targeted nanosystem for oncogene silencing therapy [59]. In this work the authors focused their strategy in silencing the oncogenic microRNA miR-155, known to upregulate many human cancers. The loading of the therapeutic anti-miR-155 siRNA (5′ACCCCUAUCACGAUUAGCAUUAA3′) was performed onto amino-modified MSNs (ca. 125 nm) with regular mesopores (2.8 nm), which were further coated with a PDA layer suitable to graft the AS1411 nucleolin-targeted aptamer. The resulting system was able to effectively repress the expression of miR-155 and NF-κB in SW480 cells and accumulate within the tumor in vivo after 72 h (during the first 48 h it was mainly accumulated in the lungs). This treatment, especially in combination with 5-fluoruracil (5-FU) made it possible to slow down tumor growth, although did not achieve a complete growth arrest. A gene silencing effect was demonstrated upon an effective downregulation of the PGP protein which sensitized treated mice against 5-FU.

Another interesting targeted nanosystem aimed at anticancer gene silencing was reported by Ngamcherdtrakul, Sangvanich and coworkers [60]. These authors developed a nanoconjugate based on MSNs (50 nm) coated with cross-linked PEI, a PEG spacer and at the outermost layer the trastuzumab monoclonal antibody (MAB). The resulting MSN@PEI@PEG-MAB nanodevice was suitable to bind siRNAs throughout a multilayered assembly. The system was first evaluated against luciferase activity using a siLUC siRNA (5′ CGGAUUACCAGGGAUUUC-Att 3′) and then with siHER2 (5′ CACGUUUGAGUC-CAUGCCCAAUU 3′) RNAs against the BT474 cell line. The system achieved a reduction in cell viability close to ca. 80%, which set an interesting starting point for the development of more advanced therapies. Beyond the interest provided by the potential of this nanosystem, it is also remarkable the freeze–thaw processing of MSN@PEI particles; as typically the combination of PEI and silica usually leads to high degradation rates in aqueous media. Herein, the use of a cytoprotectant makes it possible to maintain their structural integrity of MSN@PEI@PEG-MAB for long periods of time and their further combination with siRNA on demand.

To conclude with examples dealing with anticancer siRNA delivery, i is worth citing the work by Heidari et al., which proposed a chitosan-coated, dual-targeted nanosystem to reduce the anticancer multidrug-resistant pump MDR1 [61]. This system employed amino-modified MSNs (70 nm with 5 nm mesopores) were loaded with the anti-MDR1 siRNA and a later a chitosan coating layer responsible for retaining the cargo serve as anchoring point for targeting groups. The functionalization of such chitosan layer with a Maleimide-PEG3000-NHS bifunctional linker made it possible to covalently bind NH_2 groups of chitosan to thiol groups from cysteine groups at TAT peptides to thiol modified folic acid moieties. The overall performance of this nanosystem was evaluated against HeLa and

EPG85.257 cells with MDR1 overexpression. As a result, a significant reduction in MDR1 levels was achieved without affecting viability, which suggests that this kind of nanosystem could be employed as chemotherapeutic adjuvants by reducing the chemotherapeutic outflux from cancerous cells.

Beyond the use of siRNA delivery against cancer, there are also interesting approaches for the treatment of other genetic diseases. For instance, Hosseinpur et al. employed miRNA delivery to enhance osteogenicity. These authors employed large pore sized (40 nm) core-cone silica nanoparticles (ca. 200 nm) to load the rattus norvegicus (rno) rno-miRNA-26a-5p with the aid of PEI [62]. The system proved to have acceptable biocompatibility before miRNA loading and once loaded proved to boost the expression of seven genes associated with osteogenesis (Runx-2, OCN, Col1α1, IBSP, GSK3β, ALPL, and BMP-2) together with the production of collagen and alkaline phosphatase (ALP) which are associated with bone formation.

Another interesting example, reported by Niu, Yang and coworkers, focused on the treatment of diabetes. These authors created a nanosystem able to transfect two plasmids: the fibroblast growth factor 21 (FGF-21, known to improve glucose metabolism and insulin resistance) and Liraglutide, an analog of the Glucagon-like peptide-1 (GLP-1) with a hypoglycemic effect [63]. The nanosystem, based on hierarchically porous silica nanospheres [85], was able to load and deliver both therapeutics, and in particular the FGF-21-containing plasmid with higher efficacy than Lipofectamine 2000. In vitro, treated mice improved glucose tolerance, lowered blood glucose levels and reduce body weight without significant toxicity or side effects; which suggest a potent long-lasting effect caused by the effective delivery of both plasmids to cells.

Apart from the cited diseases, during the last years cancer immunomodulation has also become a very active field of research. Many of the reported examples employ an immunostimulant protein, typically chicken ovalbumin (OVA), to trigger an immune response against such xenoprotein. The direct consequence is the generation of an immune response against the tumor-forming cells and thus achieve anticancer vaccination. For an in-depth discussion on the topic, please refer to the following section and to reference [86]. Unfortunately, the continuous delivery of xenoproteins intratumorally is not an easy task, for which reason research groups have explored another possibility: plasmid transfection. Ideally, this will produce a sustained immune response at the tumoral region. Along this line two recent examples by Nguyen et al. and Song et al. reported the use of mesoporous silica for the delivery of OVA encoding plasmid [64,65]. In the first example, the authors employed Rambutan-like MSNs [64], a particular kind of particle with a unique spiky surface that, as highlighted above, shows superior nucleotide delivery efficiency, whereas, in the second example the authors employed a branched PEI (60 KDa) to bind the OVA pDNA into a polyplex that was latter supported on rod-shaped mesoporous silica microparticles [65]. With obvious differences on delivery kinetics consequence of particle sizes, both nanosystems were demonstrated to induce effective immune activation as demonstrated by the overexpression of typical markers: CD86, TNF-α, IFN-γ, IL-12, and IL-4. In vivo assays on mice demonstrated that plasmid-loaded, Rambutan-like MSNs, accumulate within the tumor, which in the absence of further results, suggests the possibility of achieving effective antitumor vaccination.

In addition to previous examples, in which nanosystems are designed for intravenous or intraperitoneal dosage, there are also interesting aimed at local dosage. As a representative examples of this strategy, there could be highlighted the nanodevice prepared by Valdés-Sánchez et al. to prevent blindness caused by retinitis pigmentosa [66]. In this work, the authors prepared amino-modified, Ca-dopped silica nanoparticles (ca. 150 nm with $\approx$3.5 nm mesopores) employing a sol–gel method onto which loaded the PRPF31-GFP plasmid. Upon in vitro evaluation, the system was evaluated in vivo throughout subretinal dosage in mice. Treated animals maintained the integrity of the retinal tissue and the visual function, which suggests an effective gene delivery by the silica-based non-viral vector and thus a protective effect against retinal degenerative diseases.

Another interesting example of local administration was reported by Tian et al., who developed a dendritic silica nanosystem able to deliver an anti-*Streptococcus mutans* plasmid [67]. Employed particles were prepared with cyclohexane as pore expander and an amino-modified surface to ensure electrostatically binding of the plasmid. The integrity of the DNA was ensured by treating the system against DNase I, proving that the multilayered structure gave some protection against degradation. When tested against S. Mutans, the nanosystem was able to deliver the antisense vicR plasmid, disrupting the normal function of bacterial enzymes responsible for wall biosynthesis. Consequently, although this local treatment did not result in destruction of bacteria, it made it possible to significantly disrupt the synthesis and function of biofilm, thus sensitizing these cariogenic bacteria against conventional antibiotics.

2.2. Combined Anticancer Therapies

Despite advances made in clinical practice, to date, chemotherapy remains the main therapeutic tool for the treatment of cancer. The well-known side effects occurring as a consequence of its use, especially the appearance of multidrug resistance (MDR) in relapsing cancers, demand more selective and active treatments. Combination therapy seems to improve overall efficacy through synergistic effects, although at the expense of maintaining chemotherapy's typical side-effects. Apart from the classical combination of chemotherapy with surgery or radiotherapies, during the last years new promising nanotherapies have also emerged. Among them, (1) immunotherapy promises to recruit the patient's immune system to destroy cancer cells, while (2) gene therapy claims to either overcome drug resistance by disrupting active immortality pathways of cancerous cells. In fact, both strategies have been successfully employed in the clinic and in particular the combination of chemotherapeutic drugs and siRNAs has proven to be a valuable combination against cancer [87].

However, profiting from this combined anticancer effect is not easy, as the stability of siRNA-based therapeutics is very short unless efficiently delivered. In addition, secondly, the differences on pharmacokinetics of both chemotherapeutic drug and siRNA makes it extraordinarily difficult to achieve a simultaneous action. For these reasons, nanoparticles have arisen as promising platforms to deliver both agents simultaneously and overcome one of the limitations of combination therapy. Since our first review on this topic [88], the number of formulations has grown significantly. In fact, the current state of the art is split between (1) the clinician's point of view, focused on ultra-specialized applications of particular drug-siRNA combinations, as can be guessed from reviews with titles such as Overcoming doxorubicin resistance in cancer: siRNA-loaded nanoarchitectures for cancer gene therapy [89] and Employing siRNA tool and its delivery platforms in suppressing cisplatin resistance: Approaching to a new era of cancer chemotherapy [90]; and (2) the materials scientist's point of view, which is more focused on the development of strategies and understanding mechanisms of action of such nanosystems, like the excellent review by Paris and Vallet-Regí [91].

Among the strategies reported, multiple chemotherapeutic agents, oligonucleotide sequences, and combinations of both can also be found. For instance, Zheng et al. reported the use of MSNs to deliver Sorafenib, a kinase inhibitor in combination with a Cy5-labeled Vascular Endothelial Growth Factor (VEGF) siRNA (sense: 5′-GGAGUACCCUGAUGAGAUC dTdT-3′) [68]. Their system was built employing Sorafenib-loaded, amino-capped MSNs onto which the siRNA was electrostatically deposited and the Lactobionic acid (4-O-β-D-galactopyranosyl-D-gluconic acid, LA) grafted. The resulting nanosystem showed good antiproliferative effect on four typical cancer cell lines (Huh7, HepG2, HeLa and A549), showing a synergistic behavior when both anticancer compounds were combined. In the best cases, the authors reached more than 80% reduction on viability at 60 μM concentration of particles. Another interesting example was reported by Dilnawaz and Sahoo [69]. These authors evaluated the behavior of single-drugs vs. drug combinations, both delivered by MSNs or in their free form, against healthy HEK-293 and lung adenocarcinoma A549 cell lines. According to their results, MSN-loaded combinations of Docetaxel or Etoposide with Carfilzomib showed the best effect. In a second step, they evolved their nanosystem by including a Survivin siRNA (Sense: 5′-GGACCACCGCAUCUCUACAdTdT-3′) throughout a sequential PEI-oligonucleotide coating. The evaluation of the resulting combined drug nanosystem made it possible to reach cell apoptosis up to ca. 50% due to the simultaneous knockout of Survivin and dual action of drugs. Unfortunately, no further evaluation was done in vivo to fully prove their suitability as therapeutic models.

Apart from the previous examples, interesting efforts have been also made on improving carrier biocompatibilities and release rates. For instance, Zhang et al. reported the use of a biodegradable nanosystem based on reducible coating and silica matrix to co-deliver Doxorubicin (DOX) together with an anti Bcl-2 siRNA (sense 5′-CGGGAGAUAGUGAUGAAGd TdT-3′) [70]. Prepared nanoparticles contained disulfide bridges along their structure, which are quickly cleaved in intracellular environments. The advantages of these particles go beyond fast degradation, as they show also quick cargo delivery and may induce deregulation of homeostasis throughout depletion of the cytoprotective Glutathione, which may also have an additional adjuvant effect. To build the system, the authors employed a layer-by-layer assembly strategy using hyaluronic acid as the first coating layer and PEI as a cationic compound. Data shown suggested that the nanosystem designed for DOX/Bcl-2 siRNA co-delivery led to higher rates of late apoptosis in MCF-7 cancer cell line (>99%) than those yielded with carried DOX (59%) or siBcl-2 (10%) alone, which suggest an effective gene/chemo-synergic therapy. Another interesting example was reported by Choi et al. To improve biocompatibility these authors employed Ca2+ doped MSNs to create an amino-free nanosystem able to co-deliver the anti Bcl-2 siRNA (sense: 5′-GUACAUCCAUUAUAAGCUGdTdT-3′) directly bound to MSNs′ surfaces due to the presence of Ca cations and Chloroquine as chemotherapeutic [71]. The resulting system showed a sustained but pH-dependent release of siRNA and an effective downregulation of Bcl-2 expression on SKOV3 cell lines, yielding similar values as those reported with the lipofectamine reference. The presence of Chloroquine in the nanosystem also helped in reducing cell viability, reaching up to 30% of reduction, which although not outstanding from a therapeutic point of view, confirms the efficacy of this nanoformulation.

In addition to gene silencing, several research groups have also explored combined therapies which typical chemotherapeutics and plasmids, being the p53 tumor suppressor protein and Survivin, inhibitor of caspases′ pathways, two of the most promising candidates for reverting tumoral profile. Along this line, Rong et al. and Zhou et al. reported the combination of a p53 plasmid with typical chemotherapeutic drugs to obtain double action nanotherapeutics. In the system by Zhou et al., the authors managed to load the hydrophilic drug 5-fluoruracil within the mesopores of silica nanoparticles [73]. Due to the difficulty of this process, the chemotherapeutic drug was entrapped within the mesopores with the aid of a cationic, acid-cleavable block copolymer based on 2-(dimethylamino)ethyl methacrylate (DMAEMA). The resulting nanosystem proved to have an outstanding biocompatibility with a slight reduction of viability up to 200 $\mu g \cdot mL^{-1}$ in MCF-7 cell line and hemolysis rates below 5%. This model only yielded a modest cell viability reduction (about ca. 50%) despite the dual effect and the success of p53 plasmid transfection. On the other hand, the nanosystem reported by Rong et al. employed Bortezomib (BTZ) as cargo. Chosen core–shell upconversion-silica nanohybrids were modified through a grafting step with Histone 2A as targeting moiety and finally coated with the p53 plasmid electrostatically [72]. In this case, the protein, apart from favoring migration to nucleus [92] showed additional features: (1) increased biocompatibility (augmented cell viability of empty carrier); (2) reduced hemolysis rate; and (3) enabling of the subsequent interaction with the p53 plasmid. The complete system was able to efficiently transfect p53 into cells′ nucleus, and in combination with BTZ, made it possible to reduce HeLa and NCI-H1299 cells viabilities about ca. 65% and 80% respectively, demonstrating that nuclei targeting must be considered too for the development of future nanoformulations.

In addition to previous examples, there are other combinations of plasmids and chemotherapeutic drugs reported in the literature. For instance, Tsai, Wang and coworkers [74] reported a system based on the combination of Survivin plasmid with Camptothecin, while Alibolandi and coworkers tested a combination of HNF4α plasmid and Cisplatin [75]. In the first example, amino-modified, rod-shaped MSNs were employed to create targeted nanoparticles upon grafting of the AS1411 aptamer. Then, onto the surface, the remaining unfunctionalized amino made it possible to undergo electrostatic deposition of the negatively charged plasmid to provide a transfecting system with nuclear-targeting properties. The combination of both substances showed a clear synergistic effect against C26 cancerous cells (ca. 80% viability reduction), while healthy CHO cells were not affected in this manner (ca. 30% viability reduction). Further in vivo experiments also demonstrated that the targeting agent was crucial to obtain an adequate intratumor distribution, while in its absence bioaccumulation occurred preferentially in liver and kidney. In the second example, the absence of any targeting element was justified because the final purpose of such nanosystem was a liver disease. In this model, negatively charged phosphonate-capped cisplatin-loaded MSNs were coated with a 2 KDa PEI and the corresponding plasmid. This system was able to reduce the tumorigenic capacity of Huh7 cells according to data shown, achieving even effective tumor volume reduction when both therapeutic agents were combined in the nanodevice.

In addition to plasmid transfection, the use of RNAs also offers broad potential for the development of temporary therapies. The use of micro RNAs has been reported alone and in combination with other cytotoxic compounds. For instance, Li et al. reported the use of lipid-coated MSNs to deliver the miR-137 RNA (sense: 5′-UUAUUGCUUAAGAAUACGC GUAG-3′) in vitro and in vivo [76]. Herein, to improve biocompatibility and efficacy, the authors employed three complementary strategies on their design: (1) the external coating was fabricated from red blood cells membranes′, which were (2) modified with RDG targeting moieties to enable targeting and (3) included the indocyanine green photosensitizer within the porous silica structure. As a result, the system was able to reach adequate intratumor accumulation due to the RDG targeting and the advanced immune cloaking. In vivo experiments showed that combined photothermal and gene therapy made it possible to treat solid tumors so efficiently that even tumor reduction was observed. A similar design employing RGD-targeted, lipid-coated MSNs was also reported by Oupický and coworkers, who employed a silica-based nanodevice to deliver the siPlk-1 anticancer RNA (sense: 5′-CAACCAAAGUCGAAUAUGAUU-3′) in combination with a microRNA-200c mimic (sense: 5′-UAAUACUGCCGGGUAAUGAUGGA3′) [77]. Their system, which also contained indocyanine green as a photosensitizer, was also able to destroy tumors in vivo, achieving complete tumor remission in mice and an antimetastatic activity in an orthotopic breast cancer model highlighting the potential of gene therapy in combination therapies.

As discussed above, the combination of therapeutic RNAs makes it possible to develop highly interesting combined therapies; although for such purposes the use of chimeric RNAs [93] would be of interest as combining several effects in a single nucleotide strand would facilitate delivery processes. However, the combination of different RNA sequences may be of interest too if different dosages are required. Along this line, Ahir et al. explored the potential therapeutic effect of the hsa-miR-34a-5p miRNA (sense: 5′- UGGCAGUGUCUUAGCUGGUUGU-3′) together with the anti-sequence miR-10b-5p (5′-CACAAAUUCGGUUCUACAGGGUA-3′) [78]. This system [78] was prepared from trimethylammonium-modified MSNs, which were able to coordinate both RNA strands via electrostatic interactions. Then, to conclude the system was coated with a pore-blocking PLGA layer and hyaluronic acid sequences with targeting capabilities. The complete nanosystem was able to interact with the CD44 receptor, internalize into cancer cells and promote an early apoptotic state. The external polymeric layer made it possible to accomplish in vivo studies in both mice and chicken embryo with satisfactory results, due to the great biocompatibility provided. However, upon tumor delivery, cell uptake and system disassembly, both therapeutic RNAs made it possible to achieve significant tumor reduction, although in this case, the polycationic nanoparticle may have membrane-lytic effect too and may have a contribution on the overall antitumor effect.

2.3. Gene-Based Therapies against Bone Diseases

As outlined previously, gene editing allows the treatment of many diseases of genetic origin in addition to cancer; as reviewed by Knežević and coworkers [94]. Among such diseases, bone ones have recently attracted great attention; especially those in combination with ceramic materials such as silica, which offer great osteocompatibility and interesting properties as filling material. On this topic, He and coworkers explored the antiosteoporosis effect of a plasmid encoding the bone morphogenetic protein 2 (BMP-2) (pIRES2-ZsGreen1-BMP2) in combination with the dexamethasone glucocorticoid [79]. This nanodevice employed PEI-PLL as a connecting layer due to the better biocompatibility and higher transfection efficacy than pure 25 KDa PEI. Moreover, to enable cell targeting, the outermost surface of the resulting carrier was modified with the RGD motif and thus promote cell internalization. Drug and plasmid were integrated within the nanosystem throughout sequential pore loading and DNA surface electrostatic deposition onto the positively charged polymer-coated particle. Cell cultures on Bone Marrow Mesenchymal Stem Cells (BMSCs) showed effective transfection and drug release, which yielded an overexpression of the BMP-2 protein along with other typical bone formation markers such as Alkaline phosphatase (ALP) and the osteo-related genes RUNX2, OPN, Col 1 and OCN in vitro, demonstrating the effective action of both components.

The use of miRNAs also has positive effects, as reported by Yan et al. [80] and Hosseinpour et al. [62]. These authors explored the efficacy of silica-carried 26A miRNA in the osteogenic differentiation of rat BMSCs, although using complementary delivery strategies. In the first example, the RNA was deposited onto positively charged MSNs@PEI, while in the second, the RNA was loaded within the porous structure of MSNs. As a result, both nanosystems were able to deliver the miRNA intracellularly, although the first model, modified with the cell-penetrating KALA peptide, showed better performance and improved biocompatibility. However, both nanoformulations were able to easily achieve 3-fold increases in overexpression of osteo-related genes; although the in-pore-loaded miRNA nanosystem showed higher BMP-2 expression (ca. 10-fold vs. 3.5-fold) due to the enhanced protection offered by the silica matrix in these for single-drug biopolymer delivery.

In addition to previous examples, Manzano, Vallet-Regí and coworkers also explored an osteoporosis treatment based on targeted nanotherapeutic able to deliver simultaneously an osteogenic peptide and an anti-osteoporotic siRNA. In their first formulation, the authors evaluated the codelivery of siRNA and peptide combination [81], while in a later contribution, they incorporated the alendronate bone-specific targeting moiety to improve efficacy [82]. Both formulations were prepared employing MCM-41-like MSNs, which upon peptide loading were reacted with PEI and the anti-SOST siRNA to obtain the final device. The effective silencing of SOST expression made it possible to obtain upregulation of both ALP (Alkaline phosphatase) and RunX2 proteins, responsible for bone regeneration both in vitro and in vivo. Unfortunately, their first model was only effective upon highly invasive bone injection; although in the evolved model, an alendronate-targeted nanosystem achieving satisfactory antiosteoporosis effect in female ovariectomized mice could also be achieved, but with subcutaneous dosage. In this nanoformulation, the bisphosphonate-containing alendronate was linked to the silica matrix through a PEG connector, also allowing the loading of osteostatin and siRNA, as previously described.

2.4. Other Gene-Based Therapies Employing Silica Particles as Nucleotide Carriers

In addition to the discussed examples, nucleotide-based therapy using silica-based nanocarriers could be also employed for the treatment of other genetic diseases. However, as consequence of the limited number of reported examples, these must be considered as proofs of concept rather than as potential therapeutics. For instance, Wu, Chen and coworkers reported a nanosystem able to prevent neuron damage caused by chronic inflammation in neurodegenerative disease. For so, the authors designed a multifunctional nanosystem composed by an antioxidant (curcumin) and a neurite growth promoter (RhoG containing plasmid) and the TAT cell-penetrating peptide (AYGRKKRRQRRR) to favor cell uptake [83]. The resulting device was able to transfect Na2 cells favored by the positively charged targeting peptide, as demonstrated by the transfection of a red fluorescent protein and induce neurite growth, and once internalized, induce cell growth throughout a dual action of the plasmid and the curcumin. In fact, both effects were determined separately employing Paraquat, a known generator of superoxide anion radicals, and the non-therapeutic GFP plasmid.

Another example of gene-originated disorders that could be treated with silica-based nanomaterials was reported by Tao et al., who designed a nanosystem able to deliver a microRNA-33 antagomir to prevent lipid metabolic disorders [84]. Herein, cholesterol-targeted amino-modified MSNs were loaded with an antagonist of the miRNA-33 (sense: 5'UGCAAUGCAACUACAAUGCAC-3'). As a result, the system could internalize into hepatic cells and promote lipid metabolism, thus reverting fatty liver condition. In vivo studies showed that treated mice had an accelerated lipid metabolism, as determined by the lower weight gain observed. In fact, body weights of treated mice showed similar values to lovastatin treated mice, pointing out an effect of similar magnitude to that obtained by conventional antilipidemic drugs.

In conclusion, plasmid delivery has also been employed to treat bacterial infections. Through a downregulation of extracellular polysaccharides in Streptococcus Mutans colonies, Tian et al. were able to reduce the cariogenic potential of this type of cocci [67]. Their model was built by electrostatic deposition of the negatively charged anti-VicR plasmid onto dendritic amino-modified MSNs. This particular structure of silica showed to be of capital importance to obtain adequate protection against DNA degradation and transfection into bacteria. Genetically disrupted cocci showed a reduced formation of extracellular matrices which led to a weaker and disordered biofilm, making them highly sensitive to conventional antibiotics and reducing their potential carcinogenicity.

3. Protein- and Peptide-Based Therapeutics Employing Silica-Based Nanosystems

As discussed previously, proteins have capital roles in many physiological processes. However, oppositely to nucleic acids, any possible therapeutic effect of proteins relies on how they reach their destination, as denaturalization may occur. For that reason, it is important to develop carriers that are able to keep their structural integrity during trafficking [95,96]. Because of the need for having such protective features, protein delivery has become a popular topic during the recent years. However, associated with the intrinsic complexity of such delivery, clear strategies to accomplish successful dosages are still unavailable and thus many authors still explore the possibilities of porous silica. For interested readers, the following reviews and articles deal with the insights of protein loading and delivery with MSNs [97–99]. On the other hand, peptides have attracted great attention too due their simplified structures that can partially retain functional features. Such biological activities in combination with a very convenient chemical robustness is of interest to produce low-cost therapeutics for regulating metabolic cycles and signaling processes.

Due to the number of physiological tasks developed by proteins and peptides [100], the following section will be structured in several epigraphs, covering the following aspects: (1) anticancer and pro-apoptotic proteins and peptides [101–103], (2) antibacterial proteins and peptide delivery [104,105], (3) immunostimulant nanoformulations, (4) nanocarriers for enzymes, and (5) delivery of growth factors as outlined in Figure 3 and detailed in Table 2. In this review, non-therapeutic applications of peptides will be excluded. For instance, cellular targeting [106] on both eukaryotic [107] and prokaryotic [108] cells, combined therapies using non-based silica carriers [109] and peptide-based stimuli responsive materials [47,110] will not be covered, despite their relevance.

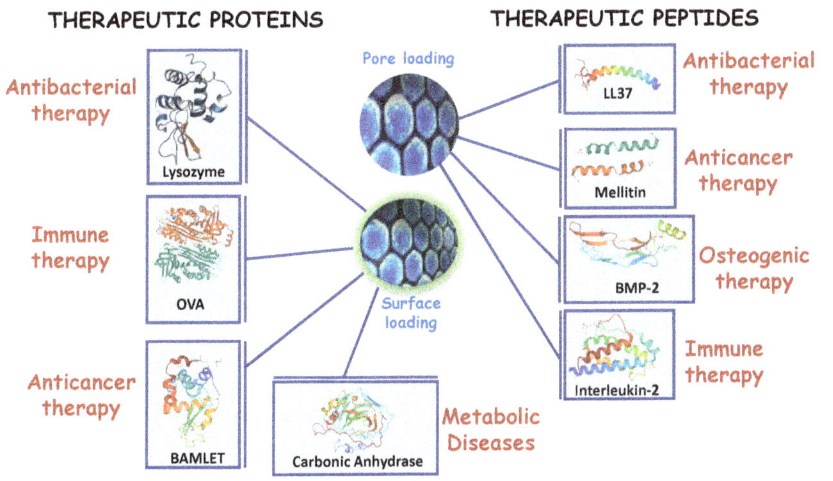

Figure 3. Delivery of proteins and peptides with mesoporous silica nanoparticles makes it possible to create nanotherapeutics for a wide range of diseases.

3.1. Anticancer Therapies

The difficulty of handling proteins, which can easily undergo denaturalization and lose their function, makes it difficult to develop interesting examples on anticancer protein delivery. Nevertheless, the adequacy of MSNs for hosting, transporting and delivering proteins was described by Lin and coworkers decades ago. They pioneered pore-enlarged MSNs to facilitate protein hosting and thus promote uptake of proteins through impermeable membranes. These authors focused on Cytochrome C (Cyt c), a relatively small protein with pro-apoptotic effect triggered through the caspase pathway [111]. Herein, unaltered MSNs with 5.4 nm mesopores could accommodate this globular protein to produce intracellular delivery, albeit without control on protein release. More recently, Griebenow and coworkers revisited Cyt c delivery with MSNs bearing thiol-modified mesopores [112]. This evolved system had cleavable bonds able to retain Cyt c within the mesopores and trigger its release upon increased levels of glutathione such as those in cytoplasmatic environments. Similarly, Shang et al. employed Cyt c as a representative example for MSN-based delivery [113]. Herein, the authors examined the loading capacity and protein activity in relation to nanoparticle sizes, rather than evaluating the overall delivery. Their findings showed that larger diameters and flatter surfaces allowed the adsorption of more protein units and yielded higher overall activities. In another interesting contribution to the topic, Davis and coworkers optimized the connecting linker used to bind proteins to particle surfaces. Employing Cyt c as model, they systematically evaluated several custom-made linkers against their most relevant criteria that affected protein delivery [114]. Among them, surface charge (>40 mV on ζ-potential), ionic charge at acidic pH, endosomal escape properties and surface retention were studied. According to their results, 1 mol% functionalization of primary amines made it possible to accomplish all those tasks successfully, being the best basis for further surface grafting strategies.

In addition to Cyt c, a broad number of anticancer proteins were also reported as payloads in nanoparticle-based delivery [96] although when dealing with MSNs, the number of examples significantly decreased.

Table 2. Examples of therapeutic proteins and peptides delivered by silica-based nanocarriers. For an illustrative review on the coupling protocols employed for linking peptides onto nanocarriers, please check reference [115].

Nanosystem	Assembly Strategy	Protein or Peptide	Secondary Therapeutic	Biomolecule Release Stimulus	Action Mechanism	Biological Evaluation	Ref.
Delivery of peptides and proteins in anticancer strategies							
MSNs	Pore loading	Cytochrome C	None	Pore release	Cell apoptosis promoter	In vitro: HeLa	[111,112]
MSNs	Surface adsorption		None	Electrostatic detachment		None	[113]
MSNs	Surface grafting		None	None		In vitro: HeLa	[114]
MSNs	Surface grafting	Concanavalin A	None	None	Targeting plus upregulation of metalloproteinases	In vitro: MC3T3-E1, HOS.	[116]
MSNs	Surface grafting	BAMLET	Docetaxel	None	Cell apoptosis promoter plus chemotheraputic	In vitro: COS7, U87 MG In vivo: Zebrafish In vivo: Balb/c Mice	[117]
MSNs	Surface grafting	K_8-Citraconate $K_8(RGD)_2$	Doxorubicin	None	Membrane disruption plus chemotherapy	In vitro: COS7, U87 MG In vivo: Mice	[118]
MSNs	Surface grafting	TPP-K-$(KLAKLAK)_2$-	Topotecan	None	Mitochondrial membrane disruption plus chemotherapy	In vitro: KB	[119]
MSNs	Surface grafting	C-$GRK_2R_2QR_3P_2Q$-RGDS C-GKGG-$D(KLAKLAK)_2$	Doxorubicin	None	Membrane disruption plus chemotherapy	In vitro: HeLa, COS7	[120]
MSNs	Surface grafting	$(RGDWWW)_2KC$	Doxorubicin	None	DNA-intercalation plus chemotherapy	In vitro: COS7, U87 MG	[121]
MSNs	Surface grafting	$(KLAKLAK)_2$	Doxorubicin	None	Membrane disruption plus chemotherapy	In vitro: HeLa	[122]
MSNs	Surface grafting and pore loading	ε-poly-*L*-lysine (surface)	C9h (pore)	Enzymatic degradation (Pore release)	Membrane disruption plus propapoptotic induction	In vitro: HeLa	[123]
MSNs	Pore loading	RDG-Hylin a1	None	Pore release	Targeted cytolytic peptide	In vitro: HeLa Hep2 In vivo: Mice	[124]
HMSNs	Cavity loading	Pepstatin A	None	Release from cavity	Inhibition of Aspartyl protease	In vitro: MCF-7	[125]
MSNs (SBA-16)	Pore loading	Alamandine	None	Pore release	Unknown	In vitro: 4T1, A549, HEK-293	[126]
MSNs	Pore loading	Fabatin	None	Pore release	Mitochondrial disfunction	In vitro: MDA-MB-23, MCF-10A In vivo: Mice	[127]
SPION@MSNs	Pore loading	Mellitin	None	Thermosensitive–hydrolytic cleavage	Apoptosis induction and suppression of angiogenesis (VEGF)	In vitro: PANC-1 In vivo: Mice	[128]

Table 2. *Cont.*

Nanosystem	Assembly Strategy	Protein or Peptide	Secondary Therapeutic	Biomolecule Release Stimulus	Action Mechanism	Biological Evaluation	Ref.
Antibacterial therapies							
MSNs	Adsorption	Lysozyme	None	Electrostatic detachment	Bacterial wall hydrolysis	In vitro: *E. coli* In vitro: HEK-293, LO2 In vivo: Mice	[129]
MSNs	Pore loading	Lysozyme	None	Pore release	Bacterial wall hydrolysis	In vitro: *E. coli*	[130]
HMSNs	Surface adsorption	Lysozyme	None	Electrostatic detachment	Bacterial wall hydrolysis	In vitro: *E. coli* In vivo: Mice	[131]
HMSNs	Cavity loading	Lysozyme	None	Cavity release	Bacterial wall hydrolysis	In vitro: *E. coli*	[132]
MSNs	Surface grafting	Concanavalin A	Levofloxacin	None	Glycopeptide targeting	In vitro: *E. coli*	[133]
MSNs	Pore loading	Bactofencin A	None	Pore release	Defensin-like peptide	In vitro: *S. aureus* In vitro: HEK-293	[134]
MSNs	Pore loading	β-defensin-2	None	Pore release	Defensin-like peptide	In vitro: *Clavibacter michiganensis*	[135]
Solid SiO_2 vs MSNs	Surface adsorption vs. pore loading	LL37	None	Surface adsorption vs. pore release	Transmembrane pore formation (α-helical shape)	In vitro: *E. coli*	[136]
RSNs	Surface deposition	LL37	None	Release from rough surface	Transmembrane pore formation (α-helical shape)	In vitro: *E. coli*	[137]
MSNs	Pore loading	NZX	None	Pore release	Antituberculotic peptide	In vitro: *M. tuberculosis* In vivo: Mice	[138]
MSNs	Surface grafting	Ovotransferrin	Gentamicin	None	Membrane lysis	In vitro: *E. coli* In vivo: Mice	[139]
MSNs (phosphonate vs. raw MCM-41)	Pore loading	Pexiganan, Indolicidin, or [I^5,R^8] Mastoparan (Electrostatically located at the surface)	*trans*-chalcone, curcumin, quercetin or berberine chloride	Electrostatic detachment	Sortase A inhibition (small molecules) plus antibiotic peptide. Pexiganan (cationic), Indolicidin (filamentator), Mastoparan (α-helix pore formation)	In vitro: *S. aureus*, *S. aureus (Methilin resistant)*, *E. coli*, *P. aeruginosa*	[140]
Osteogenic therapies							
MSNs	Pore loading	bFGF	None	Pore release	Growth factor (fibroblast)	In vitro: HUVEC	[141]
MSNs	Surface adsorption	BMP-2	None	Electrostatic detachment	Bone morphogenetic protein	In vitro: bMSCs In vivo: Mice	[142]
MSN@SPION	Pore loading	BMP-2	None	Pore release	Bone morphogenetic protein	In vitro: bMSCs	[143]
MSNs	Pore loading	Osteostatin	None	Pore release	Hormone-related peptide	In vitro: MC3T3-E1 In vivo: Rabbit	[144–146]

Table 2. *Cont.*

Nanosystem	Assembly Strategy	Protein or Peptide	Secondary Therapeutic	Biomolecule Release Stimulus	Action Mechanism	Biological Evaluation	Ref.
MSNs MS-HANs	Pore loading	OGP	None	Pore release	Osteogenic growth peptide (Unknown mechanism)	In vitro: Mesenchymal stem cells	[147]
MSNs	Surface adsorption	BMP-2 derived peptide	None	Pore release	Bone morphogenetic protein derived peptide	In vitro: BMSCs In vivo: Rat	[148]
MCaSiNs	Pore loading	GL13K	Sr^{+2} doped matrix	Pore release (peptide)	Antibiotic peptide Osteoclastogenetic promoter	In vitro: HBMSCs In vitro: *S. aureus*	[149]
				Immunotherapy			
HMSNs	Cavity loading	IgG	None	Cavity release	Proof of concept	In vitro: HeLa	[150]
RSNs	Interparticle loading	Cyt c, IgG, Anti-pAkt	None	Cavity release	Proof of concept	None	[151]
HMSNs	Pore loading	OVA	None	Pore release	Xenoprotein induced immunostimulation	In vitro: NIH3T3 In vivo: Mice	[152,153]
DMOHS	Pore loading	OVA	None	Pore release	Xenoprotein induced immunostimulation	In vivo: Mice	[154]
MSNs	Pore loading	OVA	CpG	Pore release	Xenoprotein induced immunostimulation plus a Toll-like Receptor agonist (CpG)	In vitro: RAW264.7 In vivo: Mice	[155]
HMSNs	Cavity-pore loading	IL2	Retinoic acid Doxorubicin	Lipid layer detachment. Multiple release	Immunostimulant protein (IL2) Chemotherapeutic (DOX) Apoptotic promoter (RA)	In vitro: L929 In vivo: Mice	[156]
MSNs	Pore loading	IL13	None	Pore release	Immunostimulant protein (IL13)	In vitro: BMDMs In vivo: Mice	[157]
HMSNs	Surface adsorption	ORF2	None	Electrostatic detachment	Anti circovirus vaccine	In vitro: PK15 In vivo: Mice	[158]
MSNs	Surface adsorption	SWAP	None	Electrostatic detachment	Anti-parasite vaccine	In vivo: Mice	[159]
MSNs	Surface adsorption	HSP70	None	Electrostatic detachment	Anti-mycoplasma vaccine	In vivo: Mice	[160]
MSNs	Surface adsorption	EspA	None	Electrostatic detachment	Anti *E. coli* vaccine	In vivo: Mice	[161]
MSNs	Surface adsorption	rPb27	None	Electrostatic detachment	Anti-fungi vaccine	In vitro: HEK-293 In vivo: Mice	[162]
HMSNs	Cavity loading	TRP2 (cavity)	HGP100 (pores)	Lipid layer detachment. Dual release	Dual antitumor immunostimulants	In vitro: BMDCs	[163]
MSNs	Surface	Hexahistidine	Chlorogenic acid	None	Ni scavenging (histidine) Antiinflamatory (Chlorogenic acid)	In vitro: BJ	[164]

Table 2. *Cont.*

Nanosystem	Assembly Strategy	Protein or Peptide	Secondary Therapeutic	Biomolecule Release Stimulus	Action Mechanism	Biological Evaluation	Ref.
				Enzymes			
MSNs	In-pore grafting	CA or HPR	None	None	Proof of concept	None	[165]
MSNs	In-pore grafting	CA	None	None	Proof of concept	In vitro: HeLa	[166]
MSNs	Pore loading	β-Galactosidase	None	None	Treatment of Morquio B syndrome	In vitro: N2a	[167]
MSNs	Surface grafting	SOD	None	None	Antioxidant effect (reduction of Reactive Oxygen Species)	In vitro: HeLa	[168]
MSNs	Surface grafting	SOD or GPx	None	None	Antioxidant effect (reduction of Reactive Oxygen Species)	In vitro: HeLa	[169]
MSNs	Surface grafting	Proteasomes	None	None	Anti Tau-protein aggregation	In vitro: HEK-293, HeLa	[170]
UC@MSNs	Surface adsorption	RNAase	Cisplatin prodrug	Electrostatic detachment	Combined anti-protein and chemotherapy	In vitro: HepG2, L929 In vivo: Mice	[171]

Abbreviations: BAMLET: Bovine α-lactalbumin made lethal to tumors; bFGF: basic fibroblast growth factor; BMDCs: murine Bone-marrow derived dendritic cells; BMDMs: murine bone marrow-derived macrophages; BMP-2: Bone morphogenetic protein 2; BMSCs: rat bone mesenchymal stem cells; bMSCs: Murine bone mesenchymal stem cells; CA: Carbonic anhydrase; cAMP: cyclic adenosine monophosphate; MCaSiNs: Calcium Silicate Mesoporous Particles; CpG@OVA: Ovoalbumin loaded CpG oligodeoxynucleotide; DMOHS: Dendritic Mesoporous Organosilica Hollow Spheres; DOX: doxorubicin; EspA: an immunogenic protein from enterohaemorrhagic *Escherichia coli*; GPx: Glutathione peroxidase; HBMSCs: Human bone marrow derived mesenchymal stem cells; HMSNs: Hollow Mesoporous silica Nanoparticles; HRP: Horseradish peroxidase; IgG: Immunoglobulin G; IL2: Interleukin 2; MS-HANs: Mesoporous silica-hydroxyapatite nanoparticles; OGP: Osteogenic growth peptide; ORF2: open reading frame from Porcine Circovirus Type 2; OVA: Chicken Ovoalbumin; RA: Retinoic acid (all trans); RSNs: Rough (non-porous spiky) silica nanoparticles; SOD: Superoxide dismutase; SPIONs: Superparamagnetic Iron Oxide Nanoparticles; UC@MSNs: Mesoporous silica coated up conversion nanoparticles.

The most relevant example was reported by Vallet-Regí and coworkers, who reported the effect of Concanavalin A, a lectin with anticancer and antibacterial properties [116], against murine preosteoblast (MC3T3-E1) and human osteosarcoma (HOS) cells. In this system, doxorubicin (DOX), a common chemotherapeutic, was also incorporated to the MSNs before the polymeric layer which was attached to the silica surface using pH-sensitive linkers. Finally, amide bonds were used to graft the ConA onto the surface. Only when the pH decreased enough was the bis-acetal linker cleaved, enabling the release of DOX. Although not studied, the antiproliferative effect of ConA was ensured when both species were administered simultaneously, showing a synergistic effect higher than the obtained with DOX alone.

In addition to previous examples, another promising anticancer substance is the bovine α-lactalbumin made lethal against tumor cells (BAMLET), which has also arisen as an interesting possibility for silica-based delivery. BAMLET is an emerging nanotherapeutic, which comes out of conjugating bovine α-lactalbumin (BLA) with oleic acid and acts as apoptosis promoter. With this information in mind, Pei et al. designed a nanosystem able to co-deliver a typical cytotoxic drug such as docetaxel (DTX) with BAMLET [117]. Their design was prepared from oleic acid modified MSNs, which upon loading with DTX and treatment with BLA, resulted in the formation of silica particles with a protein coating. This BAMLET corona performed several functions, as it allowed great colloidal stability and biocompatibility together with an effective diffusion barrier for the loaded cytotoxic drug. This protein-coated system was able to exert a relevant proapoptotic effect against several cancerous cell lines such as MCF-7, RBL-2H3 and HeLa, which could be increased in the presence of the cytotoxic drug. In vivo assays on mice showed very promising results, too, as the individuals treated with the DTX-MSN-BAMLET nanosystem showed complete tumor growth arrest and survival after 30 days.

Unlike proteins, whose anticancer effect is promoted by triggering biological effects, peptides can exert anticancer effects throughout two different mechanisms: (1) disruption of membrane's normal function; and (2) the activation of pro-apoptotic pathways [172]. The first mechanism operates when the peptides are enriched in cationic amino acids: Lysine (Lys, K; amine) Arginine (Arg, R; guanidine) and Histidine (Hys, H; imidazole) and is similar to the effect shown by other polycationic species [173]. Even though this mechanism is not fully understood, recent investigations on nanosystems based on TAT [174] and related cell-membrane-penetrating peptides [175] suggest a membrane-lytic effect too.

Regarding chemical reactivity, peptides bind much more easily to nanoparticles than proteins, as their relatively small size reduces the possibility of linkages on active site surroundings and permit the use of les mild conditions. Nevertheless, despite this advantage, typically polycationic peptides are known to increase the risk of vascular damage like hemolysis and embolisms during trafficking. For these reasons, masking its positive charge could be a relevant strategy in nanomedicine's design. This was explored by Zhang and coworkers on three different approaches. In the first example, they transformed a custom-targeted cationic peptide $K_8(RGD)_2$ into a negatively charged peptide using citraconic anhydride without affecting its targeting capacity [118]. On later investigations these authors reported the use of glutathione-cleavable anionic coatings for cloaking triphenyl phosphonium modified KLA peptides (KLAKLAKKLAKLAK) [119] and membrane disrupting sequences specific to mitochondria (C-GKGG-DKLAKKLAKLAK) and membranes (C-GRKKRRQRRRPPQ-RGDS) [120]. In these three contributions the authors managed to obtain significantly reduced membrane-lytic effects and demonstrated the importance of enhancing biocompatibility of nanocarriers.

Following another strategy, Zhang's research group developed a drug delivery system for DOX employing a membrane-targeted therapeutic peptide. Herein, the tryptophan-rich peptide ($(RGDWWW)_2KC$) was bound to MSNs through glutathione-mediated disulfide bonds [121], which made it possible to detach the peptide and exert a DNA-intercalant effect. As expected, the best results were obtained when DOX was also present, suggesting a combined effect. Feng et al. also tested this coating strategy for the simultaneous delivery of DOX and the anticancer peptide KLA. Herein, bovine serum albumin (BSA) corona was used as the final capping in this model [122]. This protein had a double effect: it was able to enable a protease/glutathione-mediated intracellular release and established a diffusion barrier for both therapeutic agents. The use of BSA in its wild-type form in the system is an intriguing feature that could result in additional cellular responses when combined with the other multi-apoptotic effects.

Previous examples were developed employing typical coupling processes throughout direct amidation processes. However, more recently cross- and orthogonal linkers has emerged as a powerful strategy for the conjugation of oligopeptides onto particles [176] and between bioactive fragments [177]; although the delivery of biologically active peptides is still based on the reversible bindings. In contrast to precedent examples, peptide pore loading throughout threading is also a highly convenient strategy for delivery as it makes it possible tothe incorporation of additional peptides onto particles' surfaces. This strategy was explored by Martnez-Máñez's group, who reported the use of polylysine as coatings on the C9h (YVETLDDIFEQWAHSEDLK) loaded pro-apoptotic peptide [123]. Herein, this polylysine layer performed two different roles, whereby its cationic nature (1) facilitated cellular uptake and (2) prevented C9h peptide degradation and leakage. The encapsulated peptide showed a better therapeutic profile when delivered from MSNs rather than in its free form. Unfortunately, tested therapies employing this peptide quickly reached a maximal effect far away from clinic desirable efficiencies, underlining the limited anticancer effects of peptides and the need for combining therapies to produce relevant results.

Other examples on proapoptotic peptides delivery were reported by Cao et al. [124] and Rahmani et al. [125], who studied the behavior of Hylin a1 peptide (IFGAILPLAL-GALKNLIK) and the Cathepsin D inhibitor peptide pepstatin A, respectively. In the first example, the authors tuned the mesoporous environment to facilitate threading and enable pH-dependent releases. Moreover, they also elongated the peptide strand to include an additional RGD recognition sequence [124]. In vitro studies for this system revealed that peptide encapsulation reduce hemolysis showed by the free peptide, while maintained a potent cytotoxic effect against HeLa and Hep2 cells. In the second example, the authors developed a silica-based formulation to increase peptide dosage. They evaluated both mesoporous and hollow mesoporous materials to determine the best delivery profile. Unexpectedly, these authors discovered two diverging behaviors: HMSNs were able to load less peptide than typical large-pore MSNs. However, they also showed a more sustained release (longer therapeutic effect) compared to pore-expended MSNs, which is of high interest for future developments.

In addition to the previous examples, anticancer research interests have also focused on the repurposing of bioactive substances to help fighting the disease. For example, Ferreira Soares and coworkers studied the effect of alamandine, a heptapeptide related to angiotensin [126]. According to their results, this peptide showed an improved therapeutic effect against tumoral 4T1 (murine breast cancer) and A549 (human pulmonary cancer) when loaded within the mesopores and a very low cytotoxic effect on HEK-293 cells probably consequence of the protective role exerted by the silica carrier. Other interesting examples were reported by Ramya et al. [127] and Lin et al. [128], who respectively employed Fabatin, a phytodefensin peptide obtained from the seeds of *Vicia faba*, and Mellitin, a water-soluble cationic peptide present on bee venom, in anticancer nanodevices. In the first example, Ramya et al. loaded the Fabatin peptide (LLGRCKVKSNRFNGPCLT-DTHCSTVCRGEGYKGGDCHGLRRRCMCLC) into unmodified MSNs [127]. This simple formulation was able to improve cytotoxicity of free Fabatin and drop viability of triple negative breast cancer MDA-MB-231 cells to values close to those obtained with DOX. These values were also maintained when the Fabatin-loaded nanosystem was tested in mice. Indeed, apoptosis levels on other tissues were reduced 5-fold compared with the values for chemotherapy, highlighting the potential of peptide nanoformulations. In the second example, Mellitin peptide (GIGAVLKVLTTGLPALISWIKRKRQQ) was delivered from MSNs through a sequential magnetic-enzymatic double triggering stimulus-responsive nanosystem. When the peptide was retained within the pores, great hemocompatibility and low cytotoxicity were obtained. However, when the peptide was released, viability reductions above 80% were possible to obtain in PANC-1 cells. Evaluation in mice with the combined peptide and thermal therapies, the nanosystem was able to promote tumor growth reduction of about 75% in comparison with the control. Considering that both Fabatin and Mellitin were the only therapeutic agents within these nanosystems, it seems clear that repurposing peptides with the aid of nanotechnology could open the door to new generation alternatives to conventional chemotherapies.

3.2. Antibacterial Therapies

In addition to the anticancer examples, proteins may also exert antibacterial effects too [178]. The widely known lysozyme, a 14.4 kDa enzyme, has been widely employed to destroy Gram-positive bacterial walls by cleaving the −1,4 bonds between *N*-acetylmuramic acid and *N*-acetyl-D-glucosamine. In the first example reported on the delivery of antibacterial proteins with silica, negatively charged MCM-41 MSNs were employed to electrostatically bind lysozyme units [129]. The resulting enzymatic activity on the bacterial microenvironment made it possible to destroy both Gram-positive and Gram-negative bacteria. Subsequent investigations aimed at increasing this bactericidal effect led to the development of different nanosystems. Among them, dendritic pore MSNs [130] and a set of fancy, rough, non-porous spiky nanoparticles [131] with very interesting protective features were also employed as support for lysozyme. The comparison between different models suggests that the antibacterial effect of lysozyme is mainly achieved when quick releases occur, although in the case of the spiky particles they seem to exert an additional unknown effect. Unfortunately, all these systems were tested only against planktonic bacteria, which are the most sensitive and less pathogenic state. In a more recent contribution, Ye and coworkers investigated if enlarged pore, hollow MSNs with large loading capacities (up to 350 $mg \cdot g^{-1}$) were able to destroy preformed biofilms [132]. In their work the authors found an activity threshold (400 $g \cdot mL^{-1}$) for lysozyme over which no added therapeutic effect could be obtained. In this case, the use of the carrier raised was able to obtain more prolonged effects than free lysozyme due to a more sustained release.

In addition to lysozyme, Concanavalin A (ConA) has also demonstrated certain antibacterial effects [133]. Vallet-Regí and coworkers profited from the antibiotic properties of such lectin to prepare an antibacterial nanosystem in combination with levofloxacin. Their model, based on the chemical grafting on ConA at the final step onto carboxylate-modified MCM-41 MSNs made it possible to successfully prepare a nanodevice able to deliver these two antibiotic species alike. As a result, the dual action of this nanosystem made it possible to destroy *E. coli* biofilms at a concentration of 10 mg MSNs per mL, demonstrating that combination therapy outperforms single-agent therapies in antibacterial treatments too.

Along this line, Bactofencin A also displayed an intriguing antibacterial action when delivered with SBA-15 particles. In the contribution by Hudson and coworkers a clear antibacterial effect of such protein was demonstrated against *S. aureus* [134]. In this case, the system did not include diffusion barriers, which led to a rapid outflow of the protein and thus, an almost total release within 5 h. As a result, the remaining bacteria were able to reform the biofilm quickly and return to its pathogenic state.

To conclude with the silica-based delivery of antibacterial proteins, it is noteworthy to highlight, too, the potential of human proteins for antibacterial therapies on other species. Along this line, Marcelino-Pérez et al. reported the effect of a recombinant human β-defensin 2 onto mesoporous silica nanoparticles as phytosanitary against *Clavibacter michiganensis* subspecies *michiganensis* [135]. In this work the authors obtained the human β-defensin 2 upon transfection of the engineered plasmid in *E. coli.* and loaded the protein directly into pore-enlarged silica particles. The nanosystem proved to be highly efficient against *C. michiganensis* as it was able to reduce the number of colony-forming units in plants by 10-fold. Moreover, the use of a human protein as phytosanitary is expected to have negligible effect in public health due to its human origin.

Peptides are another research line with broader potential to develop antibacterial treatments, as they promise to overcome bacterial resistance throughout non-conventional mechanisms. In fact, the integration of such antimicrobial peptides into nanocarriers has also received attention during last years [179,180]. The first examples on the topic were focused mainly on methodological aspects and on peptides able to undergo membrane-disrupting effects. However, as introduced before, recent contributions also focus on therapeutically active peptides. For example, the studies by Braun et al. on the optimal loading and particle composition (non-porous, calcinated mesoporous, and amino-capped mesoporous silicas) employed the LL37 (LLGDFFRKSKEKIGKEFKRIVQRIKDFLRNLVPRTES) antibacterial peptide [136]. The most negatively charged nanoparticles (calcined MSNs) showed the best loading profile and thus antibacterial effect against *E. coli*. The authors suggested that this effect was consequence of an optimal charge balance between both counterparts. Moreover, the appropriate protective environment of porous silica also helped in decreasing hemo- and proteolysis caused by LL37. In a later work, these authors also investigated the effect of the porous structure. By comparing enlarged pore HMSNs with typical sized-MSNs they discovered that the HMSNs showed a higher antibiotic profile consequence of a more continuous release [181], in agreement with data reported by Rahmani et al. [125]. This LL37 peptide was also successfully employed as antibiotic on silica carriers with virus-like topographies. In this contribution, reported by Häffner et al. [137], spiky nanoparticles had a crucial role, they were able to open bacterial and favor internalization of LL37 into bacteria. As a result, the complete nanosystem was able to reduce *E. coli* colonies in more than 80%, and also prevent the formation of new biofilms and bacterial aggregates.

Another promising research line for antibacterial nanodevices is the treatment of infected cells. Along this line, Tenland et al. employed MSNs to deliver to infected macrophages the NZX [138] (GFGCNGPWSEDDIQCHNHCKSIKGYKGGYCARGGFVCK CY) anti-tuberculosis peptide [182]. This MSN loaded nanosystem proved to internalized by macrophages and produce an effective peptide release able to destroy infected mycobacterium without significantly affecting host macrophages. Moreover, this particular formulation made it possible to obtain a longer therapeutic effects than free. The authors also suggested a possible operating mechanism based on the accumulation of nanocarriers within intracellular vacuoles which acted as drug reservoirs by the prevention of peptide digestion within the silica matrix.

In recent years, other emerging antibacterial peptides have also been nanoformulated using porous nanosilica as carriers. For example, Ma et al. employed an ovotransferrin-derived peptide to create a multipurpose nanotherapeutic able to act as antibiotic drug delivery system [139]. These authors demonstrated that the surface functionalization of MSNs with the OVTp12 (AGLAPYKLKPIA) peptide alone was able to induce an interesting antibiotic effect against *E. coli*, although this effect was clearly enhanced when the mesopores were loaded with gentamicin. In vivo studies of this system showed very low hemolysis rates (<10%) and a potent anti-infective effect in mice, reaching up to ca. 70% survival of individuals after 10 days.

Another interesting example on combination therapy was reported by Alharthi et al. who studied the antibiotic effect of different combined Sortase A inhibitors [183] (*trans*-Chalcone, Curcumin, Quercetin and Berberine chloride) together with polycationic antimicrobial peptides: Pexiganan (GIGKFLKKAKKFGKAFVKILKK-NH_2), Indolicidin (ILPWKW PWWPWRR-NH_2) and Mastoganan (INLKILARLAKKIL-NH_2) [140]. All examples employed inhibitor-loaded raw or phosphonate-modified MSNs which were further coated with the corresponding peptide throughout electrostatic deposition. As expected, obtained results showed that all combinations of inhibitors and peptides displayed different degrees of antibacterial effects. However, for certain combinations, the overall effect of combined therapy clearly improved the minimum inhibitory concentration values obtained for single-loaded therapeutics, suggesting that the repurposing of low-effect but cheap therapeutics could have importance on the development of new generation of nanomedicines and help in the fight against bacterial resistance.

In another interesting contribution to the topic, Zhang et al. reported the use of SiO_2 nanoparticles to create skin-inspired fibrous membranes with wound healing properties. Herein, silica particles were chemically modified to behave as the outermost hydrophobic coating of a multilayered skin-like structures [184]. Membranes were assembled throughout sequential deposition of hydrophilic and hydrophobicity components such as poly(vinyl alcohol) and poly(vinylidene fluoride) along four layers. Additionally, a quinolone antibiotic (Ciprofloxacin), together with an antioxidant (Astaxanthin), were embedded within the structure to promote the proliferation of new fibroblasts and prevent bacterial growth. As a result, this silica-containing composite was able to be implanted onto massive wounds and herein induce new tissue regeneration and complete wound healing within 2 weeks. In addition to skin regeneration, silicon-based ceramics have also demonstrated antibacterial features. As demonstrated by He and coworkers [185], polypropylene-Si_3N_4 composites showed bacteriostatic effects against *E. coli* and *S. aureus* in non-woven filtering materials such as facemasks. Despite these examples not containing therapeutic biopolymers, they show that silicon-containing ceramics are also highly convenient materials for the preparation of a broad number of healthcare materials.

3.3. Osteogenic Therapies

Growth factors, as promoters of cellular proliferation and differentiation are crucial for the development of functional vasculature, bones and wound healing. During the last few years, many studies on aging and frailty have pointed out that osteoporosis and related bone diseases could be successfully treated with both hormonal or nanomedical therapies, with this opening the way to a new generation of bone healing nanodevices [186,187]. This topic, pioneered by Zhang et al. more than a decade ago [141], has now become a hot research area in the uses of silica in nanomedicine. This first contribution, in which ad hoc MSNs were designed to host the 18 kDa basic fibroblast growth factor (bFGF) [141] was quickly followed by two interesting contributions by Gan et al., who studied two complimentary strategies to deliver the bone morphogenetic protein 2 (BMP-2). In the first contribution, these authors explored the behavior of dexamethasone-loaded, chitosan-coated MSNs to deliver such proteins attached onto particles' surfaces [142]. whereas in the second model, a pore protecting strategy was explored. In this work, the SBA-15 mesopores were capped with iron oxide nanoparticles [143] to reduce the excessive osteogenesis obtained in their first example.

Some peptides also have anabolic properties, for which reason they have also been employed as payloads in bone repairing nanodevices, especially those that focus on local tissue regeneration. Along this line, Esbrit and Vallet-Regí studied bone remineralization triggered by Osteostatin, a peptide derived from the parathyroid hormone-related protein ($PTHrP_{107-111}$) [144]. In their two contributions these authors employed SBA-15 matrices to ensure peptide preservation and sustained release. In fact, the longer-lasting effect of loading allowed these materials to induce bone regeneration in both cavitary defects [145] and osteoporotic defects [146] successfully, as demonstrated by the significant increments on bone mass obtained. Similar strategies were also reported by Mendes et al. [147] and He and coworkers [148], who reported the successful combination of therapeutic peptides with other chemical species such as (1) the Osteogenic Growth Peptide (OGP, ALKRQGRTLYGFGG) and apatite onto SBA-15 matrices and (2) the BMP-2 derived peptide (KIPKASSVPTELSAISTLYL), together with the synthetic glucocorticoid dexamethasone employing MCM-41 nanoparticles.

In more recent studies, the combination of Osteostatin with CaP containing bioactive glasses [188,189] has emerged as a powerful tool for the development of new bone therapies promoted by local delivery of calcium and phosphorus. Additionally, in recent examples, new elements have been incorporated to bioactive glasses to enhance their therapeutic effects and properties. For instance, Jiménez-Holguín et al. reported the use of Cu for bone mineralization with additional antimicrobial effect [190,191]. Mutreja et al., also reported the use of porous Sr-doped calcium silicate glasses to deliver the GL13K (GKIIKLKASLKLL) peptide. This peptide, derived from the human salivary parotid secretory protein, was chosen because its antimicrobial and osteogenic properties (reduces osteoclast differentiation). As a result, peptide delivery was able to promote bone mineralization in combination with an anti-infective effect against *Staphylococcus aureus* [149]. Data also pointed out effective increments of typical proteins associated with bone formation, demonstrating the potential of this multipurpose peptide for treating and preventing bone diseases.

3.4. Immunostimulant Proteins and Peptides

Immunotherapy is one of the most promising strategies for improving cancer therapies. The general idea is to recruit patient's immune system to attack tumors too. Nevertheless, any indiscriminate or uncontrolled activations may have disastrous consequences on patients' health. For this reason, the scientific community has shown growing interest in the development of novel therapeutics for local applications. In cancer, the enhanced retention and permeation effect (EPR) may help achieving this goal by favoring accumulation of immune stimulants within tumoral area. Then, in an ideal case, the local activation of the immune system would produce a response able to recognize and destroy cancerous cells in a vaccination-like strategy. For interested readers, a very recent review focused on the most convenient materials, aspects and limitations of this strategy [192]. In this section, we also remark that MSN-based therapies have come to be of great importance because of their features as immune adjutants [193–195].

According to published examples, there are several strategies that allow the incorporation of antibodies into silica carriers. In addition to already-discussed surface deposition and grafting and pore loading, new materials have been also developed for such purposes. One of these examples, reported by Lim et al., unique hollow MSNs showing unusual perforations were employed to load immunoglobulin (IgG) to HeLa cells [150]. Unfortunately, despite effective intracellular delivery, no immune response was achieved. Similarly, Niu et al. proposed different core–shell nanoassemblies between solid silica particles of different sizes. The resulting rough surfaces were able to host proteins within interparticle voids [151]. In particular, these authors reported nanostructures with different roughness (14, 21 and 38 nm) which could host and preserve different proteins such as Cyt c, monoclonal rabbit immunoglobulin and bigger antibody-containing proteins (HRP-linked anti-rabbit IgG antibody). In vitro studies with 38 nm hydrophobically modified RSNs showed effective deliveries of anti-pAkt antibodies into MCF-7 breast cancer cells yielding a considerable impact on cell proliferation and the downregulation of the anti-apoptotic Bcl-2 protein.

Apart from those, silica nanoparticles have also been employed for the delivery of immunostimulant proteins. Along this line, Wang, Ito, Tsuji, and colleagues, started a very prolific line of research on evaluating the stimulant effect of Hollow MSNs loaded with a chicken ovalbumin (OVA) in vivo [152]. The treatment of mice with this nanosystem produced a significant immune activation through increases of CD4+ and CD8+ T-cell populations, on the upregulation of Th1 and Th2 cytokines and of the increase in immunoglobulin antibody levels [153]. Apart from OVA, these authors also determined that employed MSNs were able to exert an immune stimulant effect, which enhanced the overall effect. Similar results were also reported by Yang et al., who reported the delivery of OVA in combination with tumor antigens from B16F10 cells. To ensure adequate and simultaneous release of OVA and antigens, rapidly degradable multi-shelled dendritic mesoporous organosilica hollow spheres (DMOHS) were chosen [154]. As a result, upregulation of CD4+ and CD8+ T-cells in combination with overexpression of Th1, interleukin-12 (IL12), gamma interferon (γ-IFN), and alpha tumor necrosis factor (α-TNF) were produced. In addition to those examples, Cha et al. also evolved the strategy by including an additional immunostimulant substance into a silica-based carrier. In their experiment, OVA was delivered in combination with an immunological danger signal, one agonist of the

Toll-like Receptor 9 (TLR9). Both species were successively loaded into large pore MSNs (20–30 nm) [155]. As a result, this system induced a similar response that those described in previous system but in a much higher scale.

Potential anticancer vaccines have also been reported employing interleukin 2 (IL2) as payload. In the work by Kong et al., IL2 was delivered in combination with all-trans retinoic acid and doxorubicin. The system was completed employing biodegradable hollow mesoporous silica nanoparticles and a lipid coating to prevent leakage and to favor biocompatibility [156]. This combination of drugs produced a general antitumor effect in mice and a significant reduction on melanoma metastasis potential, although complete tumor regression could not be achieved.

Apart from anticancer vaccination, delivery of immunoproteins is also of interest for the development of nanomedicines against other diseases. For example, Park et al. reported an Interleukin 13 (IL13)-loaded nanosystem for the treatment of autoimmune encephalomyelitis [157]. This system employed large pore MSNs (ca. 63 nm) to load the protein within the silica matrix. The intracellular delivery of IL13 induced the differentiation of bone marrow-derived macrophages to M2 macrophages, as demonstrated by the increment of CD11b, CD206 markers and the F4/80 antibody. In vivo, this nanosystem improved prognosis of the experimental autoimmune encephalomyelitis after only two intranasal injections.

In addition to previous examples non-viral-based vaccination could be also targeted for other common diseases [196,197]. The main idea is based on the delivery of antigens to immune cells without triggering a whole immune response. Along this line, several examples employing silica nanoparticles could be found in the literature. For example, Guo et al. employed HMSNs as carrier for the porcine circovirus type 2 (ORF2) protein [158]. In this nanosystem the protein was directly adsorbed onto raw nanoparticles without an additional protective layer. Despite the possible degradation, this nanoformulation achieved effective immunological activation in mice, as demonstrated by the overexpression of CD4, CD8, γ-IFN and Th1 immunoproteins. During the following years, other non-viral-based nanovaccines against (1) *Schistosoma mansoni,* by employing homogenates from the parasite [159]; (2) porcine enzootic pneumonia, by using a recombinant HSP70 antigen fragment ($HSP70_{212\text{-}600}$) from *Mycoplasma hyopneumoniae* [160]; (3) enterohaemorrhagic *E. coli*, by using a recombinant fragment of filamentous EspA protein [161]; and (4) *Paracoccidioides brasiliensis* employing the rPb27 antigen [162] were developed following similar strategies. Therapeutically, all of these nanosystems were able to immunize mice against pathogens of different nature reaching pharmacological profiles similar to those reported for free antigens and other pharmaceutical formulations with much lower macrophage's internalization rates.

Despite their activity usually being lower than those of proteins, peptides could also be employed to induce immune responses. This is of especial interest in the case of peptides that retain biological aspects from their parent proteins. For example, Xie et al. reported the use of hollow mesoporous silica carriers to deliver two melanoma-derived antigen peptides: the hydrophobic $TRP2_{180\text{-}188}$ (SVYDFFVWL) and the hydrophilic $HGP100_{25\text{-}33}$ (KVPRNQWL) peptides [163]. In order to load both peptides, the authors modified the nanocarrier with amino groups at the internal cavity and with carboxylate groups at the mesopores to enable a sequential differential double loading. To preserve both peptides and to provide enough colloidal stability, the nanosystem was further coated with a lipid bilayer in which there was incorporated an additional therapeutic compound, the toll-like receptor 4 (TLR4) agonist monophosphoryl lipid A. According to reported data, this nanosystem promoted immune cell maturation as evidenced by the upregulation of CD86, α-TNF, γ-IFN, IL12, and IL4 proteins. In vivo, vaccinated animals showed a promising tumor growth reduction on melanoma tumors together with a lower development of metastatic lymph nodes, being one few nanosystems able to reach such anticancer efficacy.

In addition to the delivery of immunostimulant compounds, recent investigations have also focused on recruiting immune cells to create advanced delivery systems. Such is the case of the article reported by Lei and Tang, who employed mesoporous silica microparticles to develop a T-cell responsive drug delivery system [198]. This system was based on a double strand DNA able to block mesopores [39]. To achieve so, one of the strands was linked to the silica while the complementary strand was bound to an anti-CD3 antibody. As a result, the hybridized DNA was able to keep Gemcitabine loaded within the pores in the absence of T-cells. However, in the presence of CD3 presenting T-cells, antibody-antigen affinity was able to dehybridize the double-stranded DNA-producing pore opening and drug outflux. The results obtained with this proof of concept suggested that this drug delivery strategy could be of interest for the development of nanoparticle-based therapies for certain lymphomas.

Apart from the presented examples aimed at the activation of the immune system to treat diseases, combinations of peptides and MSNs have been also employed to reduce immune responses. The example reported by Wang et al. was able to trap Nickel and thus reduce allergy levels [164]. To do so, the authors profited from the extraordinary coordination ability of histidine groups towards Ni cations. Through adequate chemical modification, hexahistidine peptides were grafted onto MSNs, yielding a nanosystem able to scavenge Ni and reduce allergic responses to this metal. In addition to the peptide, this nanosystem was also loaded with chlorogenic acid, an emerging cosmetic component [199] with claimed anti-inflammatory and wound healing properties in order to enhance the antiallergic performance.

3.5. Enzymatic Therapy

The primary cause of metabolic disorders is the dysregulation of enzymatic functions. This situation may lead to toxic syndromes if it remains untreated. In cancer, the dysregulation of homeostasis is also a side effect of the aberrant enzyme expression. For both reasons, enzymatic therapy has also become a promising field of research in nanomedicine, as well as when considering the lability of enzymes. Current enzymatic therapies consist of repetitive injections of enzymes, which usually leads to low-adherence treatments. In order to improve current treatments, enzymes must be modified to increase their lifespan [200]. Such are the cases of collagenase capsules, Villegas et al. [201] and anticancer p53 nanocapsules reported by Zhao et al. [202], whose degradable protective shells preserve unaltered the enzymatic function during trafficking.

However, when dealing with porous silica nanomaterials, the enzymes could be located either at the surface or within the mesopores. This second strategy is preferable for preventing enzymatic activity loss, although at the expense of being a considerable diffusion barrier for bigger substrates. Governed by simplicity, most of reported MSN-based carriers locate the proteins at the surface, as this facilitates the flux of both substrates and products. One of the most widely employed enzymes in the literature is Carbonic Anhydrase (CA), which, despite having no therapeutic impact, is widely employed as a model [165,166], especially due to the ease of measuring carbon dioxide and bicarbonate.

Apart from previous proofs of concept, MSNs have also successfully been employed as support and carriers for therapeutic enzymes, especially those aimed at the reduction of toxic syndromes. For instance, Xu et al. reported a promising model to reduce the severity of Morquio type B syndrome [167]. This disease occurs when Galactosidase enzymes fail to break the glycoside bond, thus causing bioaccumulation of oligosaccharides. In their model, the authors hosted the enzyme within the mesopores rather than on particles' surfaces, increasing the lifespan of the enzyme due to the improved protection offered by the silica matrix. It is important to remark that the small sizes and good solubility of substrate oligosaccharides made it possible to implement this configuration due to the facility of diffusion to and from mesopores. In this example ultra-large pore MSNs were required to host the 119 kDa enzyme; being core-cone shaped MSNs with dahlia-like mesopores the chosen structures. These particles were effectively internalized into N2a cells and once up taken, were able to maintain its catalytic activity much longer than the free enzyme.

Superoxide dismutase (SOD) and glutathione peroxidase (GPx) are two antioxidant enzymes widely employed to reduce Reactive Oxygen Species (ROS) intracellularly. Nevertheless, the different nature of ROS requires from nanosystems the ability to adequately expose enzymes to exert an effective antioxidant function. The system reported by Mou and colleagues for SOD delivery employed a nitrilotriacetic acid (Ni-NTA) modified silica to bind a TAT-enzyme conjugates throughout Ni^{+2} chelation [168]. To prevent extracellular ROS depletion promoted by Paraquat, the authors denaturalized the enzyme with urea, a well-known superoxide anion generator. However, when the particles were internalized by HeLa cells, the enzyme was refolded into its active state and produced the desired antioxidant effect. More recently, Mou and coworkers studied the effect of delivering two antioxidant enzymes. For so they prepared two differently loaded nanodevices—one with SOD and the other with GPx [169]. In their investigations, these authors discovered a complementing synergistic effect with this combination, where the co-delivery of both antioxidant proteins enhanced the impact of individual therapies.

Focusing on Alzheimer's disease, Han et al. provided an interesting example able to prevent aggregation of tau proteins through the introduction of human proteasomes [170]. These authors employed Mou's strategy to bind histidine moieties from active human 26S proteasomes isolated from the HEK293 cell line. The resulting system demonstrated that these enzymes could be distributed without significantly increases of proteotoxicities. Indeed, the authors asserted that their approach might be of interest for delivering proteins throughout impermeable barriers, although this system did not have potential to prevent Alzheimer's condition in vivo. In fact, for in vivo applications, all nanoparticle-based drug delivery system must be able to cross into the brain through biological barriers (blood–brain, blood–brain tumor, nose-to-brain, etc.), which require specific strategies. For an in-depth discussion on the topic, please refer to the review presented in [203].

In conclusion, enzymatic therapy may also be employed to treat cancer diseases. This was demonstrated by Teng et al., who developed a nanosystem able to deliver RNAase intracellularly. This enzyme was able to disrupt the abnormal expression of proteins within cancerous cells and thus reduce its aggressivity [171]. In their model, up-conversion nanoparticles coated with a mesoporous silica layer (UC@MSNs) were loaded with a cisplatin prodrug and coated with the RNAase to promote downregulation of proteins. As a result, the combined effect of both species boosted HepG2 apoptosis levels to almost 50% according to flow cytometry data. In vivo evaluation in mice of this nanosystem permitted stopping tumor growth and maintained tumor volume constant for almost two weeks, although these tumors could not reach complete remission.

4. Conclusions

As mentioned throughout the manuscript, the capacity and versatility of nanometric porous silica materials to develop potential therapeutic systems is enormous. This is partially a consequence of their feasibility in the development of novel strategies in nanomedical approaches. Moreover, silica offers one of the simplest methods to develop nanosystems for combination therapies, which is clearly demonstrated by the great number of systems with therapeutic potential that have been reported up to date. In addition to the delivery of oligonucleotides, peptides and proteins reviewed herein, silica particles also offer many other possibilities for creating nanotherapeutics such as targeted nanomedicines and stimuli responsive systems. However, as a consequence of their easy tunable porous structure, it is also possible to house a significant number of small and medium-sized molecules of practically any known chemical nature within the particles' matrices.

However, the use of mesoporous silica for nanomedical developments also has severe limitations. For instance, even though they are biodegradable, their slow dissolution rates and sturdiness cause them to accumulate relatively easily within tissues and during trafficking. Indeed, if inadequate surface modifications are performed, these particles show a certain tendency to aggregate and therefore produce thrombosis and embolisms. On the other hand, the delivery of functional biomacromolecules with silica-based transporters presents important limitations too, the main one being the possible immune response of patients. In fact, many nanomedical designs fail when they are tested in immunocompetent models. Fortunately, the technology needed to overcome these problems is known and although the use of biopolymers such as nucleotides, proteins and peptides can alleviate these side effects to a large extent, this technological development is still in its infancy.

In addition to previous issues, the development of oligonucleotide drug delivery technologies also led to other limitations: the use of positively charged supports and polymers. This reduces overall biocompatibility in comparison with neutral or anionic systems due to the higher hemolysis rates. Although some research groups have tried to solve this problem by using fewer positive charges or by masking polycations, the truth is that it seems to be a rather difficult problem to solve. Proof of this is that all new-generation COVID-19 vaccines have been developed using liposome technologies, which is completely different from solutions offered by oligonucleotide delivery with silica nanoparticles. Similarly, in the case of protein and enzyme delivery, there also are important limitations to be solved. In the case of in-pore loading, it is necessary to adapt pore sizes to proteins sizes and shapes while on surface location technologies that avoid the formations of protein coronas and avoid the blockage of active sites remain unsolved.

Regardless of these drawbacks, mesoporous silica nanotechnology also offers significant advantages. Among them, the main advance is the ease of developing combined therapies, which to our opinion are the future of anticancer therapies, as they have proved to be the most successful strategies to achieve complete remission of tumors. Moreover, and as outlined in this review, silica also offers interesting possibilities for the development of topical and oral therapeutics that do not require systemic injections and may favor translation into clinical practice. Even though silica nanotechnology does not reach clinical practice, we still foresee a brilliant future for this technology. In our opinion, silica has already been established as an important source for future nanomedical technologies and will remain so in the near future.

Funding: This research received no external funding.

Institutional Review Board Statement: Not applicable.

Informed Consent Statement: Not applicable.

Data Availability Statement: Not applicable.

Acknowledgments: The authors want to thank Lourdes Gude for support given.

Conflicts of Interest: The authors declare no conflict of interest.

References

1. Fus-Kujawa, A.; Prus, P.; Bajdak-Rusinek, K.; Teper, P.; Gawron, K.; Kowalczuk, A.; Sieron, A.L. An Overview of Methods and Tools for Transfection of Eukaryotic Cells in Vitro. *Front. Bioeng. Biotechnol.* **2021**, *9*, 701031. [CrossRef] [PubMed]
2. Kulkarni, J.A.; Witzigmann, D.; Thomson, S.B.; Chen, S.; Leavitt, B.R.; Cullis, P.R.; van der Meel, R. The Current Landscape of Nucleic Acid Therapeutics. *Nat. Nanotechnol.* **2021**, *16*, 630–643. [CrossRef] [PubMed]
3. Feng, R.; Patil, S.; Zhao, X.; Miao, Z.; Qian, A. RNA Therapeutics—Research and Clinical Advancements. *Front. Mol. Biosci.* **2021**, *8*, 710738. [CrossRef] [PubMed]
4. Berraondo, P.; Sanmamed, M.F.; Ochoa, M.C.; Etxeberria, I.; Aznar, M.A.; Pérez-Gracia, J.L.; Rodríguez-Ruiz, M.E.; Ponz-Sarvise, M.; Castañón, E.; Melero, I. Cytokines in Clinical Cancer Immunotherapy. *Br. J. Cancer* **2019**, *120*, 6–15. [CrossRef] [PubMed]
5. Cid, R.; Bolívar, J. Platforms for Production of Protein-Based Vaccines: From Classical to Next-Generation Strategies. *Biomolecules* **2021**, *11*, 1072. [CrossRef] [PubMed]
6. Karpiński, T.; Adamczak, A. Anticancer Activity of Bacterial Proteins and Peptides. *Pharmaceutics* **2018**, *10*, 54. [CrossRef] [PubMed]
7. Koo, H.B.; Seo, J. Antimicrobial Peptides under Clinical Investigation. *Pept. Sci.* **2019**, *111*, e24122. [CrossRef]
8. Dijksteel, G.S.; Ulrich, M.M.W.; Middelkoop, E.; Boekema, B.K.H.L. Review: Lessons Learned from Clinical Trials Using Antimicrobial Peptides (AMPs). *Front. Microbiol.* **2021**, *12*, 616979. [CrossRef]
9. Tian, Y.; van Tirrell, M.; LaBelle, J.L. Harnessing the Therapeutic Potential of Biomacromolecules through Intracellular Delivery of Nucleic Acids, Peptides, and Proteins. *Adv. Healthc. Mater.* **2022**, *11*, 2102600. [CrossRef]

10. Morais, P.; Adachi, H.; Yu, Y.-T. The Critical Contribution of Pseudouridine to MRNA COVID-19 Vaccines. *Front. Cell Dev. Biol.* **2021**, *9*, 78942. [CrossRef]
11. Xia, X. Detailed Dissection and Critical Evaluation of the Pfizer/BioNTech and Moderna MRNA Vaccines. *Vaccines* **2021**, *9*, 734. [CrossRef] [PubMed]
12. Shi, Y.; van der Meel, R.; Chen, X.; Lammers, T. The EPR Effect and beyond: Strategies to Improve Tumor Targeting and Cancer Nanomedicine Treatment Efficacy. *Theranostics* **2020**, *10*, 7921–7924. [CrossRef]
13. Zolnik, B.S.; González-Fernández, A.; Sadrieh, N.; Dobrovolskaia, M.A. Nanoparticles and the Immune System. *Endocrinology* **2010**, *151*, 458–465. [CrossRef] [PubMed]
14. Amoozgar, Z.; Yeo, Y. Recent Advances in Stealth Coating of Nanoparticle Drug Delivery Systems. *WIREs Nanomed. Nanobiotechnol.* **2012**, *4*, 219–233. [CrossRef] [PubMed]
15. Fam, S.Y.; Chee, C.F.; Yong, C.Y.; Ho, K.L.; Mariatulqabtiah, A.R.; Tan, W.S. Stealth Coating of Nanoparticles in Drug-Delivery Systems. *Nanomaterials* **2020**, *10*, 787. [CrossRef]
16. Mazayen, Z.M.; Ghoneim, A.M.; Elbatanony, R.S.; Basalious, E.B.; Bendas, E.R. Pharmaceutical Nanotechnology: From the Bench to the Market. *Futur. J. Pharm. Sci.* **2022**, *8*, 12. [CrossRef]
17. Sargazi, S.; Laraib, U.; Barani, M.; Rahdar, A.; Fatima, I.; Bilal, M.; Pandey, S.; Sharma, R.K.; Kyzas, G.Z. Recent Trends in Mesoporous Silica Nanoparticles of Rode-like Morphology for Cancer Theranostics: A Review. *J. Mol. Struct.* **2022**, *1261*, 132922. [CrossRef]
18. Crommelin, D.J.A.; van Hoogevest, P.; Storm, G. The Role of Liposomes in Clinical Nanomedicine Development. What Now? Now What? *J. Control. Release* **2020**, *318*, 256–263. [CrossRef]
19. Anselmo, A.C.; Mitragotri, S. Nanoparticles in the Clinic: An Update. *Bioeng. Transl. Med.* **2019**, *4*, e10143. [CrossRef]
20. Anselmo, A.C.; Mitragotri, S. Nanoparticles in the Clinic: An Update Post COVID-19 Vaccines. *Bioeng. Transl. Med.* **2021**, *6*, e10246. [CrossRef]
21. Hussein, H.A.; Nazir, M.S.; Azra, N.; Qamar, Z.; Seeni, A.; Tengku Din, T.A.D.A.-A.; Abdullah, M.A. Novel Drug and Gene Delivery System and Imaging Agent Based on Marine Diatom Biosilica Nanoparticles. *Mar. Drugs* **2022**, *20*, 480. [CrossRef] [PubMed]
22. Farjadian, F.; Roointan, A.; Mohammadi-Samani, S.; Hosseini, M. Mesoporous Silica Nanoparticles: Synthesis, Pharmaceutical Applications, Biodistribution, and Biosafety Assessment. *Chem. Eng. J.* **2019**, *359*, 684–705. [CrossRef]
23. Hosseinpour, S.; Walsh, L.J.; Xu, C. Biomedical Application of Mesoporous Silica Nanoparticles as Delivery Systems: A Biological Safety Perspective. *J. Mater. Chem. B* **2020**, *8*, 9863–9876. [CrossRef]
24. Sun, L.; Sogo, Y.; Wang, X.; Ito, A. Biosafety of Mesoporous Silica Nanoparticles: A Combined Experimental and Literature Study. *J. Mater. Sci. Mater. Med.* **2021**, *32*, 102. [CrossRef] [PubMed]
25. Mitchell, M.J.; Billingsley, M.M.; Haley, R.M.; Wechsler, M.E.; Peppas, N.A.; Langer, R. Engineering Precision Nanoparticles for Drug Delivery. *Nat. Rev. Drug Discov.* **2021**, *20*, 101–124. [CrossRef]
26. Lee, N.K.; Kim, S.-N.; Park, C.G. Immune Cell Targeting Nanoparticles: A Review. *Biomater. Res.* **2021**, *25*, 44. [CrossRef]
27. Rosenblum, D.; Joshi, N.; Tao, W.; Karp, J.M.; Peer, D. Progress and Challenges towards Targeted Delivery of Cancer Therapeutics. *Nat. Commun.* **2018**, *9*, 1410. [CrossRef]
28. Moodley, T.; Singh, M. Current Stimuli-Responsive Mesoporous Silica Nanoparticles for Cancer Therapy. *Pharmaceutics* **2021**, *13*, 71. [CrossRef]
29. Castillo, R.R.; Lozano, D.; González, B.; Manzano, M.; Izquierdo-Barba, I.; Vallet-Regí, M. Advances in Mesoporous Silica Nanoparticles for Targeted Stimuli-Responsive Drug Delivery: An Update. *Expert Opin. Drug Deliv.* **2019**, *16*, 415–439. [CrossRef]
30. Shen, H.; Huang, X.; Min, J.; Le, S.; Wang, Q.; Wang, X.; Dogan, A.A.; Liu, X.; Zhang, P.; Draz, M.S.; et al. Nanoparticle Delivery Systems for DNA/RNA and Their Potential Applications in Nanomedicine. *Curr. Top. Med. Chem.* **2019**, *19*, 2507–2523. [CrossRef]
31. Piperno, A.; Sciortino, M.T.; Giusto, E.; Montesi, M.; Panseri, S.; Scala, A. Recent Advances and Challenges in Gene Delivery Mediated by Polyester-Based Nanoparticles. *Int. J. Nanomed.* **2021**, *16*, 5981–6002. [CrossRef] [PubMed]
32. Qin, S.; Tang, X.; Chen, Y.; Chen, K.; Fan, N.; Xiao, W.; Zheng, Q.; Li, G.; Teng, Y.; Wu, M.; et al. MRNA-Based Therapeutics: Powerful and Versatile Tools to Combat Diseases. *Signal. Transduct. Target Ther.* **2022**, *7*, 166. [CrossRef] [PubMed]
33. Janowski, M.; Andrzejewska, A. The Legacy of MRNA Engineering: A Lineup of Pioneers for the Nobel Prize. *Mol. Ther. Nucleic Acids* **2022**, *29*, 272–284. [CrossRef] [PubMed]
34. Dong, Y.; Siegwart, D.J.; Anderson, D.G. Strategies, Design, and Chemistry in SiRNA Delivery Systems. *Adv. Drug Deliv. Rev.* **2019**, *144*, 133–147. [CrossRef] [PubMed]
35. Hattab, D.; Gazzali, A.M.; Bakhtiar, A. Clinical Advances of SiRNA-Based Nanotherapeutics for Cancer Treatment. *Pharmaceutics* **2021**, *13*, 1009. [CrossRef]
36. Duan, L.; Ouyang, K.; Xu, X.; Xu, L.; Wen, C.; Zhou, X.; Qin, Z.; Xu, Z.; Sun, W.; Liang, Y. Nanoparticle Delivery of CRISPR/Cas9 for Genome Editing. *Front. Genet.* **2021**, *12*, 673286. [CrossRef]
37. Chen, F.; Alphonse, M.; Liu, Q. Strategies for Nonviral Nanoparticle-based Delivery of CRISPR/Cas9 Therapeutics. *WIREs Nanomed. Nanobiotechnol.* **2020**, *12*, e1609. [CrossRef]
38. Radu, D.R.; Lai, C.-Y.; Jeftinija, K.; Rowe, E.W.; Jeftinija, S.; Lin, V.S.-Y. A Polyamidoamine Dendrimer-Capped Mesoporous Silica Nanosphere-Based Gene Transfection Reagent. *J. Am. Chem. Soc.* **2004**, *126*, 13216–13217. [CrossRef]

39. Castillo, R.R.; Baeza, A.; Vallet-Regí, M. Recent Applications of the Combination of Mesoporous Silica Nanoparticles with Nucleic Acids: Development of Bioresponsive Devices, Carriers and Sensors. *Biomater. Sci.* **2017**, *5*, 353–377. [CrossRef]
40. Kim, T.; Nam, K.; Kim, Y.M.; Yang, K.; Roh, Y.H. DNA-Assisted Smart Nanocarriers: Progress, Challenges, and Opportunities. *ACS Nano* **2021**, *15*, 1942–1951. [CrossRef]
41. Liu, M.; Wang, L.; Lo, Y.; Shiu, S.C.-C.; Kinghorn, A.B.; Tanner, J.A. Aptamer-Enabled Nanomaterials for Therapeutics, Drug Targeting and Imaging. *Cells* **2022**, *11*, 159. [CrossRef] [PubMed]
42. Abu-Dief, A.M.; Alsehli, M.; Al-Enizi, A.; Nafady, A. Recent Advances in Mesoporous Silica Nanoparticles for Targeted Drug Delivery Applications. *Curr. Drug Deliv.* **2022**, *19*, 436–450. [CrossRef] [PubMed]
43. Colilla, M.; Vallet-Regí, M. Chapter 13. Chemoresponsive Mesoporous Silica Nanoparticles for Targeted Drug Delivery in Cancer Therapy. In *Chemoresponsive Materials: Smart Materials for Chemical and Biological Stimulation*, 2nd ed.; Royal Society of Chemistry: London, UK, 2022; pp. 451–498.
44. Barui, S.; Cauda, V. Multimodal Decorations of Mesoporous Silica Nanoparticles for Improved Cancer Therapy. *Pharmaceutics* **2020**, *12*, 527. [CrossRef]
45. Yoo, H.; Jo, H.; Oh, S.S. Detection and beyond: Challenges and Advances in Aptamer-Based Biosensors. *Mater. Adv.* **2020**, *1*, 2663–2687. [CrossRef]
46. Aznar, E.; Oroval, M.; Pascual, L.; Murguía, J.R.; Martínez-Máñez, R.; Sancenón, F. Gated Materials for On-Command Release of Guest Molecules. *Chem. Rev.* **2016**, *116*, 561–718. [CrossRef] [PubMed]
47. Devika, V.; Sreelekshmi, P.J.; Rajeev, N.; Lakshmi, A.S.; Chandran, A.; Goutami, G.B.; Sadanandan, S. Recent Advances in Peptides-Based Stimuli-Responsive Materials for Biomedical and Therapeutic Applications: A Review. *Mol. Pharm.* **2022**, *19*, 1999–2021. [CrossRef]
48. Castillo, R.R.; Lozano, D.; Vallet-Regí, M. Mesoporous Silica Nanoparticles as Carriers for Therapeutic Biomolecules. *Pharmaceutics* **2020**, *12*, 432. [CrossRef]
49. Höbel, S.; Aigner, A. Polyethylenimines for SiRNA and MiRNA Delivery in Vivo. *WIREs Nanomed. Nanobiotechnol.* **2013**, *5*, 484–501. [CrossRef]
50. Wang, Y.; Song, H.; Yu, M.; Xu, C.; Liu, Y.; Tang, J.; Yang, Y.; Yu, C. Room Temperature Synthesis of Dendritic Mesoporous Silica Nanoparticles with Small Sizes and Enhanced MRNA Delivery Performance. *J. Mater. Chem. B* **2018**, *6*, 4089–4095. [CrossRef]
51. Chang, Q.; Liu, C.; Xie, Z.; Shu, Q.; Xie, Y.; Su, Q.; Deng, X. Polyethylenimine Functionalized Ultrasmall Mesoporous Silica Nanoparticles for SiRNA Delivery. *ChemNanoMat.* **2022**, *8*, e202100453. [CrossRef]
52. Lin, X.; Wu, W.; Fu, J.; Yang, Y.; Guo, B.; Yu, C.; Song, H. Asymmetric Silica Nanoparticles with Tailored Spiky Coverage Derived from Silica-Polymer Cooperative Assembly for Enhanced Hemocompatibility and Gene Delivery. *ACS Appl. Mater. Interfaces* **2021**, *13*, 50695–50704. [CrossRef] [PubMed]
53. Taranejoo, S.; Liu, J.; Verma, P.; Hourigan, K. A Review of the Developments of Characteristics of PEI Derivatives for Gene Delivery Applications. *J. Appl. Polym. Sci.* **2015**, *132*, 42096. [CrossRef]
54. Nhavene, E.; Andrade, G.; Faria, J.; Gomes, D.; Sousa, E. Biodegradable Polymers Grafted onto Multifunctional Mesoporous Silica Nanoparticles for Gene Delivery. *ChemEngineering* **2018**, *2*, 24. [CrossRef]
55. Slita, A.; Egorova, A.; Casals, E.; Kiselev, A.; Rosenholm, J.M. Characterization of Modified Mesoporous Silica Nanoparticles as Vectors for SiRNA Delivery. *Asian J. Pharm. Sci.* **2018**, *13*, 592–599. [CrossRef] [PubMed]
56. Khan, M.A.; Kiser, M.R.; Moradipour, M.; Nadeau, E.A.; Ghanim, R.W.; Webb, B.A.; Rankin, S.E.; Knutson, B.L. Effect of Confinement in Nanopores on RNA Interactions with Functionalized Mesoporous Silica Nanoparticles. *J. Phys. Chem. B* **2020**, *124*, 8549–8561. [CrossRef] [PubMed]
57. Du, M.; Chen, Y.; Tu, J.; Liufu, C.; Yu, J.; Yuan, Z.; Gong, X.; Chen, Z. Ultrasound Responsive Magnetic Mesoporous Silica Nanoparticle-Loaded Microbubbles for Efficient Gene Delivery. *ACS Biomater. Sci. Eng.* **2020**, *6*, 2904–2912. [CrossRef] [PubMed]
58. Li, X.; Chen, Y.; Wang, M.; Ma, Y.; Xia, W.; Gu, H. A Mesoporous Silica Nanoparticle–PEI–Fusogenic Peptide System for SiRNA Delivery in Cancer Therapy. *Biomaterials* **2013**, *34*, 1391–1401. [CrossRef]
59. Li, Y.; Duo, Y.; Zhai, P.; He, L.; Zhong, K.; Zhang, Y.; Huang, K.; Luo, J.; Zhang, H.; Yu, X. Dual Targeting Delivery of MiR-328 by Functionalized Mesoporous Silica Nanoparticles for Colorectal Cancer Therapy. *Nanomedicine* **2018**, *13*, 1753–1772. [CrossRef]
60. Ngamcherdtrakul, W.; Sangvanich, T.; Reda, M.; Gu, S.; Bejan, D.; Yantasee, W. Lyophilization and Stability of Antibody-Conjugated Mesoporous Silica Nanoparticle with Cationic Polymer and PEG for SiRNA Delivery. *Int. J. Nanomed.* **2018**, *13*, 4015–4027. [CrossRef]
61. Heidari, R.; Khosravian, P.; Mirzaei, S.A.; Elahian, F. SiRNA Delivery Using Intelligent Chitosan-Capped Mesoporous Silica Nanoparticles for Overcoming Multidrug Resistance in Malignant Carcinoma Cells. *Sci. Rep.* **2021**, *11*, 20531. [CrossRef]
62. Hosseinpour, S.; Cao, Y.; Liu, J.; Xu, C.; Walsh, L.J. Efficient Transfection and Long-Term Stability of Rno-MiRNA-26a-5p for Osteogenic Differentiation by Large Pore Sized Mesoporous Silica Nanoparticles. *J. Mater. Chem. B* **2021**, *9*, 2275–2284. [CrossRef] [PubMed]
63. Geng, S.; Qin, L.; He, Y.; Li, X.; Yang, M.; Li, L.; Liu, D.; Li, Y.; Niu, D.; Yang, G. Effective and Safe Delivery of GLP-1AR and FGF-21 Plasmids Using Amino-Functionalized Dual-Mesoporous Silica Nanoparticles in Vitro and in Vivo. *Biomaterials* **2021**, *271*, 120763. [CrossRef] [PubMed]
64. Song, H.; Yang, Y.; Tang, J.; Gu, Z.; Wang, Y.; Zhang, M.; Yu, C. DNA Vaccine Mediated by Rambutan-Like Mesoporous Silica Nanoparticles. *Adv. Ther.* **2020**, *3*, 1900154. [CrossRef]

65. Nguyen, T.L.; Yin, Y.; Choi, Y.; Jeong, J.H.; Kim, J. Enhanced Cancer DNA Vaccine via Direct Transfection to Host Dendritic Cells Recruited in Injectable Scaffolds. *ACS Nano* **2020**, *14*, 11623–11636. [CrossRef]
66. Valdés-Sánchez, L.; Borrego-González, S.; Montero-Sánchez, A.; Massalini, S.; de la Cerda, B.; Díaz-Cuenca, A.; Díaz-Corrales, F.J. Mesoporous Silica-Based Nanoparticles as Non-Viral Gene Delivery Platform for Treating Retinitis Pigmentosa. *J. Clin. Med.* **2022**, *11*, 2170. [CrossRef]
67. Tian, Y.; Zhang, Y.; Zhang, M.; Chen, X.; Lei, L.; Hu, T. Antisense *vicR*-Loaded Dendritic Mesoporous Silica Nanoparticles Regulate the Biofilm Organization and Cariogenicity of *Streptococcus mutans*. *Int. J. Nanomed.* **2022**, *17*, 1255–1272. [CrossRef]
68. Zheng, G.; Zhao, R.; Xu, A.; Shen, Z.; Chen, X.; Shao, J. Co-Delivery of Sorafenib and SiVEGF Based on Mesoporous Silica Nanoparticles for ASGPR Mediated Targeted HCC Therapy. *Eur. J. Pharm. Sci.* **2018**, *111*, 492–502. [CrossRef]
69. Dilnawaz, F.; Sahoo, S.K. Augmented Anticancer Efficacy by Si-RNA Complexed Drug-Loaded Mesoporous Silica Nanoparticles in Lung Cancer Therapy. *ACS Appl. Nano Mater.* **2018**, *1*, 730–740. [CrossRef]
70. Zhang, B.; Liu, Q.; Liu, M.; Shi, P.; Zhu, L.; Zhang, L.; Li, R. Biodegradable Hybrid Mesoporous Silica Nanoparticles for Gene/Chemo-Synergetic Therapy of Breast Cancer. *J. Biomater. Appl.* **2019**, *33*, 1382–1393. [CrossRef]
71. Choi, E.; Lim, D.-K.; Kim, S. Calcium-Doped Mesoporous Silica Nanoparticles as a Lysosomolytic Nanocarrier for Amine-Free Loading and Cytosolic Delivery of SiRNA. *J. Ind. Eng. Chem.* **2020**, *81*, 71–80. [CrossRef]
72. Rong, J.; Li, P.; Ge, Y.; Chen, H.; Wu, J.; Zhang, R.; Lao, J.; Lou, D.; Zhang, Y. Histone H2A-Peptide-Hybrided Upconversion Mesoporous Silica Nanoparticles for Bortezomib/P53 Delivery and Apoptosis Induction. *Colloids Surf. B Biointerfaces* **2020**, *186*, 110674. [CrossRef] [PubMed]
73. Zhou, S.; Ding, C.; Wang, C.; Fu, J. UV-Light Cross-Linked and PH de-Cross-Linked Coumarin-Decorated Cationic Copolymer Grafted Mesoporous Silica Nanoparticles for Drug and Gene Co-Delivery in Vitro. *Mater. Sci. Eng. C* **2020**, *108*, 110469. [CrossRef] [PubMed]
74. Tsai, P.H.; Wang, M.L.; Chang, J.H.; Yarmishyn, A.A.; Nhi Nguyen, P.N.; Chen, W.; Chien, Y.; Huo, T.I.; Mou, C.Y.; Chiou, S.H. Dual Delivery of HNF4α and Cisplatin by Mesoporous Silica Nanoparticles Inhibits Cancer Pluripotency and Tumorigenicity in Hepatoma-Derived CD133-Expressing Stem Cells. *ACS Appl. Mater. Interfaces* **2019**, *11*, 19808–19818. [CrossRef] [PubMed]
75. Babaei, M.; Abnous, K.; Taghdisi, S.M.; Taghavi, S.; Saljooghi, A.S.; Ramezani, M.; Alibolandi, M. Targeted Rod-Shaped Mesoporous Silica Nanoparticles for the Co-Delivery of Camptothecin and Survivin ShRNA into Colon Adenocarcinoma in Vitro and in Vivo. *Eur. J. Pharm. Biopharm.* **2020**, *156*, 84–96. [CrossRef] [PubMed]
76. Li, M.; Cui, X.; Wei, F.; Li, C.; Han, X. RGD Peptide Modified Erythrocyte Membrane/Porous Nanoparticles Loading Mir-137 for NIR-Stimulated Theranostics of Glioblastomas. *Nanomaterials* **2022**, *12*, 1464. [CrossRef]
77. Wang, Y.; Xie, Y.; Kilchrist, K.V.; Li, J.; Duvall, C.L.; Oupický, D. Endosomolytic and Tumor-Penetrating Mesoporous Silica Nanoparticles for SiRNA/MiRNA Combination Cancer Therapy. *ACS Appl. Mater. Interfaces* **2020**, *12*, 4308–4322. [CrossRef]
78. Ahir, M.; Upadhyay, P.; Ghosh, A.; Sarker, S.; Bhattacharya, S.; Gupta, P.; Ghosh, S.; Chattopadhyay, S.; Adhikary, A. Delivery of Dual MiRNA through CD44-Targeted Mesoporous Silica Nanoparticles for Enhanced and Effective Triple-Negative Breast Cancer Therapy. *Biomater. Sci.* **2020**, *8*, 2939–2954. [CrossRef]
79. Zhou, X.; Zhang, Q.; Chen, L.; Nie, W.; Wang, W.; Wang, H.; Mo, X.; He, C. Versatile Nanocarrier Based on Functionalized Mesoporous Silica Nanoparticles to Codeliver Osteogenic Gene and Drug for Enhanced Osteodifferentiation. *ACS Biomater. Sci. Eng.* **2019**, *5*, 710–723. [CrossRef]
80. Yan, J.; Lu, X.; Zhu, X.; Hu, X.; Wang, L.; Qian, J.; Zhang, F.; Liu, M. Effects of MiR-26a on Osteogenic Differentiation of Bone Marrow Mesenchymal Stem Cells by a Mesoporous Silica Nanoparticle–PEI–Peptide System. *Int. J. Nanomed.* **2020**, *15*, 497–511. [CrossRef]
81. Mora-Raimundo, P.; Lozano, D.; Manzano, M.; Vallet-Regí, M. Nanoparticles to Knockdown Osteoporosis-Related Gene and Promote Osteogenic Marker Expression for Osteoporosis Treatment. *ACS Nano* **2019**, *13*, 5451–5464. [CrossRef]
82. Mora-Raimundo, P.; Lozano, D.; Benito, M.; Mulero, F.; Manzano, M.; Vallet-Regí, M. Osteoporosis Remission and New Bone Formation with Mesoporous Silica Nanoparticles. *Adv. Sci.* **2021**, *8*, 2101107. [CrossRef] [PubMed]
83. Cheng, C.-S.; Liu, T.-P.; Chien, F.-C.; Mou, C.-Y.; Wu, S.-H.; Chen, Y.-P. Codelivery of Plasmid and Curcumin with Mesoporous Silica Nanoparticles for Promoting Neurite Outgrowth. *ACS Appl. Mater. Interfaces* **2019**, *11*, 15322–15331. [CrossRef] [PubMed]
84. Tao, Y.; Xu, S.; Wang, J.; Xu, L.; Zhang, C.; Chen, K.; Lian, Z.; Zhou, J.; Xie, H.; Zheng, S.; et al. Delivery of MicroRNA-33 Antagomirs by Mesoporous Silica Nanoparticles to Ameliorate Lipid Metabolic Disorders. *Front. Pharmacol.* **2020**, *11*, 921. [CrossRef]
85. Li, N.; Niu, D.; Jiang, Y.; Xu, C.; Pan, S.; He, J.; Chen, J.; Zhang, L.; Li, Y. Morphology Evolution and Spatially Selective Functionalization of Hierarchically Porous Silica Nanospheres for Improved Multidrug Delivery. *Chem. Mater.* **2017**, *29*, 10377–10385. [CrossRef]
86. Castillo, R.R.; Vallet-Regí, M. Emerging Strategies in Anticancer Combination Therapy Employing Silica-Based Nanosystems. *Biotechnol. J.* **2021**, *16*, 1900438. [CrossRef]
87. Kumar, K.; Rani, V.; Mishra, M.; Chawla, R. New Paradigm in Combination Therapy of SiRNA with Chemotherapeutic Drugs for Effective Cancer Therapy. *Curr. Res. Pharmacol. Drug Discov.* **2022**, *3*, 100103. [CrossRef]
88. Castillo, R.R.; Colilla, M.; Vallet-Regí, M. Advances in Mesoporous Silica-Based Nanocarriers for Co-Delivery and Combination Therapy against Cancer. *Expert Opin. Drug Deliv.* **2017**, *14*, 229–243. [CrossRef]

89. Paskeh, M.D.A.; Saebfar, H.; Mahabady, M.K.; Orouei, S.; Hushmandi, K.; Entezari, M.; Hashemi, M.; Aref, A.R.; Hamblin, M.R.; Ang, H.L.; et al. Overcoming Doxorubicin Resistance in Cancer: SiRNA-Loaded Nanoarchitectures for Cancer Gene Therapy. *Life Sci.* **2022**, *298*, 120463. [CrossRef]
90. Mirzaei, S.; Gholami, M.H.; Hashemi, F.; Zabolian, A.; Hushmandi, K.; Rahmanian, V.; Entezari, M.; Girish, Y.R.; Sharath Kumar, K.S.; Aref, A.R.; et al. Employing SiRNA Tool and Its Delivery Platforms in Suppressing Cisplatin Resistance: Approaching to a New Era of Cancer Chemotherapy. *Life Sci.* **2021**, *277*, 119430. [CrossRef]
91. Paris, J.L.; Vallet-Regí, M. Mesoporous Silica Nanoparticles for Co-Delivery of Drugs and Nucleic Acids in Oncology: A Review. *Pharmaceutics* **2020**, *12*, 526. [CrossRef]
92. Mosammaparast, N.; Jackson, K.R.; Guo, Y.; Brame, C.J.; Shabanowitz, J.; Hunt, D.F.; Pemberton, L.F. Nuclear Import of Histone H2a and H2b Is Mediated by a Network of Karyopherins. *J. Cell Biol.* **2001**, *153*, 251–262. [CrossRef] [PubMed]
93. Shi, X.; Singh, S.; Lin, E.; Li, H. Chimeric RNAs in Cancer. *Adv. Clin. Chem.* **2021**, *100*, 1–35. [PubMed]
94. Živojević, K.; Mladenović, M.; Djisalov, M.; Mundzic, M.; Ruiz-Hernandez, E.; Gadjanski, I.; Knežević, N. Advanced Mesoporous Silica Nanocarriers in Cancer Theranostics and Gene Editing Applications. *J. Control. Release* **2021**, *337*, 193–211. [CrossRef] [PubMed]
95. Liu, X.; Wu, F.; Ji, Y.; Yin, L. Recent Advances in Anti-Cancer Protein/Peptide Delivery. *Bioconjug. Chem.* **2019**, *30*, 305–324. [CrossRef] [PubMed]
96. Yau, A.; Lee, J.; Chen, Y. Nanomaterials for Protein Delivery in Anticancer Applications. *Pharmaceutics* **2021**, *13*, 155. [CrossRef] [PubMed]
97. Liu, H.-J.; Xu, P. Smart Mesoporous Silica Nanoparticles for Protein Delivery. *Nanomaterials* **2019**, *9*, 511. [CrossRef] [PubMed]
98. Xu, C.; Lei, C.; Yu, C. Mesoporous Silica Nanoparticles for Protein Protection and Delivery. *Front. Chem.* **2019**, *7*, 290. [CrossRef]
99. Liu, B.; Ejaz, W.; Gong, S.; Kurbanov, M.; Canakci, M.; Anson, F.; Thayumanavan, S. Engineered Interactions with Mesoporous Silica Facilitate Intracellular Delivery of Proteins and Gene Editing. *Nano Lett.* **2020**, *20*, 4014–4021. [CrossRef]
100. Lau, J.L.; Dunn, M.K. Therapeutic Peptides: Historical Perspectives, Current Development Trends, and Future Directions. *Bioorg. Med. Chem.* **2018**, *26*, 2700–2707. [CrossRef]
101. Marqus, S.; Pirogova, E.; Piva, T.J. Evaluation of the Use of Therapeutic Peptides for Cancer Treatment. *J. Biomed. Sci.* **2017**, *24*, 1–15. [CrossRef]
102. Kurrikoff, K.; Aphkhazava, D.; Langel, Ü. The Future of Peptides in Cancer Treatment. *Curr. Opin. Pharmacol.* **2019**, *47*, 27–32. [CrossRef] [PubMed]
103. Wang, Y.; Gao, Y.N.; Bai, R.; Chen, H.Y.; Wu, Y.Y.; Shang, J.J.; Bao, D.P. Identification of a Novel Anti-Cancer Protein, FIP-Bbo, from *Botryobasidium botryosum* and Protein Structure Analysis Using Molecular Dynamic Simulation. *Sci. Rep.* **2019**, *9*, 5818. [CrossRef] [PubMed]
104. Gaspar, D.; Salomé Veiga, A.; Castanho, M.A.R.B. From Antimicrobial to Anticancer Peptides. A Review. *Front. Microbiol.* **2013**, *4*, 294. [CrossRef] [PubMed]
105. Mahlapuu, M.; Håkansson, J.; Ringstad, L.; Björn, C. Antimicrobial Peptides: An Emerging Category of Therapeutic Agents. *Front. Cell Infect. Microbiol.* **2016**, *6*, 194. [CrossRef] [PubMed]
106. Khan, M.M.; Filipczak, N.; Torchilin, V.P. Cell Penetrating Peptides: A Versatile Vector for Co-Delivery of Drug and Genes in Cancer. *J. Control. Release* **2021**, *330*, 1220–1228. [CrossRef]
107. Kalmouni, M.; Al-Hosani, S.; Magzoub, M. Cancer Targeting Peptides. *Cell. Mol. Life Sci.* **2019**, *76*, 2171–2183. [CrossRef]
108. Robinson, J.A. Folded Synthetic Peptides and Other Molecules Targeting Outer Membrane Protein Complexes in Gram-Negative Bacteria. *Front. Chem.* **2019**, *7*, 45. [CrossRef]
109. Cheng, Z.; Li, Y.; Zhao, D.; Zhao, W.; Wu, M.; Zhang, W.; Cui, Y.; Zhang, P.; Zhang, Z. Nanocarriers for Intracellular Co-Delivery of Proteins and Small-Molecule Drugs for Cancer Therapy. *Front. Bioeng. Biotechnol.* **2022**, *10*, 994655. [CrossRef]
110. Xu, J.; Li, G.; Zhang, Y.; Kang, Y.; Xu, S.; Liu, H. A Dual-Responsive Drug Delivery System Based on Mesoporous Silica Nanoparticles Covered with Zipper-Type Peptide for Intracellular Transport/Release. *Colloids Surf. A Physicochem. Eng. Asp.* **2021**, *631*, 127672. [CrossRef]
111. Slowing, I.I.; Trewyn, B.G.; Lin, V.S.-Y. Mesoporous Silica Nanoparticles for Intracellular Delivery of Membrane-Impermeable Proteins. *J. Am. Chem. Soc.* **2007**, *129*, 8845–8849. [CrossRef]
112. Méndez, J.; Morales Cruz, M.; Delgado, Y.; Figueroa, C.M.; Orellano, E.A.; Morales, M.; Monteagudo, A.; Griebenow, K. Delivery of Chemically Glycosylated Cytochrome c Immobilized in Mesoporous Silica Nanoparticles Induces Apoptosis in HeLa Cancer Cells. *Mol. Pharm.* **2014**, *11*, 102–111. [CrossRef] [PubMed]
113. Shang, W.; Nuffer, J.H.; Muñiz-Papandrea, V.A.; Colón, W.; Siegel, R.W.; Dordick, J.S. Cytochrome c on Silica Nanoparticles: Influence of Nanoparticle Size on Protein Structure, Stability, and Activity. *Small* **2009**, *5*, 470–476. [CrossRef] [PubMed]
114. Huang, W.-Y.; Davies, G.-L.; Davis, J.J. Engineering Cytochrome-Modified Silica Nanoparticles to Induce Programmed Cell Death. *Chem. Eur. J.* **2013**, *19*, 17891–17898. [CrossRef]
115. Du, A.W.; Stenzel, M.H. Drug Carriers for the Delivery of Therapeutic Peptides. *Biomacromolecules* **2014**, *15*, 1097–1114. [CrossRef] [PubMed]
116. Martínez-Carmona, M.; Lozano, D.; Colilla, M.; Vallet-Regí, M. Lectin-Conjugated PH-Responsive Mesoporous Silica Nanoparticles for Targeted Bone Cancer Treatment. *Acta Biomater.* **2018**, *65*, 393–404. [CrossRef]

117. Pei, W.; Cai, L.; Gong, X.; Zhang, L.; Zhang, J.; Zhu, P.; Jiang, H.; Wang, C.; Wang, S.; Chen, J. Drug-Loaded Oleic-Acid Grafted Mesoporous Silica Nanoparticles Conjugated with α-Lactalbumin Resembling BAMLET-like Anticancer Agent with Improved Biocompatibility and Therapeutic Efficacy. *Mater. Today Bio* **2022**, *15*, 100272. [CrossRef] [PubMed]
118. Luo, G.-F.; Chen, W.-H.; Liu, Y.; Zhang, J.; Cheng, S.-X.; Zhuo, R.-X.; Zhang, X.-Z. Charge-Reversal Plug Gate Nanovalves on Peptide-Functionalized Mesoporous Silica Nanoparticles for Targeted Drug Delivery. *J. Mater. Chem. B* **2013**, *1*, 5723–5732. [CrossRef]
119. Luo, G.F.; Chen, W.H.; Liu, Y.; Lei, Q.; Zhuo, R.X.; Zhang, X.Z. Multifunctional Enveloped Mesoporous Silica Nanoparticles for Subcellular Co-Delivery of Drug and Therapeutic Peptide. *Sci. Rep.* **2014**, *4*, 6064. [CrossRef]
120. Cheng, Y.J.; Zeng, X.; Cheng, D.B.; Xu, X.D.; Zhang, X.Z.; Zhuo, R.X.; He, F. Functional Mesoporous Silica Nanoparticles (MSNs) for Highly Controllable Drug Release and Synergistic Therapy. *Colloids Surf. B Biointerfaces* **2016**, *145*, 217–225. [CrossRef]
121. Xiao, D.; Hu, J.-J.; Zhu, J.-Y.; Wang, S.-B.; Zhuo, R.-X.; Zhang, X.-Z. A Redox-Responsive Mesoporous Silica Nanoparticle with a Therapeutic Peptide Shell for Tumor Targeting Synergistic Therapy. *Nanoscale* **2016**, *8*, 16702–16709. [CrossRef]
122. Zhang, J.; Shen, B.; Chen, L.; Chen, L.; Meng, Y.; Feng, J. A Dual-Sensitive Mesoporous Silica Nanoparticle Based Drug Carrier for Cancer Synergetic Therapy. *Colloids Surf. B Biointerfaces* **2019**, *175*, 65–72. [CrossRef] [PubMed]
123. de la Torre, C.; Domínguez-Berrocal, L.; Murguía, J.R.; Marcos, M.D.; Martínez-Máñez, R.; Bravo, J.; Sancenón, F. ϵ-Polylysine-Capped Mesoporous Silica Nanoparticles as Carrier of the C9h Peptide to Induce Apoptosis in Cancer Cells. *Chem. Eur. J.* **2018**, *24*, 1890–1897. [CrossRef] [PubMed]
124. Cao, J.; Zhang, Y.; Shan, Y.; Wang, J.; Liu, F.; Liu, H.; Xing, G.; Lei, J.; Zhou, J. A PH-Dependent Antibacterial Peptide Release Nano-System Blocks Tumor Growth in Vivo without Toxicity. *Sci. Rep.* **2017**, *7*, 11242. [CrossRef] [PubMed]
125. Rahmani, S.; Budimir, J.; Sejalon, M.; Daurat, M.; Aggad, D.; Vivès, E.; Raehm, L.; Garcia, M.; Lichon, L.; Gary-Bobo, M.; et al. Large Pore Mesoporous Silica and Organosilica Nanoparticles for Pepstatin a Delivery in Breast Cancer Cells. *Molecules* **2019**, *24*, 332. [CrossRef] [PubMed]
126. Leão, N.M.; Silva Borges, B.R.; Verano-Braga, T.; Morais, J.P.; Souza Santos, R.A.; Ferreira Soares, D.C. Mesoporous Silica Nanoparticles Loaded with Alamandine as a Potential New Therapy against Cancer. *J. Drug Deliv. Sci. Technol.* **2021**, *61*, 102216. [CrossRef]
127. Ramya, S.; Paulpandi, M.; Kavithaa, K.; Saranya, T.; Winster, H.; Balachandar, V.; Narayanasamy, A. Fabatin-Loaded Silica Nanoparticle-Induced Apoptosis via Mitochondrial Dysfunction: Targeting the PI3K/AKT Molecular Pathway as a Therapeutic Implication against Triple Negative Breast Cancer. *New J. Chem.* **2021**, *45*, 17847–17861. [CrossRef]
128. Lin, F.-C.; Yu, Q.; Zink, J.I. Self-Contained Nanocapsules Carrying Anticancer Peptides for Magnetically Activated and Enzyme-Cleaved Drug Delivery. *ACS Appl. Nano Mater.* **2021**, *4*, 10771–10783. [CrossRef]
129. Li, L.; Wang, H. Enzyme-Coated Mesoporous Silica Nanoparticles as Efficient Antibacterial Agents in Vivo. *Adv. Healthc. Mater.* **2013**, *2*, 1351–1360. [CrossRef]
130. Wang, Y.; Nor, Y.A.; Song, H.; Yang, Y.; Xu, C.; Yu, M.; Yu, C. Small-Sized and Large-Pore Dendritic Mesoporous Silica Nanoparticles Enhance Antimicrobial Enzyme Delivery. *J. Mater. Chem. B* **2016**, *4*, 2646–2653. [CrossRef]
131. Song, H.; Ahmad Nor, Y.; Yu, M.; Yang, Y.; Zhang, J.; Zhang, H.; Xu, C.; Mitter, N.; Yu, C. Silica Nanopollens Enhance Adhesion for Long-Term Bacterial Inhibition. *J. Am. Chem. Soc.* **2016**, *138*, 6455–6462. [CrossRef]
132. Xu, C.; He, Y.; Li, Z.; Ahmad Nor, Y.; Ye, Q. Nanoengineered Hollow Mesoporous Silica Nanoparticles for the Delivery of Antimicrobial Proteins into Biofilms. *J. Mater. Chem. B* **2018**, *6*, 1899–1902. [CrossRef] [PubMed]
133. Martínez-Carmona, M.; Izquierdo-Barba, I.; Colilla, M.; Vallet-Regí, M. Concanavalin A-Targeted Mesoporous Silica Nanoparticles for Infection Treatment. *Acta Biomater.* **2019**, *96*, 547–556. [CrossRef] [PubMed]
134. Durack, E.; Mallen, S.; O'Connor, P.M.; Rea, M.C.; Ross, R.P.; Hill, C.; Hudson, S. Protecting Bactofencin A to Enable Its Antimicrobial Activity Using Mesoporous Matrices. *Int. J. Pharm.* **2019**, *558*, 9–17. [CrossRef] [PubMed]
135. Marcelino-Pérez, G.; Ruiz-Medrano, R.; Gallardo-Hernández, S.; Xoconostle-Cázares, B. Adsorption of Recombinant Human β-Defensin 2 and Two Mutants on Mesoporous Silica Nanoparticles and Its Effect against *Clavibacter michiganensis* Subsp. michiganensis. *Nanomaterials* **2021**, *11*, 2144. [CrossRef]
136. Braun, K.; Pochert, A.; Lindén, M.; Davoudi, M.; Schmidtchen, A.; Nordström, R.; Malmsten, M. Membrane Interactions of Mesoporous Silica Nanoparticles as Carriers of Antimicrobial Peptides. *J. Colloid Interface Sci.* **2016**, *475*, 161–170. [CrossRef]
137. Häffner, S.M.; Parra-Ortiz, E.; Browning, K.L.; Jørgensen, E.; Skoda, M.W.A.; Montis, C.; Li, X.; Berti, D.; Zhao, D.; Malmsten, M. Membrane Interactions of Virus-like Mesoporous Silica Nanoparticles. *ACS Nano* **2021**, *15*, 6787–6800. [CrossRef]
138. Tenland, E.; Pochert, A.; Krishnan, N.; Rao, K.U.; Kalsum, S.; Braun, K.; Glegola-Madejska, I.; Lerm, M.; Robertson, B.D.; Lindén, M.; et al. Effective Delivery of the Anti-Mycobacterial Peptide NZX in Mesoporous Silica Nanoparticles. *PLoS ONE* **2019**, *14*, e0212858. [CrossRef]
139. Ma, B.; Chen, Y.; Hu, G.; Zeng, Q.; Lv, X.; Oh, D.H.; Fu, X.; Jin, Y. Ovotransferrin Antibacterial Peptide Coupling Mesoporous Silica Nanoparticle as an Effective Antibiotic Delivery System for Treating Bacterial Infection In Vivo. *ACS Biomater. Sci. Eng.* **2022**, *8*, 109–118. [CrossRef]
140. Alharthi, S.; Ziora, Z.M.; Janjua, T.; Popat, A.; Moyle, P.M. Formulation and Biological Evaluation of Mesoporous Silica Nanoparticles Loaded with Combinations of Sortase A Inhibitors and Antimicrobial Peptides. *Pharmaceutics* **2022**, *14*, 986. [CrossRef]

141. Zhang, J.; Postovit, L.M.; Wang, D.; Gardiner, R.B.; Harris, R.; Abdul, M.M.; Thomas, A.A. In Situ Loading of Basic Fibroblast Growth Factor within Porous Silica Nanoparticles for a Prolonged Release. *Nanoscale Res. Lett.* **2009**, *4*, 1297–1302. [CrossRef]
142. Gan, Q.; Zhu, J.; Yuan, Y.; Liu, H.; Qian, J.; Li, Y.; Liu, C. A Dual-Delivery System of PH-Responsive Chitosan-Functionalized Mesoporous Silica Nanoparticles Bearing BMP-2 and Dexamethasone for Enhanced Bone Regeneration. *J. Mater. Chem. B* **2015**, *3*, 2056–2066. [CrossRef] [PubMed]
143. Gan, Q.; Zhu, J.; Yuan, Y.; Liu, C. PH-Responsive Fe3O4 Nanopartilces-Capped Mesoporous Silica Supports for Protein Delivery. *J. Nanosci. Nanotechnol.* **2016**, *16*, 5470–5479. [CrossRef] [PubMed]
144. Lozano, D.; Manzano, M.; Doadrio, J.C.; Salinas, A.J.; Vallet-Regí, M.; Gómez-Barrena, E.; Esbrit, P. Osteostatin-Loaded Bioceramics Stimulate Osteoblastic Growth and Differentiation. *Acta Biomater.* **2010**, *6*, 797–803. [CrossRef] [PubMed]
145. Trejo, C.G.; Lozano, D.; Manzano, M.; Doadrio, J.C.; Salinas, A.J.; Dapía, S.; Gómez-Barrena, E.; Vallet-Regí, M.; García-Honduvilla, N.; Buján, J.; et al. The Osteoinductive Properties of Mesoporous Silicate Coated with Osteostatin in a Rabbit Femur Cavity Defect Model. *Biomaterials* **2010**, *31*, 8564–8573. [CrossRef]
146. Lozano, D.; Trejo, C.G.; Gómez-Barrena, E.; Manzano, M.; Doadrio, J.C.; Salinas, A.J.; Vallet-Regí, M.; García-Honduvilla, N.; Esbrit, P.; Buján, J. Osteostatin-Loaded onto Mesoporous Ceramics Improves the Early Phase of Bone Regeneration in a Rabbit Osteopenia Model. *Acta Biomater.* **2012**, *8*, 2317–2323. [CrossRef]
147. Mendes, L.S.; Saska, S.; Martines, M.A.U.; Marchetto, R. Nanostructured Materials Based on Mesoporous Silica and Mesoporous Silica/Apatite as Osteogenic Growth Peptide Carriers. *Mater. Sci. Eng. C* **2013**, *33*, 4427–4434. [CrossRef]
148. Zhou, X.; Feng, W.; Qiu, K.; Chen, L.; Wang, W.; Nie, W.; Mo, X.; He, C. BMP-2 Derived Peptide and Dexamethasone Incorporated Mesoporous Silica Nanoparticles for Enhanced Osteogenic Differentiation of Bone Mesenchymal Stem Cells. *ACS Appl. Mater. Interfaces* **2015**, *7*, 15777–15789. [CrossRef]
149. Mutreja, I.; Kumar, D.; Hogan, K.; Campbell, E.; Mansky, K.; Aparicio, C. Strontium- and Peptide-Modified Silicate Nanostructures for Dual Osteogenic and Antimicrobial Activity. *Biomater. Adv.* **2022**, *135*, 212735. [CrossRef]
150. Lim, J.-S.; Lee, K.; Choi, J.-N.; Hwang, Y.-K.; Yun, M.-Y.; Kim, H.-J.; Won, Y.S.; Kim, S.-J.; Kwon, H.; Huh, S. Intracellular Protein Delivery by Hollow Mesoporous Silica Capsules with a Large Surface Hole. *Nanotechnology* **2012**, *23*, 085101. [CrossRef]
151. Niu, Y.; Yu, M.; Zhang, J.; Yang, Y.; Xu, C.; Yeh, M.; Taran, E.; Hou, J.J.C.; Gray, P.P.; Yu, C. Synthesis of Silica Nanoparticles with Controllable Surface Roughness for Therapeutic Protein Delivery. *J. Mater. Chem. B* **2015**, *3*, 8477–8485. [CrossRef]
152. Wang, X.; Li, X.; Ito, A.; Yoshiyuki, K.; Sogo, Y.; Watanabe, Y.; Yamazaki, A.; Ohno, T.; Tsuji, N.M. Silica Nanospheres: Hollow Structure Improved Anti-Cancer Immunity of Mesoporous Silica Nanospheres In Vivo. *Small* **2016**, *12*, 3510–3515. [CrossRef] [PubMed]
153. Wang, X.; Li, X.; Yoshiyuki, K.; Watanabe, Y.; Sogo, Y.; Ohno, T.; Tsuji, N.M.; Ito, A. Comprehensive Mechanism Analysis of Mesoporous-Silica-Nanoparticle-Induced Cancer Immunotherapy. *Adv. Healthc. Mater.* **2016**, *5*, 1169–1176. [CrossRef] [PubMed]
154. Yang, Y.; Lu, Y.; Abbaraju, P.L.; Zhang, J.; Zhang, M.; Xiang, G.; Yu, C. Multi-Shelled Dendritic Mesoporous Organosilica Hollow Spheres: Roles of Composition and Architecture in Cancer Immunotherapy. *Angew. Chem. Int. Ed.* **2017**, *56*, 8446–8450. [CrossRef] [PubMed]
155. Cha, B.G.; Jeong, J.H.; Kim, J. Extra-Large Pore Mesoporous Silica Nanoparticles Enabling Co-Delivery of High Amounts of Protein Antigen and Toll-like Receptor 9 Agonist for Enhanced Cancer Vaccine Efficacy. *ACS Cent. Sci.* **2018**, *4*, 484–492. [CrossRef] [PubMed]
156. Kong, M.; Tang, J.; Qiao, Q.; Wu, T.; Qi, Y.; Tan, S.; Gao, X.; Zhang, Z. Biodegradable Hollow Mesoporous Silica Nanoparticles for Regulating Tumor Microenvironment and Enhancing Antitumor Efficiency. *Theranostics* **2017**, *7*, 3276–3292. [CrossRef]
157. Park, J.; Choi, S.W.; Cha, B.G.; Kim, J.; Kang, S.J. Alternative Activation of Macrophages through Interleukin-13-Loaded Extra-Large-Pore Mesoporous Silica Nanoparticles Suppresses Experimental Autoimmune Encephalomyelitis. *ACS Biomater. Sci. Eng.* **2021**, *7*, 4446–4453. [CrossRef]
158. Guo, H.C.; Feng, X.M.; Sun, S.Q.; Wei, Y.Q.; Sun, D.H.; Liu, X.T.; Liu, Z.X.; Luo, J.X.; Yin, H. Immunization of Mice by Hollow Mesoporous Silica Nanoparticles as Carriers of Porcine Circovirus Type 2 ORF2 Protein. *Virol. J.* **2012**, *9*, 108. [CrossRef]
159. De Padua Oliveira, D.C.; de Barros, A.L.B.; Belardi, R.M.; de Goes, A.M.; de Oliveira Souza, B.K.; Soares, D.C.F. Mesoporous Silica Nanoparticles as a Potential Vaccine Adjuvant against Schistosoma Mansoni. *J. Drug Deliv. Sci. Technol.* **2016**, *35*, 234–240. [CrossRef]
160. Virginio, V.G.; Bandeira, N.C.; dos Anjos Leal, F.M.; Lancellotti, M.; Zaha, A.; Ferreira, H.B. Assessment of the Adjuvant Activity of Mesoporous Silica Nanoparticles in Recombinant *Mycoplasma hyopneumoniae* Antigen Vaccines. *Heliyon* **2017**, *3*, e00225. [CrossRef]
161. Hajizade, A.; Salmanian, A.H.; Amani, J.; Ebrahimi, F.; Arpanaei, A. EspA-Loaded Mesoporous Silica Nanoparticles Can Efficiently Protect Animal Model against Enterohaemorrhagic *E. coli* O157:H7. *Artif. Cells Nanomed. Biotechnol.* **2018**, *46*, S1067–S1075. [CrossRef]
162. Ferreira Soares, D.C.; Soares, L.M.; Miranda de Goes, A.; Melo, E.M.; Branco de Barros, A.L.; Alves Santos Bicalho, T.C.; Leao, N.M.; Tebaldi, M.L. Mesoporous SBA-16 Silica Nanoparticles as a Potential Vaccine Adjuvant against Paracoccidioides Brasiliensis. *Microporous Mesoporous Mater.* **2020**, *291*, 109676. [CrossRef]
163. Xie, J.; Yang, C.; Liu, Q.; Li, J.; Liang, R.; Shen, C.; Zhang, Y.; Wang, K.; Liu, L.; Shezad, K.; et al. Encapsulation of Hydrophilic and Hydrophobic Peptides into Hollow Mesoporous Silica Nanoparticles for Enhancement of Antitumor Immune Response. *Small* **2017**, *13*, 1701741. [CrossRef] [PubMed]

164. Wang, T.; Yin, L.; Ma, Z.; Zhang, Y. Chlorogenic Acid-Loaded Mesoporous Silica Nanoparticles Modified with Hexa-Histidine Peptides Reduce Skin Allergies by Capturing Nickel. *Molecules* **2022**, *27*, 1430. [CrossRef] [PubMed]
165. Gößl, D.; Singer, H.; Chiu, H.Y.; Schmidt, A.; Lichtnecker, M.; Engelke, H.; Bein, T. Highly Active Enzymes Immobilized in Large Pore Colloidal Mesoporous Silica Nanoparticles. *New J. Chem.* **2019**, *43*, 1671–1680. [CrossRef]
166. Méndez, J.; Monteagudo, A.; Griebenow, K. Stimulus-Responsive Controlled Release System by Covalent Immobilization of an Enzyme into Mesoporous Silica Nanoparticles. *Bioconjug. Chem.* **2012**, *23*, 698–704. [CrossRef]
167. Xu, C.; Yu, M.; Noonan, O.; Zhang, J.; Song, H.; Zhang, H.; Lei, C.; Niu, Y.; Huang, X.; Yang, Y.; et al. Core-Cone Structured Monodispersed Mesoporous Silica Nanoparticles with Ultra-Large Cavity for Protein Delivery. *Small* **2015**, *11*, 5949–5955. [CrossRef] [PubMed]
168. Chen, Y.P.; Chen, C.T.; Hung, Y.; Chou, C.M.; Liu, T.P.; Liang, M.R.; Chen, C.T.; Mou, C.Y. A New Strategy for Intracellular Delivery of Enzyme Using Mesoporous Silica Nanoparticles: Superoxide Dismutase. *J. Am. Chem. Soc.* **2013**, *135*, 1516–1523. [CrossRef]
169. Lin, Y.H.; Chen, Y.P.; Liu, T.P.; Chien, F.C.; Chou, C.M.; Chen, C.T.; Mou, C.Y. Approach to Deliver Two Antioxidant Enzymes with Mesoporous Silica Nanoparticles into Cells. *ACS Appl. Mater. Interfaces* **2016**, *8*, 17944–17954. [CrossRef]
170. Han, D.H.; Na, H.-K.; Choi, W.H.; Lee, J.H.; Kim, Y.K.; Won, C.; Lee, S.-H.; Kim, K.P.; Kuret, J.; Min, D.-H.; et al. Direct Cellular Delivery of Human Proteasomes to Delay Tau Aggregation. *Nat. Commun.* **2014**, *5*, 5633. [CrossRef]
171. Teng, B.; Ding, B.; Shao, S.; Wang, Z.; Tong, W.; Wang, S.; Cheng, Z.; Lin, J.; Ma, P. Intracellular RNA and Nuclear DNA-Dual-Targeted Tumor Therapy via Upconversion Nanoplatforms with UCL/MR Dual-Mode Bioimaging. *Chem. Eng. J.* **2021**, *405*, 126606. [CrossRef]
172. Xie, M.; Liu, D.; Yang, Y. Anti-Cancer Peptides: Classification, Mechanism of Action, Reconstruction and Modification. *Open Biol.* **2020**, *10*, 200004. [CrossRef] [PubMed]
173. Mader, J.S.; Hoskin, D.W. Cationic Antimicrobial Peptides as Novel Cytotoxic Agents for Cancer Treatment. *Expert Opin. Investig. Drugs* **2006**, *15*, 933–946. [CrossRef] [PubMed]
174. Torchilin, V.P. Tat Peptide-Mediated Intracellular Delivery of Pharmaceutical Nanocarriers. *Adv. Drug Deliv. Rev.* **2008**, *60*, 548–558. [CrossRef] [PubMed]
175. Boohaker, J.R.; Lee, W.M.; Vishnubhotla, P.; M. Perez, J.L.; R. Khaled, A. The Use of Therapeutic Peptides to Target and to Kill Cancer Cells. *Curr. Med. Chem.* **2012**, *19*, 3794–3804. [CrossRef] [PubMed]
176. Khan, M.A.; Ghanim, R.W.; Kiser, M.R.; Moradipour, M.; Rogers, D.T.; Littleton, J.M.; Bradley, L.H.; Lynn, B.C.; Rankin, S.E.; Knutson, B.L. Strategy for Conjugating Oligopeptides to Mesoporous Silica Nanoparticles Using Diazirine-Based Heterobifunctional Linkers. *Nanomaterials* **2022**, *12*, 608. [CrossRef]
177. Klabenkova, K.; Fokina, A.; Stetsenko, D. Chemistry of Peptide-Oligonucleotide Conjugates: A Review. *Molecules* **2021**, *26*, 5420. [CrossRef]
178. Castillo, R.R.; Vallet-Regí, M. Recent Advances Toward the Use of Mesoporous Silica Nanoparticles for the Treatment of Bacterial Infections. *Int. J. Nanomed.* **2021**, *16*, 4409–4430. [CrossRef]
179. Nordström, R.; Malmsten, M. Delivery Systems for Antimicrobial Peptides. *Adv. Colloid Interface Sci.* **2017**, *242*, 17–34. [CrossRef]
180. Wang, C.; Hong, T.; Cui, P.; Wang, J.; Xia, J. Antimicrobial Peptides towards Clinical Application: Delivery and Formulation. *Adv. Drug Deliv. Rev.* **2021**, *175*, 113818. [CrossRef]
181. Braun, K.; Pochert, A.; Gerber, M.; Raber, H.F.; Lindén, M. Influence of Mesopore Size and Peptide Aggregation on the Adsorption and Release of a Model Antimicrobial Peptide onto/from Mesoporous Silica Nanoparticles in Vitro. *Mol. Syst. Des. Eng.* **2017**, *2*, 393–400. [CrossRef]
182. Tenland, E.; Krishnan, N.; Rönnholm, A.; Kalsum, S.; Puthia, M.; Mörgelin, M.; Davoudi, M.; Otrocka, M.; Alaridah, N.; Glegola-Madejska, I.; et al. A Novel Derivative of the Fungal Antimicrobial Peptide Plectasin Is Active against *Mycobacterium tuberculosis*. *Tuberculosis* **2018**, *113*, 231–238. [CrossRef] [PubMed]
183. Zrelovs, N.; Kurbatska, V.; Rudevica, Z.; Leonchiks, A.; Fridmanis, D. Sorting out the Superbugs: Potential of Sortase A Inhibitors among Other Antimicrobial Strategies to Tackle the Problem of Antibiotic Resistance. *Antibiotics* **2021**, *10*, 164. [CrossRef]
184. Zhang, W.; Guan, X.; Qiu, X.; Gao, T.; Yu, W.; Zhang, M.; Song, L.; Liu, D.; Dong, J.; Jiang, Z.; et al. Bioactive Composite Janus Nanofibrous Membranes Loading Ciprofloxacin and Astaxanthin for Enhanced Healing of Full-Thickness Skin Defect Wounds. *Appl. Surf. Sci.* **2023**, *610*, 155290. [CrossRef]
185. Han, M.-C.; Cai, S.-Z.; Wang, J.; He, H.-W. Single-Side Superhydrophobicity in Si_3N_4-Doped and SiO_2-Treated Polypropylene Nonwoven Webs with Antibacterial Activity. *Polymers* **2022**, *14*, 2952. [CrossRef] [PubMed]
186. Hosseinpour, S.; Walsh, L.J.; Xu, C. Modulating Osteoimmune Responses by Mesoporous Silica Nanoparticles. *ACS Biomater. Sci. Eng.* **2022**, *8*, 4110–4122. [CrossRef] [PubMed]
187. Burdușel, A.-C.; Gherasim, O.; Andronescu, E.; Grumezescu, A.M.; Ficai, A. Inorganic Nanoparticles in Bone Healing Applications. *Pharmaceutics* **2022**, *14*, 770. [CrossRef]
188. Heras, C.; Sanchez-Salcedo, S.; Lozano, D.; Peña, J.; Esbrit, P.; Vallet-Regi, M.; Salinas, A.J. Osteostatin Potentiates the Bioactivity of Mesoporous Glass Scaffolds Containing Zn^{2+} Ions in Human Mesenchymal Stem Cells. *Acta Biomater.* **2019**, *89*, 359–371. [CrossRef]
189. Pérez, R.; Sanchez-Salcedo, S.; Lozano, D.; Heras, C.; Esbrit, P.; Vallet-Regí, M.; Salinas, A.J. Osteogenic Effect of ZnO-Mesoporous Glasses Loaded with Osteostatin. *Nanomaterials* **2018**, *8*, 592. [CrossRef]

190. Jiménez-Holguín, J.; Sánchez-Salcedo, S.; Vallet-Regí, M.; Salinas, A.J. Development and Evaluation of Copper-Containing Mesoporous Bioactive Glasses for Bone Defects Therapy. *Microporous Mesoporous Mater.* **2020**, *308*, 110454. [CrossRef]
191. Jiménez-Holguín, J.; Sánchez-Salcedo, S.; Cicuéndez, M.; Vallet-Regí, M.; Salinas, A.J. Cu-Doped Hollow Bioactive Glass Nanoparticles for Bone Infection Treatment. *Pharmaceutics* **2022**, *14*, 845. [CrossRef]
192. Viswanath, D.I.; Liu, H.-C.; Huston, D.P.; Chua, C.Y.X.; Grattoni, A. Emerging Biomaterial-Based Strategies for Personalized Therapeutic in Situ Cancer Vaccines. *Biomaterials* **2022**, *280*, 121297. [CrossRef] [PubMed]
193. Wang, X.; Li, X.; Ito, A.; Sogo, Y.; Ohno, T. Particle-Size-Dependent Toxicity and Immunogenic Activity of Mesoporous Silica-Based Adjuvants for Tumor Immunotherapy. *Acta Biomater.* **2013**, *9*, 7480–7489. [CrossRef]
194. Li, X.; Wang, X.; Sogo, Y.; Ohno, T.; Onuma, K.; Ito, A. Mesoporous Silica-Calcium Phosphate-Tuberculin Purified Protein Derivative Composites as an Effective Adjuvant for Cancer Immunotherapy. *Adv. Healthc. Mater.* **2013**, *2*, 863–871. [CrossRef] [PubMed]
195. Wang, X.; Li, X.; Ito, A.; Watanabe, Y.; Sogo, Y.; Tsuji, N.M.; Ohno, T. Stimulation of In Vivo Antitumor Immunity with Hollow Mesoporous Silica Nanospheres. *Angew. Chem. Int. Ed.* **2016**, *55*, 1899–1903. [CrossRef] [PubMed]
196. Kim, M.-G.; Park, J.Y.; Shon, Y.; Kim, G.; Shim, G.; Oh, Y.-K. Nanotechnology and Vaccine Development. *Asian J. Pharm. Sci.* **2014**, *9*, 227–235. [CrossRef]
197. Pati, R.; Shevtsov, M.; Sonawane, A. Nanoparticle Vaccines Against Infectious Diseases. *Front. Immunol.* **2018**, *9*, 2224. [CrossRef] [PubMed]
198. Lei, K.; Tang, L. T Cell Force-Responsive Delivery of Anticancer Drugs Using Mesoporous Silica Microparticles. *Mater. Horiz.* **2020**, *7*, 3196–3200. [CrossRef]
199. Saewan, N. Effect of Coffee Berry Extract on Anti-Aging for Skin and Hair—In Vitro Approach. *Cosmetics* **2022**, *9*, 66. [CrossRef]
200. Villegas, M.R.; Baeza, A.; Vallet-Regí, M. Nanotechnological Strategies for Protein Delivery. *Molecules* **2018**, *23*, 1008. [CrossRef]
201. Villegas, M.R.; Baeza, A.; Usategui, A.; Ortiz-Romero, P.L.; Pablos, J.L.; Vallet-Regí, M. Collagenase Nanocapsules: An Approach to Fibrosis Treatment. *Acta Biomater.* **2018**, *74*, 430–438. [CrossRef]
202. Zhao, M.; Liu, Y.; Hsieh, R.S.; Wang, N.; Tai, W.; Joo, K.-I.; Wang, P.; Gu, Z.; Tang, Y. Clickable Protein Nanocapsules for Targeted Delivery of Recombinant P53 Protein. *J. Am. Chem. Soc.* **2014**, *136*, 15319–15325. [CrossRef] [PubMed]
203. Li, J.; Zhao, J.; Tan, T.; Liu, M.; Zeng, Z.; Zeng, Y.; Zhang, L.; Fu, C.; Chen, D.; Xie, T. Nanoparticle Drug Delivery System for Glioma and Its Efficacy Improvement Strategies: A Comprehensive Review. *Int. J. Nanomed.* **2020**, *15*, 2563–2582. [CrossRef] [PubMed]

pharmaceutics

MDPI

Review

Biopolymer-Based Nanosystems for siRNA Drug Delivery to Solid Tumors including Breast Cancer

Md Abdus Subhan [1,*] and Vladimir P. Torchilin [2,3,*]

1 Department of Chemistry, ShahJalal University of Science and Technology, Sylhet 3114, Bangladesh
2 CPBN, Department of Pharmaceutical Sciences, North Eastern University, Boston, MA 02115, USA
3 Department of Chemical Engineering, North Eastern University, Boston, MA 02115, USA
* Correspondence: subhan-che@sust.edu (M.A.S.); v.torchilin@northeastern.edu (V.P.T.)

Abstract: Nanobiopolymers such as chitosan, gelatin, hyaluronic acid, polyglutamic acid, lipids, peptides, exosomes, etc., delivery systems have prospects to help overwhelmed physiological difficulties allied with the delivery of siRNA drugs to solid tumors, including breast cancer cells. Nanobiopolymers have favorable stimuli-responsive properties and therefore can be utilized to improve siRNA delivery platforms to undruggable MDR metastatic cancer cells. These biopolymeric siRNA drugs can shield drugs from pH degradation, extracellular trafficking, and nontargeted binding sites and are consequently suitable for drug internalization in a controlled-release fashion. In this review, the utilization of numerous biopolymeric compounds such as siRNA drug delivery systems for MDR solid tumors, including breast cancers, will be discussed.

Keywords: biopolymer; drug delivery; solid tumors; breast cancer; siRNA therapy

1. Introduction

Biopolymers are promising to apply for drug delivery as well as therapeutic agents for cancer therapy. Biopolymeric materials demonstrate enhanced biocompatibility as well as siRNA delivery competence, targeting varieties of cancer genes in solid tumors in vitro and in vivo. Both natural and synthetic biopolymers are effective vehicles for siRNA drug delivery to cancer cells. Biopolymers, such as alginate, pullulan, cellulose, polylactic acid, chitosan, and exosomes have been efficiently utilized in siRNA drug delivery to solid tumors. Further, siRNA therapeutics have been meticulously explored and have been translated to the clinic through lipid NPs [1,2]. However, there are several pitfalls in siRNA delivery proficiency and targeting. Cationic lipids have toxicity since they can interact with enzymes [3,4]. Cationic lipids can interact with RNA and enable encapsulation; however, they can also be toxic [5]. A high concentration of cationic lipids can destroy the cellular membrane, causing cell lysis and necrotic death [3,4]. To overcome these shortcomings, different polymer-based nanoparticle carriers such as biopolymers, exosomes, and peptide nanoassemblies have been utilized for efficient siRNA delivery approaches [6]. Exosomes are extracellular vesicles, and they are mediators of short- and long-distance intercellular communication in health and diseases. Exosomes are promising platforms for siRNA delivery due to their nonimmunogenic and nontoxic natures. Exosomes can be extracted from different biological sources, such as cells and tissues.

Many natural polysaccharide polymers are utilized in siRNA delivery [7]. The key advantage of polysaccharides is the presence of various functional groups suitable for functionalization to achieve structural variety and polymers [8]. The most frequently utilized polysaccharide for siRNA delivery is chitosan. The primary amino groups in chitosan make it a polycation that facilitates the association of siRNA and complex formation [9]. For increasing the solubility of chitosan, numerous alterations have been performed, and water-soluble chitololigosaccarides have been fabricated. These chitololigosaccarides have been utilized as siRNA delivery vehicles [10].

Collagen is another biocompatible, safe natural biopolymer, which is suitable for drug delivery. For one study, siRNA-loaded collagen formulations were fabricated and utilized as localized and sustained-released platforms of siRNA in vivo to solid tumors with promising outcomes [7,11].

Biopolymers such as HA or chitosan are utilized in drug delivery as siRNA vehicles. These biomaterials are manageable via numerous alterations. PEI, aminopropyl triethoxysilane (APTES), and chitosan are cationic polymers used for coating nanoparticles for packing siRNA through electrostatic attractions [12–14]. APTES is frequently utilized for amine functionalization of mesoporous silica nanoparticles (MSNs) [14]. Chitosan is an amine-rich natural polysaccharide, which is an excellent biopolymer for siRNA delivery [15]. Several anticancer drugs have been delivered by the chitosan–MSN system to tumors, including breast tumors [16–18]. One study reported on chitosan-functionalized MSN for gene delivery with EPR (enhanced permeability and retention) uptake in HeLa cells [17].

Chitosan acts as an siRNA delivery vehicle due to its exclusive benefits, cationic charge, and simple alteration [1,19]. The key drawback of using chitosan-based polymers as cancer drug carriers is the low transfection competence and endosomal release. Chitosan-based polymeric NPs for siRNA delivery demonstrated improved transfection efficiencies by controlling some features such as length, charge, etc. [20]. To enhance the transfection efficiency of chitosan NPs, an ionic gelatin method has been developed [21]. The incorporation of heparin into chitosan nanoparticles increased transfection efficiency. The efficiency of siRNA-induced silencing of VEGF of ARPE-19 cells was 25% more with heparin compared to chitosan nanoparticles without heparin [22].

Recently, the US-FDA has approved four siRNA-based drugs: patisiran, givosiran, lumasiran, and inclisiran. For this reason, siRNA drugs have received increased attention of researchers. Biopolymers are auspicious as siRNA drug delivery platforms as well as therapeutic agents for effective cancer therapy.

2. Sources of Biopolymers Utilized as Drug Delivery Vehicles

Biopolymers or green polymeric materials are promising to apply in drug delivery as well as therapeutic agents for cancer therapy [23,24]. Biopolymers including alginate, pullulan, cellulose, polylactic acid, chitosan, and exosomes have been efficiently utilized in siRNA drug delivery to cancer cells. These biopolymers can be achieved from different sources such as algae, fungi, plants, animals, microbes, etc. (Figure 1). However, exosomes are obtained from tissues, biological fluids, and conditioned media of cells in culture. Biopolymers have gained a lot of attention in the advancement of innovative anticancer drug delivery platforms. Biopolymer-based NPs can be developed in many ways. These biopolymeric NPs have been established as suitable platforms for efficient drug delivery with enhanced selectivity and low toxicity.

Biopolymers are usually composed of non-toxic, safe monomers, and they are carbon neutral for the environment. Biomass morphology is generally structured as nanomaterials. As a result, biopolymers enable the synthesis of biocompatible nanomaterials [23,24]. Chitosan, starch, dextran, and cellulose are polysaccharides derived from plants and microbial biofilms; these biopolymers are frequently utilized for the synthesis of nanomaterials. Biopolymers are unique in composition and possess various important physicochemical and physiological properties suitable for drug delivery. Further, biopolymeric nanomaterials can be formed by attaching metals, metal oxides, and metal sulfides to biopolymers [25,26]. Biopolymers can form molecular capsules by intramolecular hydrogen bonding. For example, starch can form polymeric nanocomposites by incorporating metals or metal oxides. Further, chitosan and derivatives are frequently used for producing biocompatible nanoparticles for drug delivery. Nanomaterials can be incorporated in biopolymers by both sorption and impregnation [23,24,27].

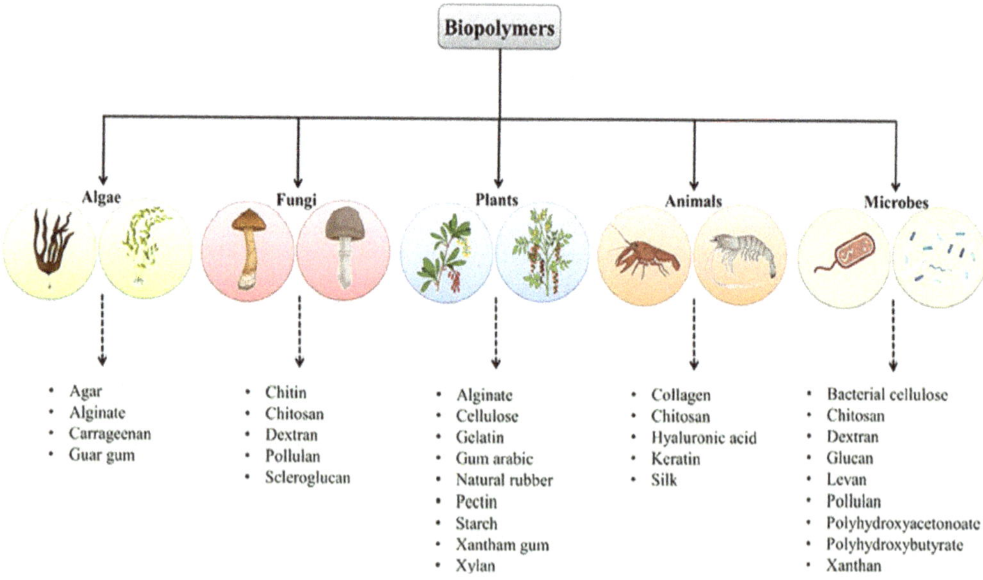

Figure 1. Biopolymers could be achieved for different sources. Adapted with permission from Ref. [23]. Copyright © 2023 Pathak, Singh, Singh, Sharma, Singh, Gupta, Mishra, Mishra and Tripathi.

3. Biopolymer-Based Nanocarrier for siRNA Drug Delivery to Solid Tumors

The importance of effective cancer therapy is emerging increasingly [28]. Due to the feasibility of polymer functionalization, enhanced targeting effects have been observed with polymeric NPs decorated with different peptides or ligands. Chitosan-based NPs were conjugated with HAD (hyaluronic acid dialdehyde) to fabricate an siRNA nanoplatform, siRNA@chitosan-HAD, for the delivery of siRNA targeting CD44 [29]. A natural biopolymer, curdlan, was modified, and represented as CuVa polymers it demonstrated enhanced biocompatibility as well as siRNA delivery competence targeting the *STAT3* gene in solid tumors in vitro [30]. Exosomes result from extracellular vesicles formed in the cells [31]. They are prospective candidates for siRNA delivery, as they are easy to handle, nonimmunogenic, and nontoxic [32,33]. Bovine milk exosomes are used as vehicles for siRNA delivery to tumors. siRNA delivery via milk exosomes silenced major oncogenes with 2- to 10-fold knockdown in several solid cancers in vitro and in vivo [33]. Exosomes with a targeted approach (e.g., CD47 targeting) have shown promising outcomes in solid tumors of deadly cancers, including GBM (glioblastoma), PDAC (pancreatic ductal adenocarcinoma), and TNBC (triple-negative breast cancer) [33–35]. The angiopep-2 (An2)-functionalized STST3siRNA-loaded exosome (Exo-An2-siRNA) is a potential drug to enhance the treatment of solid tumors. Exo-An2-siRNA demonstrated excellent blood stability, efficient cellular uptake, and effective BBB penetration ability in GBM patients. Exo-An2-siRNA exhibited enhanced in vitro anti-cancer effects, as exosomes protect siRNA, and modified An2 targets tumors effectively. Exo-An2-siRNA showed inhibition of proliferation in a xenograft model of GBM with no side effects and enhancing median survival time. The developed nanoplatform Exo-An2-siRNA is a promising, safe therapy approach for solid tumors targeting the *STAT3* gene. This approach may be effective in breast cancer therapy [35].

A method was developed to utilize serum-derived exosomes as drug delivery carriers for siRNA drugs in vivo. The study used exosomes as vehicles for siRNA delivery to lung macrophages in vivo; exosomes containing siRNA exerted functional roles in the lung in vivo. The serum-derived exosomes are non-toxic and do not activate immune responses and were found to be safe for in vivo use [36].

Further, bovine milk exosomes were utilized as an innovative siRNA delivery platform targeting specific genes including VEGF, EGFR, AKT, MAPK, BCl2, SUR, and KRAS [33]. A $5'^{-32}$ P-labeled siKRAS was used as a tracer for evaluating exosome loading with siRNAs. The study exhibited the knockdown of target gene expression ranging from 2- to 10-fold in different cancers by exosome-loaded siRNAs. Since KRAS has been correlated with different solid cancers, siKRAS delivery to different solid tumors may be a promising siRNA therapeutic approach for solid tumor therapy. In this study, siKRASG12S was used to target KRAS in lung cancer cells. Extracellular vehicles, particularly exosomes derived from milk used as carriers for siRNA-loading and delivery signify a feasible natural nanocarrier for siRNA delivery to cancer cells utilizing systemic or oral delivery platforms [33,37,38].

The cellular membrane (~5 nm) can be passed by cell-penetrating peptides (CPPs), 5–30 residue peptides [39]. Peptides are important siRNA drug carriers. In particular, CPPs are short cationic peptides that facilitate intracellular delivery of several biological cargos, including siRNA and mRNA [40].

Amphiphilic phospholipid peptide dendrimers (AmPPDs) were fabricated as an efficient siRNA delivery platform. A DSPE-KK_2-AmPPDsiRNA delivery system mediated the effective intracellular uptake and endosomal release of siRNA complexes and resulted in gene knockdown, demonstrating anticancer efficiency both in vitro and in vivo in solid tumors [41]. A peptide-based p5RHH NPs-condensing oligonucleotide was developed for siRNA delivery, which reduced the KRAS gene in solid tumors in vivo, demonstrating cancer inhibition effectively [42].

Introduction of protein and mAbs into a polymeric vehicle can advance the siRNA delivery to tumors [43–45]. A nanosystem composed of a cyclodextrin-containing polycation, an adamantane-conjugated PEG (AD-PEG), human transferrin-decorated AD-PEG, and therapeutic siRNA was fabricated. The hTf receptor is overexpressed on tumor cells. The developed nanosystem successfully delivered siRNA to tumor cells through hTf receptor-mediated endocytosis, and it downregulated the target genes in solid tumors. The nanocaplet transferred siRNA into the cytoplasm and caused gene silencing through RNAi. The targeted NPs also showed deep tissue penetration due to transferrin-stimulated transcytosis [46]. A 3D spheroid study with Hep3B cancer cells exhibited a transferrin conjugated nanosystem delivered siRNA effectively into the spheroid with a depth of 70 μm, indicating Tf-mediated deep tissue permeation, gene knockdown, and cancer cell inhibition [43,46]. In this study, the deep tissue penetration of the nanocaplet was evaluated by confocal laser scanning microscopy (CLSM, $\lambda_{ext} = 488$) using $siRNA^{A488}$ fluorescently labeled with Alexa Flour 488 in the nanodrugs. Since the brain endothelial cells overexpress Tf-receptors, the nanocaplet may have the potential to overcome the blood–brain barrier. Thus, an in vivo study utilizing the nanocaplet may be suitable for delivering siRNA to brain tissues, such as GBM.

Poly-L-arginine and siRNA were consecutively adsorbed on an anionic liposome NP, and the outermost layer was decorated with propargyl-modified NPs with 1,2-distearoyl-sn-glycero-3-phosphoethanolamine (DSPE), DSPE-PEG, and cholesterol to fabricate long blood-circulating materials [43,47]. Such coformulation of cationic polymers with lipid improved the polyplex stability and enhanced siRNA loading efficacy and steadiness [48–50].

siRNA–lipid complexes are known as lipoplexes. Cationic liposomes combined with lipids have shown excellent outcomes in siRNA drug delivery for cancer therapy [21]. However, immune responses and undesired toxicities associated with lipoplexes should be addressed. Various approaches have been developed to diminish lipoplex toxicity. Although their entrapment efficiency might be low, neutral liposomes have reduced such toxicity. Further, cationic lipoplexes have been coated with anionic polymers, including sodium salts of hyaluronic acid, and alginic acid carboxymethyl cellulose. The coated cationic liposomes showed enhanced siRNA delivery to solid tumors (Figure 2) [21].

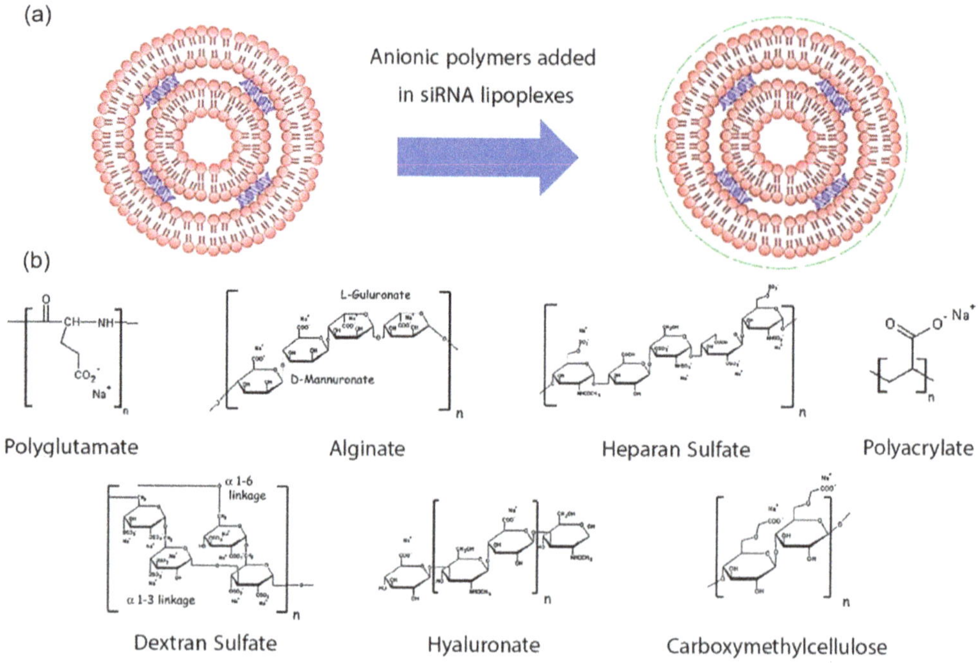

Figure 2. (**a**) Anionic polymers included in siRNA lipoplexes. (**b**) Different anionic polymers. Adapted with permission from Ref. [21]. Copyright © 2023–2022 John Wiley & Sons, Inc.

The benefit of polymeric nanocarriers is that they are not stimulators of the immune response. Electrostatic interactions between polymers and siRNA are readily achievable through linkages (thiol, maleimide, disulfide, etc.). The natural cationic polymers have led to efficient delivery approaches due to low toxicity, biodegradability, and biocompatibility [21,51–53].

The siRNA rigid structure incites weak interactions with polycation, and polyplexes are less effective in protecting siRNA against nucleases. Increasing the polycation amount could improve this pitfall while increasing toxicity, and hence an optimum polycation quantity should be utilized. Further, dendrimers have flexibility and prospects in siRNA loading [21,54].

Nanoparticles PEG-*b*-PLA or PEG-*b*-PLGA have received increased curiosity for siRNA drug delivery but are restricted by the reduced encapsulation efficacy of siRNA. To overcome this pitfall, a cationic lipid-induced NP platform known as CLAN with more siRNA encapsulation competence was fabricated [55].

Cell membrane coating has been established as an efficient approach to enhance the in vivo execution of NPs. Cell membrane decoration on NPs improves their colloidal stability and blood circulation and also benefits targeted drug delivery. Cancer cell membrane-decorated biomimetic NPs could deliver siRNA to homologous tumors in vivo [56,57]. Cell membrane-coated polyplexes are associated with problems in releasing siRNA into cytoplasm. To resolve this problem, charge-reversal polymers are utilized to facilitate siRNA release in cytoplasm [58,59]. PEI–siRNA polyplexes were coated with citraconic anhydride-grafted poly-L-lysine, which was then decorated with angiopep-2 peptide-coated red blood cell membrane (Ang-RBCm) [59]. The RBC membrane decoration offered long-term blood circulation, and targeting of angiopep-2 enhanced the accumulation of drugs into solid tumors. Such cell membrane-decorated charge reversal hybrid nanocomplexes proficiently inhibited tumor growth by knockdown of target genes via siRNA delivery.

Virus/polymer chimeric NPs permit simultaneous expression and silencing of numerous genes [60,61]. NPs comprising a Bcl-2-interacting mediator of cell death (BIM)-encoding adeno-associated virus (AAV) core and an acid-degradable polyketal (PK) shell were developed for enhanced siRNA delivery to tumors. After endocytosis of nanoparticles into the cell, the release of siRNA was facilitated by PK shell degradation. The BIM-encoding AAV was then transported into the cell nucleus for BIM expression. BIM enhanced the apoptosis of cancer cells. Thus, the polymer/virus chimeric NPs upregulated the BIM expression and simultaneously inhibited the cancer cells [43,60,61].

PLL produced by *Streptomyces albus* was utilized as a scaffold for the production of polyphenol-grafted bio-inspired polymers for siRNA delivery. Phenol, catechol, and pyrogallol were conjugated onto PLL through a simple reaction, and the cellular uptake of polyphenol-grafted polymer-siRNA complex by HeLa cells was examined [62]. A series of polyphenol-grafted low molecular weight polymer–siRNA complexes was effective in siRNA delivery inspired by natural polyphenol. The phenolic moieties enhanced the siRNA binding ability and stability of the polyplex. Among the polyphenols, polycatechols demonstrated the highest performance in siRNA binding, cellular uptake, and gene silencing [62]. The polycatechol–siRNA platform could be a highly promising bioinspired polymer for siRNA delivery in solid tumors.

Green tea-based EGCG enables siRNA condensation by low molecular-weight polymers, yielding green NPs (GNPs), showing effective siRNA-induced gene silencing [63]. Biocompatible components of GNPs such as EGCG, siRNA, and PLL demonstrated reduced toxicity on the transfected cells. Similarly, antisense oligodeoxynucleotides, DNAzymes, and peptide nucleic acids may be utilized to design bioinspired polymer-based platforms for the fabrication of GNPs for a robust system that is promising for gene/siRNA delivery (Figure 3) [63].

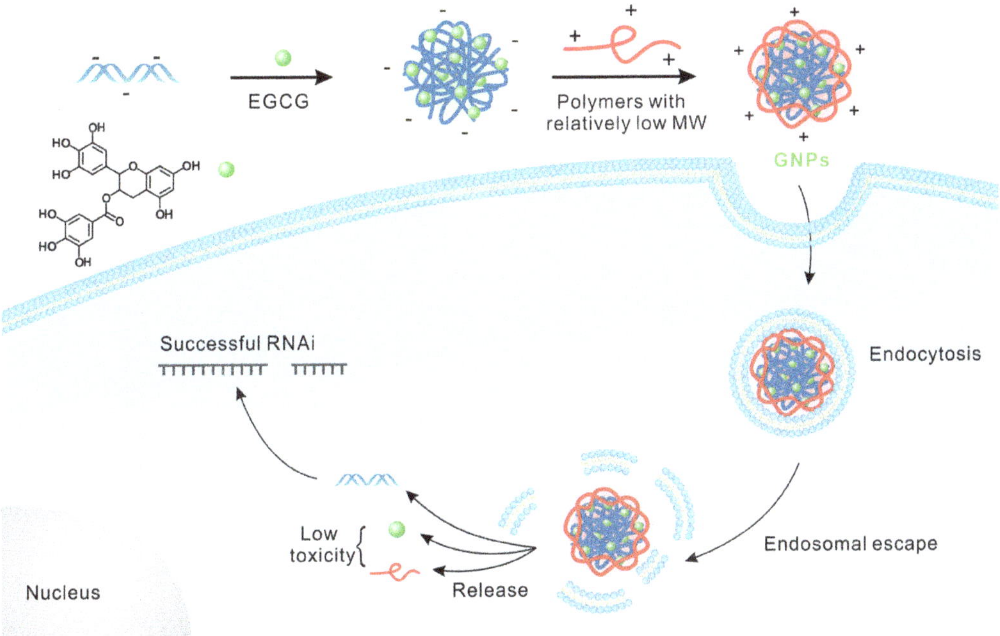

Figure 3. GNPs prepared by the supramolecular approach exhibited a gene silencing mechanism with high efficiency and low toxicity. Adapted with permission from Ref. [63]. Copyright © 2023 American Chemical Society.

Further, utilizing siRNA drugs for the treatment of hematological malignancies has been ineffective because blood cancer cells demonstrate significant resistance to standard transfection methods [64]. However, there are some studies concerning siRNA delivery to blood cancers. For example, CD20/CD44 dual targeted siRNA delivery using layer-by-layer NPs (LbL NPs, PLGA/PLA/siRNA/PLA; PLGA, poly-(D,L-lactide-co-glycolide); PLA, poly-l-arginine) with BCL-2 siRNA in B cell lymphoma and leukemia cells downregulated *BCL-2*, induced apoptosis, and reduced the blood cancer cells. Herein, the outermost layer of LbL NPs consists of hyaluronic acid (a CD44-ligand) covalently conjugated to CD20 antibodies. The CD20/CD44 dual-targeting outer layer offered precise binding to blood cancer cells, followed by receptor-mediated endocytosis of the LbL NPs [64].

In recent years, there have been many pre-clinical studies using biopolymer-induced siRNA therapeutics for the enhanced therapy of cancer. To date, several siRNA-mediated drugs have been in different phases of clinical trials or have received FDA approval for the treatment of a variety of diseases, including solid tumors, as demonstrated in Table 1 [65,66]. siRNA drugs including patisiran, givosiran, lumasiran, and inclisiran have received approval from the FDA [67]. As a result, siRNA drugs have received the enhanced attention of medical doctors and scientists. The design of apposite siRNA-assisted delivery systems is a vital issue for clinical applications in cancer therapy. These approaches could be potentially supportive in reducing cancer deaths in the near future.

Table 1. siRNA therapeutics in different stages of clinical trials.

siRNA Drugs	Delivery System	Target Gene	Disease	Company	Status/ Phase
Patisiran	LNP-siRNA	TTR	TTR-mediated amyloidosis	Alnylam	Approved, FDA 2018
Givosiran	GalNAc-siRNA	ALAS1	Acute hepatic porphyria	Alnylam	Approved, FDA 2019
Lumasiran	GalNAc-siRNA	HAO1	Primary heperoxaluria type 1	Alnylam	Approved, FDA 2020
Inclisiran	GalNAc-siRNA	PCSK9	Hypercholesterolemia	Alnylam Novartis	Approved, FDA 2021
Vutrisiran (ALN-TTRSCO2)	GalNAc-siRNA	TTR	TTR-mediated amyloidosis	Alnylam	III
Nedosiran (DCR-PHXC)	GalNAc-siRNA	LDHA	Primary heperoxaluria	Alnylam	III
Fitusiran (ALN-AT3SC)	GalNAc-siRNA	AT	Hemophilia A and B and rare blood disorder	Alnylam Sanofi Genzyme	III
Atu027	Liposomal-siRNA, AtuPLEX	PKN3	Solid tumors, advanced or metastatic pancreatic cancer	Silence Therapeutics	II completed
TMK-PLK1	LNP-siRNA	PLK1	Solid tumors, adrenal cortical carcinoma, hepatocellular carcinoma	Arbutus Biopharma	II completed
siG12D LODER	siG12D-LODER™ system and chemotherapy	G12D mutated KRAS	Pancreatic cancer, pancreatic ductal adenocarcinoma	Silenseed Ltd.	II ongoing
ALN-VSP02	LNP-siRNA	VEGF, KSP	Solid tumors	Alnylam	I completed

4. Breast Cancer Therapy with Biopolymer-Based siRNA Nanoplatform

Breast cancer is the second leading cause of death in woman after lung cancer worldwide. However, cancer death rates including breast cancer have been decreasing due to the development of advanced screening technologies and significant improvement of immunotherapy, targeted therapy, and gene therapy [68]. Natural herbal compounds or biopolymers have significant roles in inhibiting drug resistance in breast cancer, impeding breast cancer stem cells (BCSCs). BCSCs utilize different cellular mechanisms, including membrane transporters, anti-apoptotic, pro-survival, and self-renewal-linked signaling pathways. Biopolymers may disrupt these mechanisms and hence impede breast tumors, inhibiting the BCSCs, and the induction of apoptosis [69]. Therefore, this review highlighted developments in biopolymer-mediated siRNA drug delivery systems for breast tumors. Biopolymer-induced delivery of chemotherapeutics and siRNA is promising and has demonstrated synergistic effects as a single delivery therapeutic [12]. Methotrexate (MTX) is a potential drug for breast cancer treatment, effectively inhibiting the dihydrofolate reductase (DHFR) receptor, which suppresses DNA synthesis. MTX can significantly induce apoptosis by inhibiting the STAT3 in the JAK/STAT pathway in tumor cells. Inhibition of the STAT3 protein by siRNA is an effective approach to hinder breast tumors [70]. A nanoparticle delivery system for the co-delivery of MTX and STAT3siRNA was developed and investigated against MCF-7 breast cancer cells in vitro. Mesoporous silica nanoparticles (MSNs) were functionalized with chitosan by covalent grafting facilitated by APTES using glutaraldehyde as a linker. APTES and grafted chitosan offered an amine rich surface coating on MSN NPs, permitting electrostatic loading of both MTX and siRNA on MSN [14]. The co-delivery of MTX and STAT3siRNA to breast cancer cells in vitro efficiently inhibited the DHFR receptor and inhibited breast tumors. Further, the co-delivery of doxorubicin (Dox) and p53 by deionized chitosan derivative functionalized MSNs tempted noteworthy apoptosis in breast cancer cells due to the synergistic effects of Dox and the p53 gene [18].

PEG-modified chitosan was synthesized for survivin-targeted siRNA delivery to breast cancer cells to inhibit breast tumors and to prevent metastasis [71]. The outcome of the study demonstrated that siRNA loaded in PEGylated chitosan (PEG-chitosan-siRNA) delivered to breast cancer cells showed enhanced cellular uptake and inhibited 4T1 tumor cell growth significantly. The flow cytometry study and confocal scanning microscopy study demonstrated that PEG-chitosan-siRNA particles were uptaken more easily than naked siRNA. The developed PEG-chitosan-siRNA nanoparticle significantly reduced tumor growth in vivo in a xenograft mouse model. The study exhibited that the PEG-chitosan-siRNA nanoplatform is safe, effective in tumor inhibition, and efficient in preventing metastasis [71].

Gold NPs modified with trimethyl chitosan (TMC) were developed as efficient delivery carriers to enhance the stability and cellular uptake of siRNA targeting EGFR in breast tumors. EGFRsiRNA was complexed with AuNPs-TMC through electrostatic interactions to develop the nanosystem AuNPs-TMC-EGFRsiRNA using a facile synthesis process. The synthesized AuNPs-TMC-EGFRsiRNA NPs effectively delivered siRNA to breast cancer cells. Therefore, AuNPs-TMC-EGFRsiRNA appeared as a prospective and potential nanodelivery system for breast cancer therapy [72].

A breast cancer-specific gene 1-siRNA (BCSG1-siRNA) plasmid was fabricated and then encapsulated by chitosan-silicon dioxide NPs. The slow release of BCSG1-siRNA to MCF-7 cells demonstrated enhanced cytotoxicity, reduced proliferation, and increased apoptosis. BCSG1-siRNA downregulated BCSG1 and inhibited tumor growth significantly. As a result, chitosan-silicon dioxide NP-BCSG1-siRNA is an effective siRNA delivery system to inhibit breast tumors [73].

A pH-responsive nanoplatform was fabricated introducing antiPOL siRNA and guanidine-CO_2-functionalized chitosan at a neutral pH to knock down the POLR2 in TNBC tumors. The siRNA nanodrug was effectively internalized into TNBC cells harboring hemizygous TP53 deletion, overcoming intracellular trafficking at low pH. The downregulation of POLR2 caused TNBC cell death with little or no toxicity [74].

A triblock copolymer, a hollow-sphere polymersome PVPON-PDMS-PVPON was designed for siRNA delivery to breast cancer cell lines [75]. The polymersomes assembled in the presence of 1 μM PARPsiRNA-loaded siRNA inside the nanocarrier. When the developed nanovehicles were disrupted by ultrasound in vitro, the siRNA was released unaltered. siRNA-loaded polymersomes were tested against HER2+, trastuzumumab-resistant breast cancer cells in vitro, which reduced the PPARP1 protein level in the breast cancer cells and impeded MDR genes. Herein, a fluorescent dye was attached covalently to the outside of the nanocapsules. Thus, the targeting molecule was added to the nanocapsules to deliver with the polymersomes into the tumors. This nanosystem with PARP1siRNA was treated with TNBC cells in a tumor mouse model, and the delivered siRNA significantly knocked down the expression of PARP1 and increased by four-fold the survival of mice bearing TNBC tumors. PARPis are suitable for targeting TNBC tumors with defects in DNA repair and could alter the TME (tumor microenvironment) [76–83]. However, it has been challenging to combine many PAPRis with chemotherapy due to bone marrow suppression. Targeting PARP1i in the tumor may permit novel combination therapy approach. The polymersome-loaded PARP1siRNA nanosystem is a biodegradable, non-ionic polymeric nanovesicles proficient to encapsulate and deliver PARP1siRNA for knocking down PARP1 in vivo. This study provided an advanced nanopodium for the development of an effective drug delivery nanocarrier system for siRNA therapy of breast cancer, including TNBC [75].

A stable polyplex was developed using PEG-modified L-arginine oligo(-alkylaminosiloxane) grafted with PEI for siRNA delivery to breast cancer cells [84]. The polyplex was used to deliver the anti-luciferase siRNA to the luciferase-expressed TNBC cell line MDA-MB-231-Luc-D3H2LN, and anti-ABCB1 siRNA to doxorubicin-resistant MCF-7 ADR. Transfection of MDA-MB-231-Luc-D3H2LN cells with anti-luciferase siRNA demonstrated comprehensive knockdown of luciferase expression. When MCF-7 ADR cells were transfected with anti-ABCB1 siRNA, efficient downregulation of ABCB1 was observed. Therefore, this approach could be effective against breast cancer and TNBC therapy [84].

Lipid-based carriers are efficient in siRNA delivery [85]. Piperazine is frequently utilized in bioactive compounds and medicine [86,87]. A recent study reported that piperazine-containing ionizable lipid (Pi-Lipids) favorably deliver genes to immune cells in vivo without targeting the ligand [86]. Herein, the author demonstrated that Pi-A10 LNP (Pi-LNPs that contain PPZ-A10) effectively delivered mRNA/siRNA to solid tumors. LNP-10A was also formulated with siGFP and siLuc, and LNP10AsiGFP showed 25% knockdown of GFP, but no knockdown of luciferase was observed using LNP10AsiLuc. This study showed that high throughput in vivo investigations can recognize NPs with natural nonhepatocyte tropism, and the lipid with bioactive small molecule motifs can deliver mRNA/siRNA to immune cells in vivo. Further, LNP-10A was formulated with sipGP. This strategy could be applied to deliver siRNA to breast cancer cells [88].

CD44 expression is linked to the JAK2/STAT3 pathway in TNBC hostility. Incorporation of PEG–lipid complexes inside the lipid bilayers increased the sustained release of siRNA and evasion of the mononuclear phagocytic system. Cationic liposomes are widely used for siRNA delivery platforms for enhanced delivery, without enzymatic degradation, and effective cytoplasmic release [89].

Lipid-based polymeric nanocarriers are utilized for siRNA delivery to chemoresistant breast cancer. A nanoplatform formulated by hexadecylated PEI-siRNA encapsulated with lipid segments (DSPE-PEG, DSPE, cholesterol, tripalmitin) and fabricated with apolipoprotein was used for megalin-targeted siRNA delivery to chemo-treated breast cancer cells, which demonstrated enhanced delivery and a reduction of disease recurrence [89].

Calcium apatite NPs-siRNA downregulated BCL-2, EGFR, and ERBB-2 genes, resulting in the cell death-inhibiting PI3K/MAPK signaling pathway in a mouse model of breast tumors [90]. A calcium phosphate-PEG-polyanion polymer promoted the siRNA delivery targeting anti-apoptotic gene BCL-xL and BCL-2 in breast cancer cell line MCF-7. After 48 h of incubation the BCL-xL and BCL-2 genes were reduced by 49% and 23%, respectively [91].

Further, MCF-7 and MDA-MB-231 cell lines were studied to knock down SUR and BCL2 genes utilizing siSUR- and siBCL2-loaded bovine milk exosomes. siSUR was highly effective in inhibiting the breast cancer cells with ~80% knockdown of the SUR gene [33]. Bovine milk exosomes are natural, stable, safe, feasible, and promising nanocarriers for siRNA delivery to breast cancer cells.

NPs consisting of coupled PEI and folic acid to core/shell $Fe_3O_4@SiO_2$ were fabricated to deliver Notch-shRNA to TNBC cell lines, MDA-MB-231 [92]. The study demonstrated that superparamagnetic $Fe_3O_4@SiO_2$/PEI-FA/Notch-1shRNA NPs induced delivery of Notch-1shRNA to TNBC cells effectively with no toxicity. The nanodelivery cargo was capable of delivering Notch-1shRNA efficiently while protecting shRNA from degradation by exogenous DNaseI and nucleases. The preferential uptake of superparamagnetic $Fe_3O_4@SiO_2$/PEI-FA/Notch-1shRNA NPs by TNBC cells MDA-MB-231 was revealed by MRI and fluorescence microscopy. Significantly decreased expression of Notch-1 was exhibited by MDA-MB-231 cells. Thus, the $Fe_3O_4@SiO_2$/PEI-FA/Notch-1shRNA nanosystem is an effective theranostic system exploring both Notch-1 inhibition in TNBC cells and superparamagnetic property-enabled MRI and fluorescence imaging simultaneously [92].

A PEI nanocomplex was fabricated with EpCAM aptamer (EpApt) and EpCAM siRNA (siEP) to study its effects against breast cancer. The fabricated PEI-EpApt-siEp knocked down EpCAM and subdued the proliferation of breast cancer cells MCF-7. The study demonstrated that PEI-EpApt-siEp has higher binding ability to MCF-7 cells than EpApt or scrambled aptamer or PEI-ScrApt-SiEp [65,93].

A nanosystem was fabricated with lipid-NPs-siRNA coated with a Fab antibody to utilize against heparin-binding EGF-like growth factor (αHB-EGF LNP siRNA). A developed αHB-EGF LNP siRNA nanoplatform targeting PLK1 was examined for TNBC cell line MDA-MB-231 overexpressing HB-EGF on the cell surface. The result indicated the suppression of PLK1 protein, with the inhibition of the TNBC tumor revealing αHB-EGF LNP siRNA as a prospective target for TNBC tumors expressing HB-EGF [94].

A potential nanoplatform using a liposome–polycation–DNA complex (LPD) conjugated with tumor specific antibody (TLPD) was developed for siRNA delivery targeting EGFR in TNBC cells. TLPD was prepared by a self-assembled process followed by anti-EGFR Fab antibody conjugation (TLPD-FCC). The TLPD-FCC-siRNA demonstrated enhanced binding affinity and luciferase gene silencing (20%) in EGFR, overexpressing the TNBC cell line MDA-MB-231 in vitro. This study showed that TLPD-FCC has great prospects in delivering siRNA to EGFR overexpressing breast cancer cells [95].

A TNBC tumor targeting RGD-PEG-ECO/siDANCR nanosystem was developed for studying breast cancer cells. TNBC cell lines MDA-MB-231 and BT549 cells treated with the RGD-PEG-ECO/siDANCR NPs showed ~90% silencing of DANCR expression for up to 7 days, demonstrating an effective siRNA delivery approach for targeting and an enhanced therapeutic effect [96].

ICAM-1 targeted Lcn2 siRNA encapsulated liposome binds TNBC cells MDA-MB-231 significantly, and effective Lcn2 downregulation by ICAM-Lcn2-LPs led to a substantial lessening in VEGF from MDA-MB-231 cells, resulting in diminished angiogenesis, both in vitro and in vivo. The study demonstrated a promising approach for inhibiting angiogenesis and for tumor growth retardation in TNBC tumors [97].

Cationic bovine serum albumin-conjugated siS100A4 was coated with exosome membrane to develop self-assembled nanosystem CBSA/siS100A4@Exosome for enhanced drug delivery to cancer cells (Figure 4). The in vivo study demonstrated that CBSA/siS100A4@Exosome has a higher affinity towards cancer cells compared to CBSA/siS100A4@liposomes, and it exhibited significant gene silencing as well as inhibiting the growth of malignant breast cancer cells. The CBSA/siS100A4@Exosome nanoplatform exhibits a promising strategy to impede postoperative metastasis in breast cancer [98,99].

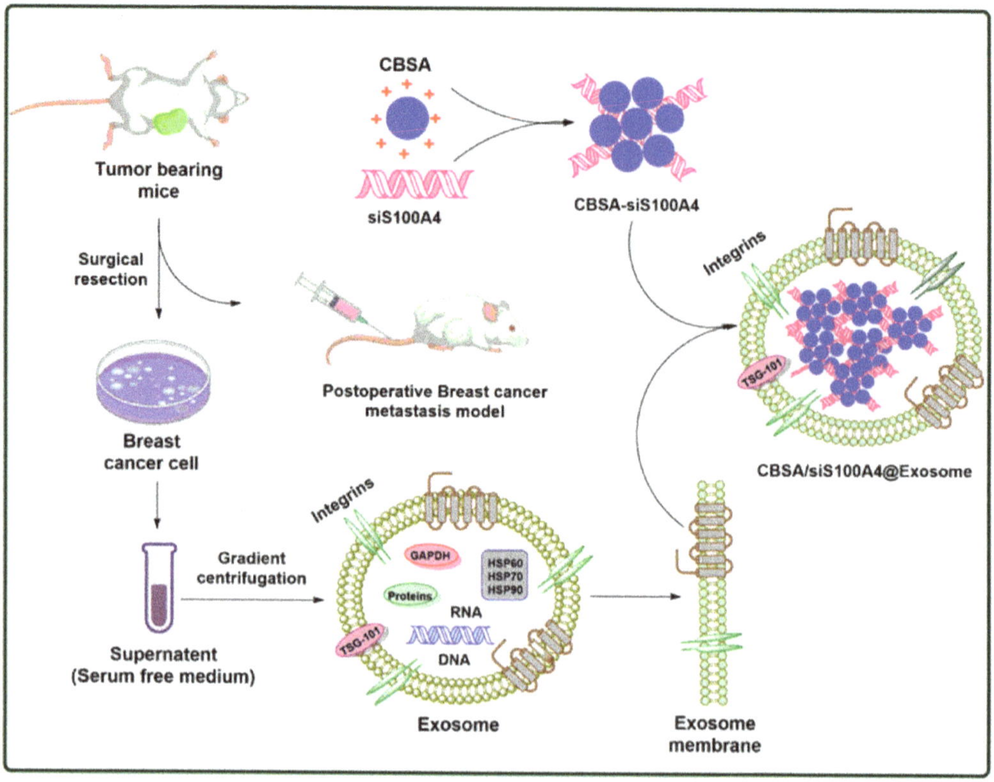

Figure 4. Post-operation metastatic inhibition in breast tumors through CBSA/siS100A4@Exosome. Adapted with permission from Ref. [99]. Copyright @2022, Goyal, Chopra, singh, Dua and Gautam.

Exosomes were collected from fibroblastic BJ cells (BJExo) to treat different cancer cells, including breast tumors [100–104]. *PIK3CA* was silenced by siPIK3CA to improve the breast tumor therapy outcome. *PIK3CA* silencing by siPIK3CA delivery reduced cell migration capacity, impeding the metastasis process. This study confirmed the PI3K/AKT/mTOR pathway and *PIK3CA* as an effective target for breast cancer therapy and introduced BJ-derived exosomes as a promising nanocarrier for siRNA delivery. The study utilized siPIK3CA to target MDA-MB-231 and MDA-MB-453 cell lines and led to the reduction in mRNA expression of *PIK3CA, AKT*, and *mTOR* [104]. The study outcome was supported by previous results [101–103].

Exosome-loaded siRNA therapeutics have been effective in downregulating BCL-2, PLK1, KRAS, and survivin anti-apoptotic proteins that induce tumor cell mitosis and cell growth factor. The exosome-mediated siRNA therapy was effective in inhibiting cell proliferation and migration. BCL-2 siRNA-loaded exosome derived from NK cells was utilized to treat ER+ breast cancer cells and led to increased apoptosis in breast cancer cells [105].

RNA-NP-modified exosome was fabricated to target three cancer cells, including breast cancer cells, simultaneously. In the structure of the RNA-NP, the pRNA of phage phi29 was extended into an arrow shape to connect with the RNA ligands, and fluorescent alexa647 was added for imaging [106]. The RNA ligand was utilized for specific targeting and exosome for effective membrane fusion. The resulting ligand-displaying exosome was capable of specific delivery of siRNA to cells and potentially inhibiting tumor growth. The exosome harboring EGFR aptamer-loaded siSurvivin NPs (EGFRapt/Exosome/siSurvivin) inhibited breast tumor in MDA-MB-468 cells orthotopic xenograft breast tumor bearing mice [106,107].

Ligands can be expressed, fusing with exosomal surface proteins by engineered exosome-producing cells, for targeting cancer cells [108]. Modified exosome-producing engineered HEK293T cells were utilized to target HER2+ breast cancer cells. The result demonstrated that HEK293T cells formed DARPin G3 on the surface of exosomes. These modified exosomes can bind to HER2/neu and deliver siRNA to SKBR3 breast cancer cells, downregulating gene expression up to 70%. This study exhibited a potential approach to facilitate gene and drug transfer to HER2+ cancer cells [108].

MCF-7 breast cancer cell-derived exosomes loaded with siRNA were used against CD44 in breast cancer cells. Exosome-siRNA targeting CD44 expression (Exos-siCD44) effectively silenced the expression. When cocultured on Exos-siCD44, breast cancer cells demonstrated diminished cell proliferation and improved sensitivity to doxorubicin. The same effect was detected in vivo in mice [109].

Nanoscale exosome-mimics (Ems) were generated from non-tumorigenic MCF-10A cells, and then siRNA was encapsulated into the EMs by the electroporation technique. The CDK4siRNA-loaded EMs significantly downregulated CDK4 mRNA protein expression in vitro and in vivo. Therefore, EMs from similar cell types may be effective as siRNA carriers for the development of innovative breast cancer therapeutics with no toxicity [110]. There have been numerous pre-clinical studies utilizing biopolymer-based siRNA delivery systems with promising results for breast cancer therapy (Table 2). Biopolymer-induced siRNA delivery is a prospective strategy to treat breast tumors to improve survival of breast cancer patients.

Table 2. Preclinical studies of selected biopolymer-induced siRNA delivery platforms for breast cancer treatment.

Target	Nanoparticle	Cell Line	Reference
RhoA	Chitosan-coated polyisohexylcyanoacrylate (PIHCA) NPs	MDA-MB-231	[111]
Survivin	Lipid substituted polymer	MDA-MB-231	[112]
HB-EGF	Fab's antibody-modified LNP	MDA-MB-231	[94]
EGFR	Fab conjugated liposomal NPs	MDA-MB-231	[95]
EGFR	PMLA-based nanobioconjugate	MDA-MB-468	[106]
eGFP	F3-targeted liposomal NPs	MDA-MB-231 MDA-MB-435	[113]
VEGF-A, VEGFR-1, VEGFR-2, and neuropilin-1	Chitosan nanocomplexes	Breast tumor induced in rat using N-nitroso-N-methylurea	[114]
Lipocalin-2	ICAM-1 conjugated liposomes	MDA-MB-231	[97]
HR	EpCAM aptamer polyethylineimine	MCF-7	[93]
EGFR	CPP-loaded nanobubbles	MDA-MB-231	[115]
CD44	Exosome	MCF-7	[109]
PIK3CA	BJExo	MDA-MB-231, MDA-MB-453	[104]
SUR, Bcl2	Bovine milk exosomes	MCF-7, MDA-MB-231	[33]
ABCB1	PEG-L-arginine oligo(-alkylaminosiloxane)-PEI polyplex	MCF-7 ADR, MDA-MB-231	[84]

5. Conclusions and Future Perspective

Biopolymers have been proficiently applied in siRNA drug delivery to cancer cells, including breast cancer. The sources of the biopolymers are algae, fungi, plants, animals, microbes, etc. Further, exosomes are obtained from tissues, biological fluids, and cells. Biopolymers are effective for the progress of innovative anticancer drug delivery platforms. Biopolymer-based NP systems have been recognized as cost-effective environmentally friendly platforms for effective drug delivery with enhanced selectivity and reduced toxicity. Nanobiopolymers have promising stimuli-responsive properties and therefore can be used to enhance siRNA delivery platforms to undruggable MDR metastatic cancer cells, including breast cancer. The biopolymeric siRNA drug formulations can shield drugs from pH degradation, extracellular trafficking, nontargeted binding sites, and they are accordingly appropriate for drug delivery and internalization into tumors in a controlled release manner. Numerous natural polysaccharide polymers are utilized in siRNA delivery for cancer therapy. The key advantage of polysaccharides is the presence of various functional groups for modification and polymerization. The most frequently applied polysaccharide for siRNA drug delivery to cancer is chitosan, which is biodegradable, biocompatible, inexpensive, and has low toxicity. For increasing the solubility of chitosan and making it suitable for endo-cytoplasmic release, various modifications have been performed. CPP-based siRNA drugs are promising siRNA delivery vehicles, as they can easily cross the cell membrane. Exosome-loaded siRNA therapeutics have been effective in downregulating cancer genes as well as anti-apoptotic and other proteins that prompt tumor cell mitosis and cell growth factors. Exosome-mediated siRNA therapy is promising and prospective in inhibiting cell proliferation and migration in solid tumors, resulting in enhanced tumor inhibition in solid tumors, including breast tumors, improving the survival of patients. However, for producing biopolymer NPs, functionalization and modification should be performed perfectly and precisely for making them pertinent nanocarriers for siRNA-based drug delivery for overcoming systemic and extra-cellular hurdles. Further, before translation of biopolymer-based siRNA drugs into the clinic, their immune activation and toxicity profiles must be addressed effectively.

Author Contributions: M.A.S. conceived the idea of research, performed the study, prepare the manuscript, and revised it. V.P.T. contributed in manuscript revisions, and supervised the whole project. All authors have read and agreed to the published version of the manuscript.

Funding: This research received no external funding.

Institutional Review Board Statement: Not applicable.

Informed Consent Statement: Not applicable.

Data Availability Statement: Not applicable.

Acknowledgments: Authors acknowledged both SUST and Northeastern University.

Conflicts of Interest: The authors declare no conflict of interest.

References

1. Byun, M.J.; Lim, J.; Kim, S.-N.; Park, D.-H.; Kim, T.-H.; Park, W.; Park, C.G. Advances in Nanoparticles for Effective Delivery of RNA Therapeutics. *BioChip J.* **2022**, *16*, 128–145. [CrossRef] [PubMed]
2. Hou, X.; Zaks, T.; Langer, R.; Dong, Y. Lipid nanoparticles for mRNA delivery. *Nat. Rev. Mater.* **2021**, *6*, 1078–1094. [CrossRef] [PubMed]
3. Xue, H.Y.; Liu, S.; Wong, H.L. Nanotoxicity: A key obstacle to clinical translation of siRNA-based nanomedicine. *Nanomedicine* **2014**, *9*, 295–312. [CrossRef] [PubMed]
4. Ho, W.; Gao, M.; Li, F.; Li, Z.; Zhang, X.Q.; Xu, X. Next-generation vaccines: Nanoparticle-mediated DNA and mRNA delivery. *Adv. Healthc. Mater.* **2021**, *10*, e2001812. [CrossRef]
5. Tenchov, R.; Bird, R.; Curtze, A.E.; Zhou, Q. Lipid nanoparticles- from liposomes to mRNA vaccine delivery, a landscape of research diversity and advancement. *ACS Nano.* **2021**, *15*, 16982–17015. [CrossRef]
6. Smith, S.A.; Selby, L.I.; Johnston, A.P.R.; Such, G.K. The Endosomal Escape of Nanoparticles: Toward More Efficient Cellular Delivery. *Bioconjugate Chem.* **2019**, *30*, 263–272. [CrossRef]
7. Şenel, B.; Büyükköroğlu, G. Applications of Lipidic and Polymeric Nanoparticles for siRNA Delivery. In *Antisense Therapy*; IntechOpen: London, UK, 2019. [CrossRef]

8. Cavallaro, G.; Sardo, C.; Craparo, E.F.; Porsio, B.; Giammona, G. Polymeric nanoparticles for siRNA delivery: Production and applications. *Int. J. Pharm.* **2017**, *525*, 313–333. [CrossRef]
9. Ragelle, H.; Vandermeulen, G.; Préat, V. Chitosan-based siRNA delivery systems. *J. Control. Release* **2013**, *172*, 207–218. [CrossRef]
10. Mao, S.; Sun, W.; Kissel, T. Chitosan-based formulations for delivery of DNA and siRNA. *Adv. Drug Deliv. Rev.* **2010**, *62*, 12–27. [CrossRef]
11. Peng, H.; Yang, H.; Song, L.; Zhou, Z.; Sun, J.; Du, Y.; Lu, K.; Li, T.; Yin, A.; Xu, J.; et al. Sustained delivery of siRNA/PEI complex from in situ forming hydrogels potently inhibits the proliferation of gastric cancer. *J. Exp. Clin. Cancer Res.* **2016**, *35*, 57. [CrossRef]
12. Shakeran, Z.; Varshosaz, J.; Keyhanfar, M.; Mohammad-Beigi, H.; Rahimi, K.; Sutherland, D.S. Co-delivery of STAT3 siRNA and methotrexate in breast cancer cells. *Artif. Cells Nanomed. Biotechnol.* **2022**, *50*, 29–39. [CrossRef] [PubMed]
13. Li, X.; Chen, Y.; Wang, M.; Ma, Y.; Xia, W.; Gu, H. A mesoporous silica nanoparticle—PEI—Fusogenic peptide system for siRNA delivery in cancer therapy. *Biomaterials* **2013**, *34*, 1391–1401. [CrossRef] [PubMed]
14. Hao, N.; Jayawardana, K.W.; Chen, X.; Yan, M. One-Step Synthesis of Amine-Functionalized Hollow Mesoporous Silica Nanoparticles as Efficient Antibacterial and Anticancer Materials. *ACS Appl. Mater. Interfaces* **2015**, *7*, 1040–1045. [CrossRef] [PubMed]
15. Tan, W.B.; Zhang, Y. Surface modification of gold and quantum dot nanoparticles with chitosan for bioapplications. *J. Biomed. Mater. Res. Part A* **2005**, *75A*, 56–62. [CrossRef]
16. Lv, G.; Qiu, L.; Liu, G.; Wang, W.; Li, K.; Zhao, X.; Lin, J. pH sensitive chitosan-mesoporous silica nanoparticles for targeted delivery of a ruthenium complex with enhanced anticancer effects. *Dalton Trans.* **2016**, *45*, 18147–18155. [CrossRef]
17. Nhavene, E.P.F.; Andrade, G.F.; Faria, J.A.Q.A.; Gomes, D.A.; de Sousa, E.M.B. Biodegradable Polymers Grafted onto Multifunctional Mesoporous Silica Nanoparticles for Gene Delivery. *Chemengineering* **2018**, *2*, 24. [CrossRef]
18. Lin, J.-T.; Liu, Z.-K.; Zhu, Q.-L.; Rong, X.-H.; Liang, C.-L.; Wang, J.; Ma, D.; Sun, J.; Wang, G.-H. Redox-responsive nanocarriers for drug and gene co-delivery based on chitosan derivatives modified mesoporous silica nanoparticles. *Colloids Surfaces B: Biointerfaces* **2017**, *155*, 41–50. [CrossRef]
19. Cao, Y.; Tan, Y.F.; Wong, Y.S.; Liew, M.W.J.; Venkatraman, S. Recent Advances in Chitosan-Based Carriers for Gene Delivery. *Mar. Drugs* **2019**, *17*, 381. [CrossRef]
20. Soliman, O.Y.; Alameh, M.G.; De Cresenzo, G.; Buschmann, M.D.; Lavertu, M. Efficiency of Chitosan/Hyaluronan-Based mRNA Delivery Systems In Vitro: Influence of Composition and Structure. *J. Pharm. Sci.* **2020**, *109*, 1581–1593. [CrossRef]
21. Sargazi, S.; Arshad, R.; Ghamari, R.; Rahdar, A.; Bakhshi, A.; Karkan, S.F.; Ajalli, N.; Bilal, M.; Díez-Pascual, A.M. siRNA-based nanotherapeutics as emerging modalities for immune-mediated diseases: A preliminary review. *Cell Biol. Int.* **2022**, *46*, 1320–1344. [CrossRef]
22. Pilipenko, I.; Korzhikov-Vlakh, V.; Sharoyko, V.; Zhang, N.; Schafer-Korting, M.; Rühl, E.; Zoschke, C.; Tennikova, T. pH-sensitive chitosan-heparin nanoparticles for effective delivery of genetic drugs into epithelial cells. *Pharmaceutics* **2019**, *11*, 317. [CrossRef] [PubMed]
23. Pathak, N.; Singh, P.; Singh, P.K.; Sharma, S.; Singh, R.P.; Gupta, A.; Mishra, R.; Mishra, V.K.; Tripathi, M. Biopolymeric nanoparticles based effective delivery of bioactive compounds toward the sustainable development of anticancerous therapeutics. *Front. Nutr.* **2022**, *9*, 963413. [CrossRef] [PubMed]
24. Sarkar, S.; Ponce, N.T.; Banerjee, A.; Bandopadhyay, R.; Rajendran, S.; Lichtfouse, E. Green polymeric nanomaterials for the photocatalytic degradation of dyes: A review. *Environ. Chem. Lett.* **2020**, *18*, 1569–1580. [CrossRef]
25. Farshchi, E.; Pirsa, S.; Roufegarinejad, L.; Alizadeh, M.; Rezazad, M. Photocatalytic/biodegradable film based on carboxymethyl cellulose, modified by gelatin and TiO2–Ag nanoparticles. *Carbohydr Polym.* **2019**, *216*, 189–196. [CrossRef] [PubMed]
26. Kolangare, I.M.; Isloor, A.M.; Karim, Z.A.; Kulal, A.; Ismail, A.F.; Inamuddin; Asiri, A. M. Antibiofouling hollow-fiber membranes for dye rejection by embedding chitosan and silver-loaded chitosan nanoparticles. *Environ. Chem. Lett.* **2019**, *17*, 581–587. [CrossRef]
27. Morin-Crini, N.; Lichtfouse, E.; Torri, G.; Crini, G. Applications of chitosan in food, pharmaceuticals, medicine, cosmetics, agriculture, textiles, pulp and paper, biotechnology, and environmental chemistry. *Environ. Chem. Lett.* **2019**, *17*, 1667–1692. [CrossRef]
28. Bae, J.; Park, S.J.; Shin, D.S.; Lee, J.; Park, S.; Kim, H.J.; Kwon, O.S. A dual functional conductive hydrogel containing titania@polypyrrolecyclodextrin hybrid nanotubes for capture and degradation of toxic chemical. *Biochip J.* **2021**, *15*, 162–170. [CrossRef]
29. Liang, Y.; Wang, Y.; Wang, L.; Liang, Z.; Li, D.; Xu, X.; Chen, Y.; Yang, X.; Zhang, H.; Niu, H. Self-crosslinkable chitosan-hyaluronic acid dialdehyde nanoparticles for CD44-targeted siRNA delivery to treat bladder cancer. *Bioact. Mater.* **2021**, *6*, 433–446. [CrossRef]
30. Erdene-Ochir, T.; Ganbold, T.; Zandan, J.; Han, S.; Borjihan, G.; Baigude, H. Alkylation enhances biocompatibility and siRNA delivery efficiency of cationic curdlan nanoparticles. *Int. J. Biol. Macromol.* **2020**, *143*, 118–125. [CrossRef]
31. O'Brien, K.; Breyne, K.; Ughetto, S.; Laurent, L.C.; Breakefield, X.O. RNA delivery by extracellular vesicles in mammalian cells and its applications. *Nat. Rev. Mol. Cell Biol.* **2020**, *21*, 585–606. [CrossRef]
32. Luan, X.; Sansanaphongpricha, K.; Myers, I.; Chen, H.; Yuan, H.; Sun, D. Engineering exosomes as refined biological nanoplatforms for drug delivery. *Acta Pharmacol. Sin.* **2017**, *38*, 754–763. [CrossRef] [PubMed]

33. Aqil, F.; Munagala, R.; Jeyabalan, J.; Agrawal, A.K.; Kyakulaga, A.-H.; Wilcher, S.A.; Gupta, R.C. Milk exosomes—Natural nanoparticles for siRNA delivery. *Cancer Lett.* **2019**, *449*, 186–195. [CrossRef] [PubMed]
34. Yang, Z.; Shi, J.; Xie, J.; Wang, Y.; Sun, J.; Liu, T.; Zhao, Y.; Zhao, X.; Wang, X.; Ma, Y.; et al. Large-scale generation of functional mRNA-encapsulating exosomes via cellular nanoporation. *Nat. Biomed. Eng.* **2020**, *4*, 69–83. [CrossRef] [PubMed]
35. Liang, S.-F.; Zuo, F.-F.; Yin, B.-C.; Ye, B.-C. Delivery of siRNA based on engineered exosomes for glioblastoma therapy by targeting STAT. *Biomater. Sci.* **2022**, *10*, 1582–1590. [CrossRef]
36. Zhang, D.; Lee, H.; Wang, X.; Rai, A.; Groot, M.; Jin, Y. Exosome-Mediated Small RNA Delivery: A Novel Therapeutic Approach for Inflammatory Lung Responses. *Mol. Ther.* **2018**, *26*, 2119–2130. [CrossRef]
37. Agrawal, A.K.; Aqil, F.; Jeyabalan, J.; Spencer, W.A.; Beck, J.; Gachuki, B.W.; Alhakeem, S.S.; Oben, K.; Munagala, R.; Bondada, S.; et al. Milk-derived exosomes for oral delivery of paclitaxel. *Nanomedicine* **2017**, *13*, 1627–1636. [CrossRef]
38. Pi, F.; Binzel, D.W.; Lee, T.J.; Li, Z.; Sun, M.; Rychahou, P.; Li, H.; Haque, F.; Wang, S.; Croce, C.M.; et al. Nanoparticle orientation to control RNA loading and ligand display on extracellular vesicles for cancer regression. *Nat. Nanotechnol.* **2018**, *13*, 82–89. [CrossRef]
39. Shoari, A.; Tooyserkani, R.; Tahmasebi, M.; Löwik, D.W.P.M. Delivery of Various Cargos into Cancer Cells and Tissues via Cell-Penetrating Peptides: A Review of the Last Decade. *Pharmaceutics* **2021**, *13*, 1391. [CrossRef]
40. Jones, S.W.; Christison, R.; Bundell, K.; Voyce, C.J.; Brockbank, S.M.V.; Newham, P.; Lindsay, M.A. Characterisation of cell-penetrating peptide-mediated peptide delivery. *Br. J. Pharmacol.* **2005**, *145*, 1093–1102. [CrossRef]
41. Dong, Y.; Chen, Y.; Zhu, D.; Shi, K.; Ma, C.; Zhang, W.; Rocchi, P.; Jiang, L.; Liu, X. Self-assembly of amphiphilic phospholipid peptide dendrimer-based nanovectors for effective delivery of siRNA therapeutics in prostate cancer therapy. *J. Control. Release* **2020**, *322*, 416–425. [CrossRef]
42. Strand, M.S.; Krasnick, B.A.; Pan, H.; Zhang, X.; Bi, Y.; Brooks, C.; Wetzel, C.; Sankpal, N.; Fleming, T.; Goedegebuure, S.P.; et al. Precision delivery of RAS-inhibiting siRNA to KRAS driven cancer via peptide-based nanoparticles. *Oncotarget* **2019**, *10*, 4761–4775. [CrossRef] [PubMed]
43. Wang, H.; Zhang, S.; Lv, J.; Cheng, Y. Design of polymers for siRNA delivery: Recent progress and challenges. *VIEW* **2021**, *2*, 20200026. [CrossRef]
44. Peer, D.; Park, E.J.; Morishita, Y.; Carman, C.V.; Shimaoka, M. Systemic Leukocyte-Directed siRNA Delivery Revealing Cyclin D1 as an Anti-Inflammatory Target. *Science* **2008**, *319*, 627–630. [CrossRef] [PubMed]
45. Lu, H.; Wang, D.; Kazane, S.; Javahishvili, T.; Tian, F.; Song, F.; Sellers, A.; Barnett, B.; Schultz, P.G. Synthesis of Bispecific Antibodies using Genetically Encoded Unnatural Amino Acids. *J. Am. Chem. Soc.* **2013**, *135*, 13885. [CrossRef]
46. Kohata, A.; Hashim, P.K.; Okuro, K.; Aida, T. Transferrin-Appended Nanocaplet for Transcellular siRNA Delivery into Deep Tissues. *J. Am. Chem. Soc.* **2019**, *141*, 2862–2866. [CrossRef]
47. Khan, O.F.; Kowalski, P.S.; Doloff, J.C.; Tsosie, J.K.; Bakthavatchalu, V.; Winn, C.B.; Haupt, J.; Jamiel, M.; Langer, R.; Anderson, D.G. Endothelial siRNA delivery in nonhuman primates using ionizable low–molecular weight polymeric nanoparticles. *Sci. Adv.* **2018**, *4*, eaar8409. [CrossRef]
48. Xu, X.; Saw, P.E.; Tao, W.; Li, Y.; Ji, X.; Yu, M.; Mahmoudi, M.; Rasmussen, J.; Ayyash, D.; Zhou, Y.; et al. Tumor Microenvironment-Responsive Multistaged Nanoplatform for Systemic RNAi and Cancer Therapy. *Nano Lett.* **2017**, *17*, 4427–4435. [CrossRef]
49. Yu, D.; Khan, O.F.; Suvà, M.L.; Dong, B.; Panek, W.K.; Xiao, T.; Wu, M.; Han, Y.; Ahmed, A.U.; Balyasnikova, I.V.; et al. Multiplexed RNAi therapy against brain tumor-initiating cells via lipopolymeric nanoparticle infusion delays glioblastoma progression. *Proc. Natl. Acad. Sci. USA* **2017**, *114*, E6147–E6156. [CrossRef]
50. Dahlman, J.E.; Kauffman, K.J.; Xing, Y.; Shaw, T.E.; Mir, F.F.; Dlott, C.C.; Langer, R.; Anderson, D.G.; Wang, E.T. Barcoded nanoparticles for high throughput in vivo discovery of targeted therapeutics. *Proc. Natl. Acad. Sci. USA* **2017**, *114*, 2060–2065. [CrossRef]
51. Srivastava, A.; Gowda, D.V.; Madhunapantula, S.V.; Shinde, C.G.; Iyer, M. Mucosal vaccines: A paradigm shift in the development of mucosal adjuvants and delivery vehicles. *APMIS* **2015**, *123*, 275–288. [CrossRef]
52. Parmar, R.G.; Poslusney, M.; Busuek, M.; Williams, J.M.; Garbaccio, R.; Leander, K.; Walsh, E.; Howell, B.; Sepp-Lorenzino, L.; Riley, S.; et al. Novel Endosomolytic Poly(amido amine) Polymer Conjugates for Systemic Delivery of siRNA to Hepatocytes in Rodents and Nonhuman Primates. *Bioconjugate Chem.* **2014**, *25*, 896–906. [CrossRef] [PubMed]
53. Farshbaf, M.; Davaran, S.; Zarebkohan, A.; Annabi, N.; Akbarzadeh, A.; Salehi, R. Significant role of cationic polymers in drug delivery systems. *Artif. Cells Nanomed. Biotechnol.* **2018**, *46*, 1872–1891. [CrossRef] [PubMed]
54. Kim, B.; Park, J.-H.; Sailor, M.J. Rekindling RNAi Therapy: Materials Design Requirements for In Vivo siRNA Delivery. *Adv. Mater.* **2019**, *31*, e1903637. [CrossRef] [PubMed]
55. Xu, C.F.; Iqbal, S.; Shen, S.; Luo, Y.L.; Yang, X.; Wang, J. Development of "CLAN" Nanomedicine for Nucleic Acid Therapeutics. *Small* **2019**, *15*, e1900055. [CrossRef] [PubMed]
56. Xu, C.; Liu, W.; Hu, Y.; Li, W.; Di, W. Bioinspired tumor-homing nanoplatform for co-delivery of paclitaxel and siRNA-E7 to HPV-related cervical malignancies for synergistic therapy. *Theranostics* **2020**, *10*, 3325–3339. [CrossRef]
57. Zhang, L.; Deng, S.; Zhang, Y.; Peng, Q.; Li, H.; Wang, P.; Fu, X.; Lei, X.; Qin, A.; Yu, X. Homotypic Targeting Delivery of siRNA with Artificial Cancer Cells. *Adv. Health Mater.* **2020**, *9*, 1900772. [CrossRef]

58. Liu, Y.; Zou, Y.; Feng, C.; Lee, A.; Yin, J.; Chung, R.; Park, J.B.; Rizos, H.; Tao, W.; Zheng, M.; et al. Charge Conversional Biomimetic Nanocomplexes as a Multifunctional Platform for Boosting Orthotopic Glioblastoma RNAi Therapy. *Nano Lett.* **2020**, *20*, 1637. [CrossRef]
59. Wang, Y.; Ji, X.; Ruan, M.; Liu, W.; Song, R.; Dai, J.; Xue, W. Worm-Like Biomimetic Nanoerythrocyte Carrying siRNA for Melanoma Gene Therapy. *Small* **2018**, *14*, e1803002. [CrossRef]
60. Cho, S.K.; Kwon, Y.J. Simultaneous gene transduction and silencing using stimuli-responsive viral/nonviral chimeric nanoparticles. *Biomaterials* **2012**, *33*, 3316–3323. [CrossRef]
61. Hong, C.A.; Cho, S.K.; Edson, J.A.; Kim, J.; Ingato, D.; Pham, B.; Chuang, A.; Fruman, D.A.; Kwon, Y.J. Viral/Nonviral Chimeric Nanoparticles To Synergistically Suppress Leukemia Proliferation *via* Simultaneous Gene Transduction and Silencing. *ACS Nano* **2016**, *10*, 8705–8714. [CrossRef]
62. Shen, W.; Wang, R.; Fan, Q.; Gao, X.; Wang, H.; Shen, Y.; Li, Y.; Cheng, Y. Natural Polyphenol Inspired Polycatechols for Efficient siRNA Delivery. *CCS Chem.* **2020**, *2*, 146–157. [CrossRef]
63. Shen, W.; Wang, Q.; Shen, Y.; Gao, X.; Li, L.; Yan, Y.; Wang, H.; Cheng; Y. Green Tea Catechin Dramatically Promotes RNAi Mediated by LowMolecular-Weight Polymers. *ACS Cent. Sci.* **2018**, *4*, 1326–1333. [CrossRef]
64. Choi, K.Y.; Correa, S.; Min, J.; Li, J.; Roy, S.; Laccetti, K.H.; Dreaden, E.; Kong, S.; Heo, R.; Roh, Y.H.; et al. Binary Targeting of siRNA to Hematologic Cancer Cells In Vivo Using Layer-by-Layer Nanoparticles. *Adv. Funct. Mater.* **2019**, *29*, 1900018. [CrossRef]
65. Hattab, D.; Bakhtiar, A. Bioengineered siRNA-Based Nanoplatforms Targeting Molecular Signaling Pathways for the Treatment of Triple Negative Breast Cancer: Preclinical and Clinical Advancements. *Pharmaceutics* **2020**, *12*, 929. [CrossRef] [PubMed]
66. Gangopadhyay, S.; Gore, K.R. Advances in siRNA therapeutics and synergistic effect on siRNA activity using emerging dual ribose modifications. *RNA Biol.* **2022**, *19*, 452–467. [CrossRef] [PubMed]
67. Cao, W.; Li, R.; Pei, X.; Chai, M.; Sun, L.; Huang, Y.; Wang, J.; Barth, S.; Yu, F.; He, H. Antibody–siRNA conjugates (ARC): Emerging siRNA drug formulation. *Med. Drug Discov.* **2022**, *15*, 100128. [CrossRef]
68. Islam, F.; Ward, E.M.; Sung, H. Annual Report to the Nation on the Status of Cancer, Part 1: National Cancer Statistics. *J. Natl. Cancer Inst.* **2021**, *113*, 1648–1669. [CrossRef]
69. Bozorgi, A.; Khazaei, S.; Khademi, A.; Khazaei, M. Natural and herbal compounds targeting breast cancer, a review based on cancer stem cells. *Iran J. Basic Med. Sci.* **2020**, *23*, 970–983. [CrossRef]
70. Hanafi-Bojd, M.Y.; Ansari, L.; Malaekeh-Nikouei, B. Codelivery of anticancer drugs and siRNA by mesoporous silica nanoparticles. *Ther. Deliv.* **2016**, *7*, 649–655. [CrossRef]
71. Sun, P.; Huang, W.; Jin, M.; Wang, Q.; Fan, B.; Kang, L.; Gao, Z. Chitosan-based nanoparticles for survivin targeted siRNA delivery in breast tumor therapy and preventing its metastasis. *Int. J. Nanomed.* **2016**, *11*, 4931–4945. [CrossRef] [PubMed]
72. Baghani, L.; Noroozi Heris, N.; Khonsari, F.; Dinarvand, S.; Dinarvand, M.; Atyabi, F. Trimethyl-Chitosan Coated Gold Nanoparticles Enhance Delivery, Cellular Uptake and Gene Silencing Effect of EGFR-siRNA in Breast Cancer Cells. *Front. Mol. Biosci.* **2022**, *9*, 871541. [PubMed]
73. Cui, L.; Zheng, R.; Lui, W.; Shen, P.; Tang, Y.; Luo, J.; Zhang, W.; Jia, G.; Wang, Y.; Zhao, S.; et al. Preparation of chitosan silicon dioxide/BCSG1 siRNA nanoparticles to enhance therapeutic efficacy in breast cancer cells. *Mol. Med. Rep.* **2018**, *17*, 436–441. [CrossRef] [PubMed]
74. Xu, J.; Liu, Y.; Li, Y.; Wang, H.; Stewart, S.; Van Der Jeught, K.; Agarwal, P.; Zhang, Y.; Liu, S.; Zhao, G.; et al. Precise targeting of POLR2A as a therapeutic strategy for human triple negative breast cancer. *Nat. Nanotechnol.* **2019**, *14*, 388–397. [CrossRef] [PubMed]
75. Yang, Y.; Kozlovskaya, V.; Zhang, Z.; Xing, C.; Zaharias, S.; Dolmat, M.; Qian, S.; Zhang, J.; Warram, J.M.; Yang, E.S.; et al. Poly(*N*-vinylpyrrolidone)-*block*-Poly(dimethylsiloxane)-*block*-Poly(*N*-vinylpyrrolidone) Triblock Copolymer Polymersomes for Delivery of PARP1 siRNA to Breast Cancers. *ACS Appl. Bio Mater.* **2022**, *5*, 1670–1682. [CrossRef] [PubMed]
76. Ashworth, A. A Synthetic Lethal Therapeutic Approach: Poly(ADP) Ribose Polymerase Inhibitors for the Treatment of Cancers Deficient in DNA Double-Strand Break Repair. *J. Clin. Oncol.* **2008**, *26*, 3785–3790. [CrossRef]
77. Telli, M.L.; Ford, J.M. Novel Treatment Approaches for Triple-Negative Breast Cancer. *Clin. Breast Cancer* **2010**, *10*, E16–E22. [CrossRef]
78. Zimmer, A.S.; Gillard, M.; Lipkowitz, S.; Lee, J.-M. Update on PARP Inhibitors in Breast Cancer. *Curr. Treat. Options Oncol.* **2018**, *19*, 21. [CrossRef]
79. Lee, A.; Djamgoz, M.B. Triple negative breast cancer: Emerging therapeutic modalities and novel combination therapies. *Cancer Treat. Rev.* **2018**, *62*, 110–122. [CrossRef]
80. Loibl, S.; O'Shaughnessy, J.; Untch, M.; Sikov, W.M.; Rugo, H.S.; McKee, M.D.; Huober, J.; Golshan, M.; von Minckwitz, G.; Maag, D.; et al. Addition of the PARP inhibitor veliparib plus carboplatin or carboplatin alone to standard neoadjuvant chemotherapy in triple-negative breast cancer (BrighTNess): A randomised, phase 3 trial. *Lancet Oncol.* **2018**, *19*, 497–509. [CrossRef]
81. Timms, K.M.; Abkevich, V.; Hughes, E.; Neff, C.; Reid, J.; Morris, B.; Kalva, S.; Potter, J.; Tran, T.V.; Chen, J.; et al. Association of BRCA1/2 Defects with Genomic Scores Predictive of DNA Damage Repair Deficiency Among Breast Cancer Subtypes. *Breast Cancer Res.* **2014**, *16*, 475. [CrossRef]
82. Garufi, G.; Palazzo, A.; Tortora, G.; Bria, E.; Carbognin, L. Homologous Repair Deficiency Status and Response to Neoadjuvant Chemotherapy for Triple-Negative Breast Cancer: The Best Current Biomarker to Select the Most Appropriate Treatment? *J. Cancer Sci. Clin. Ther.* **2021**, *5*, 124–133. [CrossRef]

83. Li, Y.; Zhan, Z.; Yin, X.; Fu, S.; Deng, X. Targeted Therapeutic Strategies for Triple-Negative Breast Cancer. *Front. Oncol.* **2021**, *11*, 4517. [CrossRef]
84. Lu, S.; Morris, V.B.; Labhasetwar, V. Effectiveness of Small Interfering RNA Delivery via Arginine-Rich Polyethylenimine-Based Polyplex in Metastatic and Doxorubicin-Resistant Breast Cancer Cells. *J. Pharmacol. Exp. Ther.* **2019**, *370*, 902–910. [CrossRef] [PubMed]
85. Rietwyk, S.; Peer, D. Next-Generation Lipids in RNA Interference Therapeutics. *ACS Nano* **2017**, *11*, 7572–7586. [CrossRef] [PubMed]
86. Magriotis, P.A. Recent progress toward the asymmetric synthesis of carbon-substituted piperazine pharmacophores and oxidative related heterocycles. *RSC Med. Chem.* **2020**, *11*, 745–759. [CrossRef] [PubMed]
87. Girase, P.S.; Dhawan, S.; Kumar, V.; Shinde, S.R.; Palkar, M.B.; Karpoormath, R. An appraisal of anti-mycobacterial activity with structure-activity relationship of piperazine and its analogues: A review. *Eur. J. Med. Chem.* **2021**, *210*, 112967. [CrossRef]
88. Ni, H.; Hatit, M.Z.C.; Zhao, K.; Loughrey, D.; Lokugamage, M.P.; Peck, H.E.; Del Cid, A.; Muralidharan, A.; Kim, Y.; Santangelo, P.J.; et al. Piperazine-derived lipid nanoparticles deliver mRNA to immune cells in vivo. *Nat. Commun.* **2022**, *13*, 4766. [CrossRef]
89. Ashique, S.; Almohaywi, B.; Haider, N.; Yasmin, S.; Hussain, A.; Mishra, N.; Garg, A. siRNA-based nanocarriers for targeted drug delivery to control breast cancer. *Adv. Cancer Biol. Metastasis* **2022**, *4*, 100047. [CrossRef]
90. Hafiz, M.M.; Mazatulikhma, M.Z.; Faiz, F.M.; Saifulaman, M.M. Targeted RNAi of the mitogen-activated protein kinase pathway genes in acute myeloid leukemia cells. *Sains Malays.* **2013**, *42*, 1131–1137.
91. Kamaruzman, N.I.; Tiash, S.; Ashaie, M.; Chowdhury, E.H. siRNAs Targeting Growth Factor Receptor and Anti-Apoptotic Genes Synergistically Kill Breast Cancer Cells through Inhibition of MAPK and PI-3 Kinase Pathways. *Biomedicines* **2018**, *6*, 73. [CrossRef] [PubMed]
92. Yang, H.; Li, Y.; Li, T.; Xu, M.; Chen, Y.; Wu, C.; Dang, X.; Liu, Y. Multifunctional Core/Shell Nanoparticles Cross-linked Polyetherimide-folic Acid as Efficient Notch-1 siRNA Carrier for Targeted Killing of Breast Cancer. *Sci. Rep.* **2014**, *4*, 7072. [CrossRef]
93. Subramanian, K.; Kanwar, J.R.; Athalya, P.K.; Janakiraman, N.; Khetan, V.; Kanwar, R.K.; Eluchuri, S.; Krishnakumar, S. EpCAM aptamer mediated cancer cell specific delivery of EpCAM siRNA using polymeric nanocomplex. *J. Biomed. Sci.* **2015**, *22*, 4. [CrossRef]
94. Okamoto, A.; Asai, T.; Hirai, Y.; Shimizu, K.; Koide, H.; Minamino, T.; Oku, N. Systemic Administration of siRNA with Anti-HB-EGF Antibody-Modified Lipid Nanoparticles for the Treatment of Triple-Negative Breast Cancer. *Mol. Pharm.* **2018**, *15*, 1495–1504. [CrossRef] [PubMed]
95. Gao, J.; Liu, W.; Xia, Y.; Sun, J.; Chen, H.; Li, B.; Zhang, D.; Qian, W.; Meng, Y.; Deng, L.; et al. The promotion of siRNA delivery to breast cancer overexpressing epidermal growth factor receptor through anti-EGFR antibody conjugation by immunoliposomes. *Biomaterials* **2011**, *32*, 3459–3470. [CrossRef] [PubMed]
96. Vaidya, A.M.; Sun, Z.; Ayat, N.; Schilb, A.; Liu, X.; Jiang, H.; Sun, D.; Scheidt, J.; Qian, V.; He, S.; et al. Systemic Delivery of Tumor-Targeting siRNA Nanoparticles against an Oncogenic LncRNA Facilitates Effective Triple-Negative Breast Cancer Therapy. *Bioconjugate Chem.* **2019**, *30*, 907–919. [CrossRef] [PubMed]
97. Guo, P.; Yang, J.; Di Jia, M.A.M.; Auguste, D.T. ICAM-1-targeted, Lcn2 siRNA-encapsulating liposomes are potent anti-angiogenic agents for triple negative breast cancer. *Theranostics* **2016**, *6*, 1–13. [CrossRef]
98. Zhao, L.; Gu, C.; Gan, Y.; Shao, L.; Chen, H.; Zhu, H. Exosome-mediated siRNA delivery to suppress postoperative breast cancer metastasis. *J. Control. Release* **2020**, *318*, 1–15. [CrossRef] [PubMed]
99. Goyal, R.; Chopra, H.; Singh, I.; Dua, K.; Gautam, R.K. Insights on prospects of nano-siRNA based approaches in treatment of Cancer. *Front. Pharmacol.* **2022**, *13*, 985670. [CrossRef]
100. Kucharzewska, P.; Christianson, H.C.; Welch, J.E.; Svensson, K.J.; Fredlund, E.; Ringnér, M.; Mörgelin, M.; Bourseau-Guilmain, E.; Bengzon, J.; Belting, M. Exosomes reflect the hypoxic status of glioma cells and mediate hypoxia-dependent activation of vascular cells during tumor development. *Proc. Natl. Acad. Sci. USA* **2013**, *110*, 7312–7317. [CrossRef]
101. Keam, B.; Kim, S.; Ahn, Y.-O.; Kim, T.M.; Lee, S.-H.; Kim, D.-W.; Heo, D.S. In vitro anticancer activity of PI3K alpha selective inhibitor BYL719 in head and neck cancer. *Anticancer. Res.* **2015**, *35*, 175–182.
102. Kim, M.J.; Lee, S.J.; Ryu, J.H.; Kim, S.H.; Kwon, I.C.; Roberts, T.M. Combination of KRAS gene silencing and PI3K inhibition for ovarian cancer treatment. *J. Control. Release* **2020**, *318*, 98–108. [CrossRef] [PubMed]
103. Yuan, Y.; Wen, W.; Yost, S.E.; Xing, Q.; Yan, J.; Han, E.S.; Mortimer, J.; Yim, J.H. Combination therapy with BYL719 and LEE011 is synergistic and causes a greater suppression of p-S6 in triple negative breast cancer. *Sci. Rep.* **2019**, *9*, 7509. [CrossRef] [PubMed]
104. Silva, R.; Ferreira, D.; Rodrigues, L.R. Exosome-based delivery of RNAi leads to breast cancer inhibition. *J. Drug Deliv. Sci. Technol.* **2022**, *78*, 103931. [CrossRef]
105. Kaban, K.; Hinterleitner, C.; Zhou, Y.; Salva, E.; Kantarci, A.; Salih, H.; Märklin, M. Therapeutic Silencing of BCL-2 Using NK Cell-Derived Exosomes as a Novel Therapeutic Approach in Breast Cancer. *Cancers* **2021**, *13*, 2397. [CrossRef] [PubMed]
106. Inoue, S.; Patil, R.; Portilla-Arias, J.; Ding, H.; Konda, B.; Espinoza, A.; Mongayt, D.; Markman, J.L.; Elramsisy, A.; Phillips, H.W.; et al. Nanobiopolymer for Direct Targeting and Inhibition of EGFR Expression in Triple Negative Breast Cancer. *PLoS ONE* **2012**, *7*, e31070. [CrossRef] [PubMed]
107. Zhang, Y.; Liu, Q.; Zhang, X.; Huang, H.; Tang, S.; Chai, Y.; Yang, C. Recent advances in exosome-mediated nucleic acid delivery for cancer therapy. *J. Nanobiotechnol.* **2022**, *20*, 279. [CrossRef] [PubMed]

108. Limoni, S.K.; Moghadam, M.F.; Moazzeni, S.M.; Gomari, H.; Salimi, F. Engineered Exosomes for Targeted Transfer of siRNA to HER2 Positive Breast Cancer Cells. *Appl. Biochem. Biotechnol.* **2019**, *187*, 352–364. [CrossRef]
109. Wang, X.; Cheng, K.; Zhang, G.; Jia, Z.; Yu, Y.; Guo, J.; Hua, Y.; Guo, F.; Li, X.; Zou, W.; et al. Enrichment of CD44 in Exosomes from Breast Cancer Cells Treated with Doxorubicin Promotes Chemoresistance. *Front. Oncol.* **2020**, *10*, 960. [CrossRef]
110. Yang, Z.; Xie, J.; Zhu, J.; Kang, C.; Chiang, C.; Wang, X.; Wang, X.; Kuang, T.; Chen, F.; Chen, Z.; et al. Functional exosome-mimic for delivery of siRNA to cancer: In vitro and in vivo evaluation. *J. Control. Release* **2016**, *243*, 160–171. [CrossRef]
111. Pillé, J.Y.; Li, H.; Blot, E.; Bertrand, J.-R.; Pritchard, L.-L.; Opolon, P.; Maksimenko, A.; Lu, H.; Vannier, J.-P.; Soria, J.; et al. Intravenous delivery of anti-RhoA small interfering RNA loaded in nanoparticles of chitosan in mice: Safety and e_cacy in xenografted aggressive breast cancer. *Hum. Gene Ther.* **2006**, *17*, 1019–1026. [CrossRef]
112. Aliabadi, H.M.; Landry, B.; Mahdipoor, P.; Uludağ, H. Induction of Apoptosis by Survivin Silencing through siRNA Delivery in a Human Breast Cancer Cell Line. *Mol. Pharm.* **2011**, *8*, 1821–1830. [CrossRef] [PubMed]
113. Da Silva, L.C.G.; Santos, A.O.; Bimbo, L.M.; Moura, V.; Ramalho, J.S.; De Lima, M.C.P.; Simões, S.; Moreira, J.N. Toward a siRNA-containing nanoparticle targeted to breast cancer cells and the tumor microenvironment. *Int. J. Pharm.* **2012**, *434*, 9–19. [CrossRef] [PubMed]
114. Salva, E.; Kabasakal, L.; Eren, F.; Özkan, N.; Çakalağaoğlu, F.; Akbuğa, J. Local delivery of chitosan/VEGF siRNA nanoplexes reduces angiogenesis and growth of breast cancer in vivo. *Nucleic Acid Ther.* **2012**, *22*, 40–48. [CrossRef] [PubMed]
115. Jing, H.; Cheng, W.; Li, S.; Wu, B.; Leng, X.; Xu, S.; Tian, J. Novel cell-penetrating peptide-loaded nanobubbles synergized with ultrasound irradiation enhance EGFR siRNA delivery for triple negative Breast cancer therapy. *Colloids Surf. B Biointerfaces* **2016**, *146*, 387–395. [CrossRef]

pharmaceutics

MDPI

Review

Polymeric Biomaterials for Topical Drug Delivery in the Oral Cavity: Advances on Devices and Manufacturing Technologies

Paula de Freitas Rosa Remiro [1], Mariana Harue Taniguchi Nagahara [1], Rafael Abboud Azoubel [2], Michelle Franz-Montan [3], Marcos Akira d'Ávila [2] and Ângela Maria Moraes [1,*]

[1] Department of Engineering of Materials and of Bioprocesses, School of Chemical Engineering, University of Campinas, Campinas 13083-852, SP, Brazil
[2] Department of Manufacturing and Materials Engineering, School of Mechanical Engineering, University of Campinas, Campinas 13083-860, SP, Brazil
[3] Department of Biosciences, Piracicaba Dental School, University of Campinas, Piracicaba 13414-903, SP, Brazil
* Correspondence: ammoraes@unicamp.br

Abstract: There are several routes of drug administration, and each one has advantages and limitations. In the case of the topical application in the oral cavity, comprising the buccal, sublingual, palatal, and gingival regions, the advantage is that it is painless, non-invasive, allows easy application of the formulation, and it is capable of avoiding the need of drug swallowing by the patient, a matter of relevance for children and the elderly. Another advantage is the high permeability of the oral mucosa, which may deliver very high amounts of medication rapidly to the bloodstream without significant damage to the stomach. This route also allows the local treatment of lesions that affect the oral cavity, as an alternative to systemic approaches involving injection-based methods and oral medications that require drug swallowing. Thus, this drug delivery route has been arousing great interest in the pharmaceutical industry. This review aims to condense information on the types of biomaterials and polymers used for this functionality, as well as on production methods and market perspectives of this topical drug delivery route.

Keywords: oral cavity; drug delivery; polymers; biomaterials; topical administration

1. Introduction

The most common route for drug delivery is the oral route [1,2], in which the drugs (tablets, capsules, syrup, solutions, suspensions, powder, emulsions, etc.) are placed in the mouth and swallowed. Then, the drug can be absorbed in the gastrointestinal tract. However, drug absorption can be affected by limited drug chemical and biological stability or by physiological barriers [2]. Another limiting factor for the use of the oral route of administration is that many people have difficulties in swallowing, especially the elderly and children [3,4]. Liquid and semi-solid formulations do not exhibit several of the problems concerning swallowing, but they often present palatability issues and can cause nausea and gastric discomfort. Added to these problems, the difficulty of local treatment of diseases affecting oral mucosa is another concern [5].

An alternative to the oral route of administration is transbuccal drug delivery, with advantages, such as bypassing the first-pass hepatic metabolism and avoiding drug degradation by gastrointestinal enzymes [6]. Moreover, topical administration is convenient and easy to access, minimally invasive, and presents higher drug bioavailability in comparison to transdermal administration [7]. Thus, the development of biomaterials capable of delivering drugs topically via the oral cavity is an attractive alternative for enhancing medication bioavailability and for drug administration in emergency situations, when intravenous application is impaired. However, despite the higher permeability of oral mucosa when compared to the skin, the epithelium consists in an efficient barrier to drug permeation and protects underlying connective tissues from damage [8–12]. Therefore, efficient transbuccal delivery remains a great challenge.

There is a range of biomaterials that can be used to deliver the drug through the oral mucosa. The drugs may or may not have systemic action and this depends on the type of device, dosage, and mucosa interaction, among other factors [13]. The biomaterials that are most widespread in the literature for this purpose are pressure tablets, fast-dissolving films and gels (oro-dissolving), mucoadhesive films and gels, and microneedles [7,14].

Biomaterials for drug delivery can be manufactured from numerous classes of materials. Nonetheless, polymers have very interesting characteristics for this purpose, and a large range of flexible, versatile, and non-toxic molecules can be selected among them for transbuccal delivery. These polymers should have a particular set of characteristics [15,16]. These characteristics include being non-toxic, non-irritating, and free of leachable substances. The polymers should not promote irritation and infection of the oral mucosa. In addition, the mechanical properties of the polymers should be compatible with the type of application, and the films formed by these polymers should have no taste. If they do, they should be palatable. Finally, the polymers should be easily accessible, of low cost, and should show adequate shelf-life and proper mucoadhesive properties.

Polymers can be roughly divided into synthetic and natural classes. The first includes those produced in chemical industries and the second, those that can be extracted from a natural source. It is also possible to have natural polymers chemically modified to have improved properties, and another subclass, called smart polymers, which represents molecules capable of responding to changes in the surrounding microenvironment [17].

Polymeric biomaterials containing drugs can be prepared in several ways, e.g., by solvent casting, direct compression, hot melting extrusion, rolling method, and 3D printing [18,19]. The best technique will depend on the nature of the polymer, the required architecture of the biomaterial, its application site, and the dosage of drug required for the treatment.

Figure 1 shows the increasing interest in polymeric oral cavity devices lately. As a result, it is reasonable to foresee that, shortly, polymeric devices will play an essential role in oral healthcare. Since polymeric formulations for drug delivery via topical application in the oral cavity represent a very promising and interesting therapeutic approach, this review addresses the types of polymeric biomaterials used for this purpose, their production methods, and the market perspectives related to this drug delivery route.

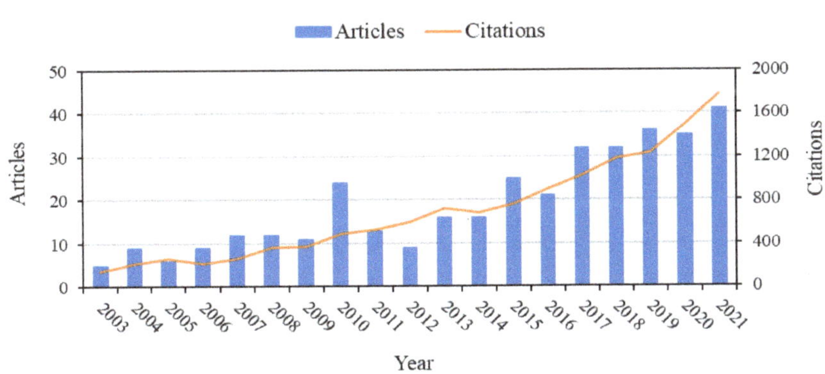

Figure 1. Number of articles in the last years addressing the evaluation of polymers used to produce formulations devices for oral cavity application and the citations associated with these publications. The literature search was based on Web of Science, with the keywords "Polymers + oral cavity".

2. Types of Polymeric Biomaterials Commonly Used in Drug Delivery in the Oral Cavity

The main types of biomaterials used for drug delivery in the oral cavity are fast-dissolving films and gels, mucoadhesive tablets, films and gels, and microneedles, which will be discussed below.

2.1. Fast-Dissolving Films (Oro-Dissolving)

An oral fast-dissolving film consists of a very thin oral strip that is simply placed on the patient's tongue or any oral mucosal tissue. The film, instantly wet by saliva, rapidly hydrates and adheres to the site of application, where it disintegrates and dissolves, releasing the medication [20,21]. Oral dissolving films have large surface areas, which facilitates their fast disintegration in the oral cavity. The drug can then be absorbed directly and reach the systemic circulation. These biomaterials do not require special storage or transport conditions as they are flexible, compact, and have a long shelf-life. These formulations can be used directly by the patient, anywhere, and at any time. They can be produced with less drug quantities, as the first-pass loss through the hepatic metabolism effect is negligible, which reduces the drug's side effects. These dosage forms are also favorable for those suffering from dysphagia, repeated vomiting, hypertension, heart attack, asthma, nausea, paralysis, and mental disorders [15,22,23]. This type of biomaterial also allows administering high drug concentrations faster when compared to conventional tablets, for example.

There is a wide variety of polymers that can be useful for producing fast-dissolving films. The use of these materials has aroused great interest in the medical and nutraceutical fields. The polymers can be used alone or combined with other materials, depending on the required attributes of the film. For this application, the polymers should be water-soluble, as they must dissolve in saliva, showing a high wettability rate, but appropriate disintegration time [15,16]. Some of the polymers and drugs used to produce oral fast-dissolving films are listed in Table 1.

Figure 2 illustrates a representative diagram of an example of one of the fabrication techniques of an oral dissolving film, up to the moment of its application.

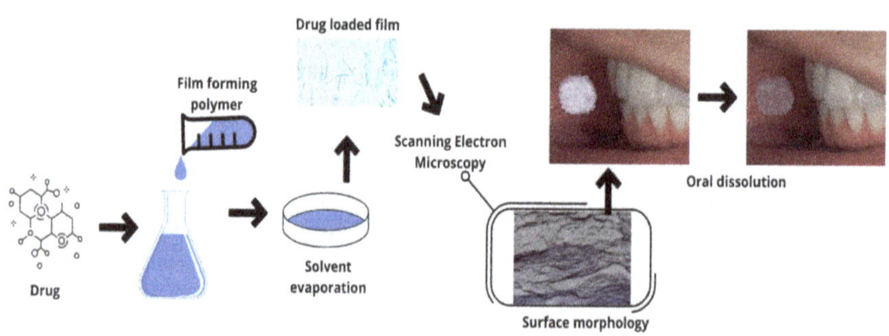

Figure 2. Schematic representation of film production by the casting technique until its application in the mouth, and its dissolution.

If fast film dissolution and rapid drug release are not desired, an alternative to prolonged drug release can be the use of mucoadhesive biomaterials, as discussed below.

Table 1. Examples of drugs and polymers used in the formulation of oral fast-dissolving films *.

Polymer	Drug	Therapeutic Use	Tested Model	References
HPMC/PVA	Levocetirizine dihydrochloride	Antihistamine	In vitro dissolution tests, and in vivo studies in rats	[24]
HPMC/PVA	Telmisartan	Hypertension	In vitro dissolution test	[25]
PVP	Paracetamol/caffeine	Analgesic and antipyretic	In vitro dissolution test	[26]
Chitosan	Metformin	Diabetes	In vitro dissolution test	[27]
PEG/400	Lercanidipine	Hypertension, and angina pectoris	In vitro Dissolution test and ex vivo drug permeation through porcine buccal mucosa	[28]
HPMC	Mirtazapine	Depression	In vitro dissolution test	[29]
HPMC/Alginate	Lidocaine	Anesthetic	In vitro dissolution test	[30]
Pullullan	Salbutamol sulfate	Asthma	In vitro dissolution tests, and in vivo studies on humans	[31]
Gelatin/Starch	Vitamin C	Assists in numerous functions	In vitro dissolution tests, and in vivo studies on humans	[32]
Chitosan/Pullullan	Aspirin	Minor aches, pains, and fever	In vitro dissolution test	[33]
Gelatin/gelatinized tapioca starch	Lidocaine	Anesthetic	In vitro dissolution test and ex vivo drug permeation through chick chorioallantoic membrane (CAM)	[34]
HPMC	Venlafaxine	Depression	In vitro dissolution test	[35]
Pectin/CMC	Paroxetin	Depression and anxiety	In vitro dissolution test and ex vivo drug permeation through chicken buccal pouch	[36]

* CMC (carboxymethyl cellulose), HPMC (hydroxypropyl methylcellulose), PEG (polyethylene glycol), PVP (polyvinyl pyrrolidone), PVA (polyvinyl acrylate).

As observed in Table 1, numerous polymers can be used to produce films with rapid degradation, and cellulose-based biomaterials stand out for this purpose. Table 1 also presents a large spectrum of drugs that can be administered through the oral cavity, minimizing the need for other invasive and often uncomfortable routes.

2.2. *Mucoadhesive Biomaterials*

Mucoadhesion can be defined as the attachment of macromolecules to mucous membranes. Mucoadhesive formulations are useful since products without these characteristics can be leached and ingested by patients, causing interruptions or reductions in drug absorption during the treatment [37–39]. Adhesion of a material to a mucous membrane or a mucus-covered surface has been employed to prolong drug contact with adsorption sites and, consequently, to improve drug absorption [40].

The mechanisms involved in mucoadhesion are not yet fully elucidated, with divergence in the methods used to quantify the interaction between polymers and mucosa [41]. Numerous polymers can be used in the oral region to achieve mucoadhesion. These polymers should have the ability to hydrate and swell when in contact with the mucus-lined epithelium [37,42]. The mucoadhesive ability of a dosage form is dependent upon a variety of factors, including the nature of the mucosal tissue and the physicochemical properties of the polymeric formulation [43]. The most common mucoadhesive biomaterials are described below.

2.2.1. Buccal Tablets

Buccal tablets are easy to prepare, as they can be made only by compression, a process easy to scale up, efficient, and an economic method for large-scale production. However, among the formulations generally used for oral purposes, it is considered the least comfortable. Despite this limitation, mucoadhesive buccal tablets can be loaded with a greater amount of drug than films and gels, being useful for treatments in which the required drug concentration is high, and slow drug release is desired in comparison to fast-dissolving films [37,39,43]. An example of a mucoadhesive buccal tablet and a summarized mechanism of how it remains adhered to the mucosa are shown in Figure 3.

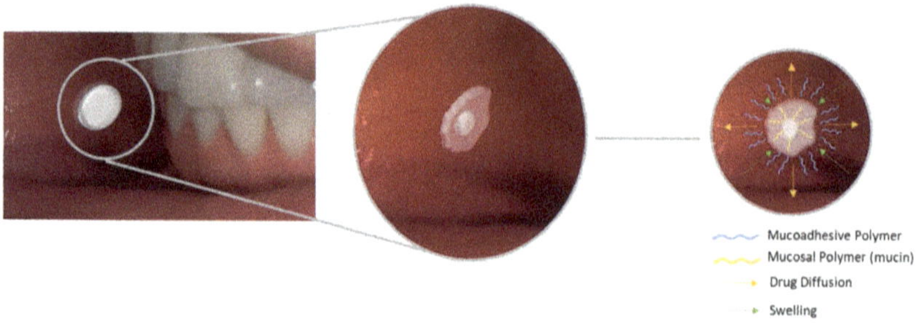

Figure 3. Factors that influence the behavior of mucoadhesive tablets and their drug release profile.

Numerous works in the literature report the use of buccal tablets for drug delivery. Koirala et al. (2021) evaluated tablets of aceclofenac for sustained drug release, to improve patient compliance for the management of different types of pain [44]. Çelik et al. (2017) developed a study to design and optimize risperidone mucoadhesive buccal tablets for systemic delivery routes and concluded that this biomaterial can be used as an alternative treatment for schizophrenia [37]. Abruzzo et al. (2015) prepared mucoadhesive tablets consisting of chitosan/gelatin microparticles compressed with propranolol hydrochloride for buccal delivery and concluded that this formulation can be effective for the treatment of hypertension, angina, atrial fibrillation, postinfarction, sinus tachycardia, arrhythmias, and obstructive cardiomyopathies [39]. Due to the route of administration, this type of biomaterial also allows bypassing the extensive hepatic first-pass, avoiding some side effects. Chandira et al. (2009) developed mucoadhesive tablets of clarithromycin, a macrolide antibiotic, which were designed to extend the gastric residence time after oral administration. The authors were able to obtain formulations capable of releasing more than 90% of the drug in 12 h, while keeping the physicochemical properties unchanged [45].

In another example, buccal adhesive tablets containing theophylline (a methylxanthine drug used in the therapy of respiratory diseases such as chronic obstructive pulmonary disease, asthma, or emphysema) were developed using direct compression [43]. The formulations were developed using a water-soluble resin with various combinations of mucoadhesive polymers. The theophylline tablets were evaluated for tensile strength, swelling capacity, and ex vivo mucoadhesion performance. The authors concluded that, in general, the majority of the developed formulations presented suitable adhesion and controlled drug release.

2.2.2. Mucoadhesive Films and Gels

Among the formulations commonly used for transbuccal drug delivery (tablets, films, gels, pastes, and sprays), mucoadhesive films offer many advantages, such as high flexibility and large surface area, which implies a greater area for drug absorption [46–48]. Another relevant advantage is the fact that they ensure a higher drug dosage when compared to gels and other formulations such as pastes and sprays, since they can be leached by saliva [46,48]. They also have the advantage of being comfortable, as they are thin, flexible, and most of the time have mechanical properties suitable for their purpose.

Gels, in turn, have the limitation of not supplying the amount of drug to the mucosa in a homogeneous way, but they have some advantages over other types of formulation, such as the relatively faster release of the incorporated drug and easy preparation. In addition, their simple administration and greater mucoadhesiveness, allowing adhesion to the mucosa in the gingival pocket, as well as rapid elimination through normal catabolic pathways, which reduces allergic reactions at the application site [49], are attractive characteristics.

The use of mucoadhesive films and gels for drug delivery in the oral cavity is well consolidated in the literature. These films are being proposed for the adjuvant treatment of oral carcinomas [5,50], chronic pain [51], for the therapy of small buccal lesions [49], to treat symptoms of migraine headaches, for pressure control [46,52,53], to treat recurrent aphthous stomatitis [54], as a topical anesthetic in dentistry [55,56], and as an anxiolytic [57], among other applications.

2.3. Microneedles (MNs)

Microneedles are three-dimensional microstructures with microscale length (usually less than 1500 μm) that can break the barrier of transdermal drug delivery. They can pierce the stratum corneum and generate transient microchannels through which external molecules can passively diffuse into the skin [58]. These biomaterials can penetrate and release drugs into the skin, which is about 10–100 times less permeable than the mucosa [59]. These biomaterials have been employed for drug delivery, vaccination, bio-sensing, and diagnostic purposes, and analysis or current literature indicates that the development of microneedles tends to grow in the near future. For instance, microneedles have been shown to have potentially good effects as transmucosal delivery systems, as recently suggested in a randomized clinical trial [60].

Microneedles can access adequate tissue depths, being able to deliver the drug deep into tissue without stimulating nerves in the underlying tissue and damaging blood vessels. Thus, the treatment with microneedles enables a minimally invasive delivery of several molecules to the tissue (skin, mucosa) in a way that overcomes the limitations of conventional methods of transdermal drug administration [58]. Another advantage of these biomaterials is their versatility, as they can be used to deliver drugs not only to the oral cavity [61], but also to the skin [62,63] and eyes [64,65].

For the reasons discussed above, this type of biomaterial will be discussed in greater detail. Microneedles can be broadly classified based on their overall shape and tip shape, which are important design and fabrication issues. Different designs of microneedles have been proposed and fabricated, such as cylindrical, conical, pyramidal, candle, spike, spear, square, pentagonal, hexagonal, octagonal, and rocket shapes [66–68]. They can be produced using metals [69–72], glasses [73], silicon [74–76], polymers [77–81], ceramics [82], borosilicate [83], and carbon nanotube-polyimide [84], among other materials. Furthermore, different techniques of drug loading and delivery may be employed [66]. Figure 4 summarizes the microneedle types, fabrication materials, and methods.

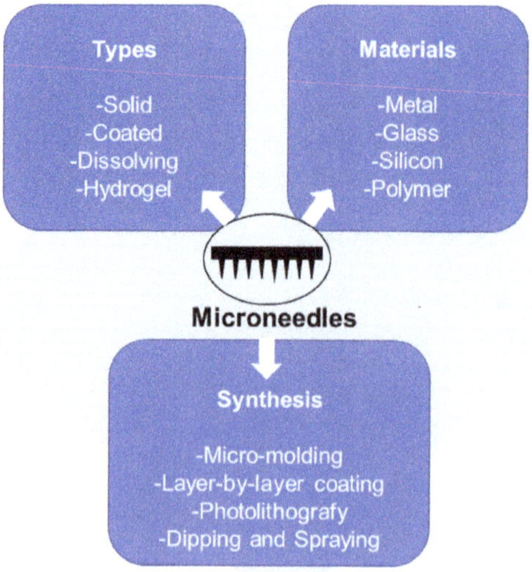

Figure 4. Diagram summarizing microneedle types, materials and production methods.

2.3.1. Types of Microneedles Concerning the Technique of Drug Loading and Delivery

Microneedles can be divided into five main classes: solid, coated, dissolving, hollow, and hydrogel-based microneedles, as illustrated in Figure 5, showing different working mechanisms.

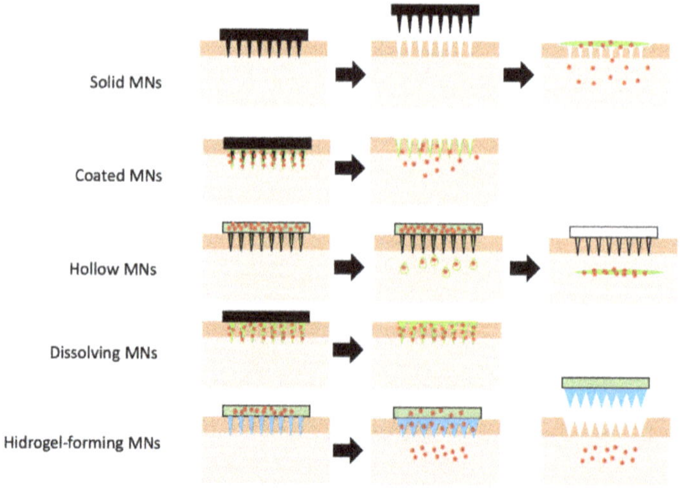

Figure 5. Schematic diagram showing the different types of microneedles and the mechanisms through which the different microneedles are used for drug delivery (adapted from [85]).

Solid microneedles (SMNs) are used to mechanically disrupt the tissues and create transient pores of micron dimensions before administration of the active pharmaceutical ingredients (APIs) from the external reservoir. Therefore, in SMNs-based drug delivery, the SMNs are first pressed against the tissue surface, followed by the application of traditional patches, or other types of dosage forms, such as the ones mentioned before, containing drug molecules. This allows for molecules of different physico-chemical properties to be delivered by using SMNs [86,87]. Solid microneedles are easy to manufacture, and usually have better mechanical properties and sharper tips when compared to hollow microneedles [88].

Coated microneedles are those in which the drug coats the microneedle surface as a solid film. When the coated microneedle is inserted into the tissue, it carries the film containing the active ingredient. Once in the tissue, within a few minutes, the coating dissolves, releasing the drug, and then it is possible to remove the microneedles and safely discard them [89].

The coated microneedle is a multifunctional system and can be used to deliver numerous molecules, such as proteins and DNA, as well as viruses [90]. One of the advantages of coated microneedles is that their mechanical properties do not vary, as occurs with soluble microparticles. The structural base of these microneedles is fairly resistant, and only the coating material and drug go through changes [89].

Dissolving microneedles (DMs) are ultra-small needles composed of water-soluble materials, usually with lengths in the micrometer range (less than 1000 μm). They create pores in the tissue and release the active principle upon microneedles dissolution [91]. These biomaterials have several advantages, such as having no risk of leaving harmful materials in the tissue, and no generation of sharp needle waste. These biomaterials present a relatively low cost, are easy to produce, and their fabrication at industrial scale is feasible [92].

Dissolving microneedles are reported in the literature for several uses, such as transdermal delivery of huperzine A for the treatment of Alzheimer's disease [93]. They can also be used for the delivery of ibuprofen, frequently used as a low molecular weight model drug demanding high dosages [94], and for the delivery of human growth hormones [95].

In fact, when it comes to delivery to the oral mucosa, soluble microneedles show a great rise lately in terms of the number of studies and publications. They can be configured as a new approach to drug delivery to the oral mucosa, effective and painless. Seon-Woo and co-workers developed a dissolving microneedle system for the oral mucosal to deliver triamcinolone acetonide to treat aphthous stomatitis [96].

Caffarell-Salvador et al. (2021) [97] designed a highly drug-loaded microneedle patch to deliver macromolecules and applied it to the buccal area, which allows faster delivery than through the skin. They successfully delivered 1-mg payloads of human insulin and human growth hormone to the buccal cavity of swine within 30 s. These are just a few examples of how dissolving microneedles can be a very useful tool for faster and more effective drug delivery.

Hollow microneedles are needles with an inner conduit for administering drugs in the tissue. These microneedles are similar to hypodermic injections, which allow pressure-actuated flow of a liquid formulation. The pressure, and consequently, the flow rate, can be modulated for fast and high dosage drug injection (bolus), slow infusion, or time-varying delivery rates [98].

Hydrogel-forming microneedles (HFMs) represent the newest form of microneedles, consisting of swellable polymers (crosslinked hydrogels) showing different performances from the microneedles mentioned above. When inserted into the tissue, HFMs swell due to the hydrophilic nature of the hydrogels, a property useful for many applications in biomedicine [99,100].

There are three types of drug loading methods in the case of hydrogel microneedles. Some microneedles only have drugs at the tips, some in patches, others have drugs in both places: needles and patches. One of the advantages of hydrogel microneedles is that the amount of drug loaded into the needles is greater than the possible amount to be loaded in solid and hollow microneedles. Another advantage of this type of microneedle is that it can be made from different types of polymers, making it possible to adjust the drug dosage according to the characteristics of the polymers used in its production. In fact, polymers and polysaccharides are very attractive for this purpose due to their high compatibility, degradability, and non-toxicity [101–103].

Zhu et al. (2022) developed a lidocaine-loaded hyaluronic acid adhesive microneedle patch for oral mucosal topical anesthesia. In this study, the authors reached the requirements for oral clinical application, i.e., rapid administration, water resistance, and adhesion, circumventing the disadvantages of the pain inflicted by direct local injection of anesthetic drugs and its ingestion, as well as the unpleasant taste of local surface anesthesia ointments [104].

Ye and coworkers (2016) developed a crosslinked hyaluronic acid microneedle patch coupled with pancreatic cells and enzymes for the delivery of insulin to regulate glucose levels, without the need for implantation, overcoming problems with the immune response and long-term efficacy of pancreatic cells therapy [105].

2.3.2. Polymeric Microneedles

As mentioned above, microneedles can be manufactured with numerous materials, such as metals, ceramics, silicon, silica glass and carbohydrates. However, in this work the focus will be on discussing polymeric microneedles.

Polymeric microneedle (PMN) systems are interesting biomaterials because they can control drug delivery, have tunable properties, and are easy and practical for patient self-administration. They have the main advantages of being biocompatible, and easily and painlessly penetrate the stratum corneum or mucosa, delivering their contents into where they can be absorbed into the systemic circulation. These biomaterials allow controlling drug release kinetics. Depending on the application site, they reach specific tissues, as well as respond to changes in the physiological conditions of the surrounding environment [106]. Furthermore, polymeric microneedles are able to deliver a greater quantity of low molecular weight molecules for biological therapies and vaccines.

In addition to the conventional requirements of biocompatibility and biodegradability, solubility and mechanical properties are relevant regarding the production of polymeric microneedles. Polymeric microneedles are produced through three main ways. The first approach refers to coating the metal microneedles with the polymer and drug to be released, giving rise to coated polymeric microneedles. The second approach, the dissolving microneedles, requires the incorporation of the drug into the matrix of soluble polymeric microneedles. In this method, delivery efficiency is determined by the rate of polymer dissolution after insertion. The third approach allows drug delivery through passive diffusion, as in hydrogel microneedles, or polymer matrix degradation. This approach can be associated with a secondary encapsulation procedure (micro or nano-formulations) which can be adjusted in a bio-responsive way [107].

Dissolvable polymeric microneedles are considered the most effective approach for drug delivery and can be used in numerous applications. The drugs to be delivered can be incorporated into dissolvable and degradable polymers [107]. One of the benefits of this type of microneedle is the fact that the biomaterial itself carries the drug, and it is not necessary to apply it at two different times, as with solid microneedles. The major challenge of polymeric microneedles is their penetration into tissues. Polymers generally have lower mechanical strength than silicon and metals, and the penetration of these microneedles into the tissue can be impaired [108,109]. Therefore, two or more polymers or additional materials can be combined to increase the mechanical strength of MNs [109].

Frequently reported polymers for the production of dissolving MNs are poly(vinylalcoh ol) (PVA) [110–112], hyaluronic acid (HAc) [113,114], hydroxypropyl methylcellulose (HPMC) [115,116], carboxymethyl cellulose (CMC) [115,116], fluorescein isothiocyanate (FITC)-dextran [117,118], sodium alginate [119,120], and other biodegradable polymers, such as chitosan [119,121], polylactic acid (PLA) [122], and polyglycolic acid (PLGA) [123], among others.

There are several ways to produce polymeric microneedles, but the most common and scalable is using molding techniques. However, this technique has some limitations as it usually involves several laborious steps, such as preparation of the polymer formulation, mold design, and fabrication and plasticization of thermoplastic polymers, which may limit the use of thermosensitive drugs [124].

Another technique that has been widely used for the manufacture of polymeric microneedles is three-dimensional (3D) printing. This strategy gives the prototyping and manufacturing methods the flexibility to produce MN patches in a one-step manner with high levels of shape complexity and reproducibility [125]. This technique offers customization, cost-efficiency, a rapid response time between design iterations, and enhanced accessibility. The increase of printing resolution, the accuracy of the features, and the accessibility of low-cost raw printing materials have stimulated the use of 3D printing for the fabrication of microneedle platforms [126].

3. Steps Involved in the Production of an Oral Dispositive to Drug Delivery

Pires et al. (2015) [127], based on the work of Ratner et al. (2013) [128], outlined some important steps to identify the need for a new biomaterial, its manufacture, testing, regulation, sale, and deployment. In the present work, a parallel can be established for the development of any biomaterial for drug delivery in the oral cavity. Thus, the steps required from designing to placing a new formulation on the market for this purpose and its use by the patient are illustrated in Figure 6.

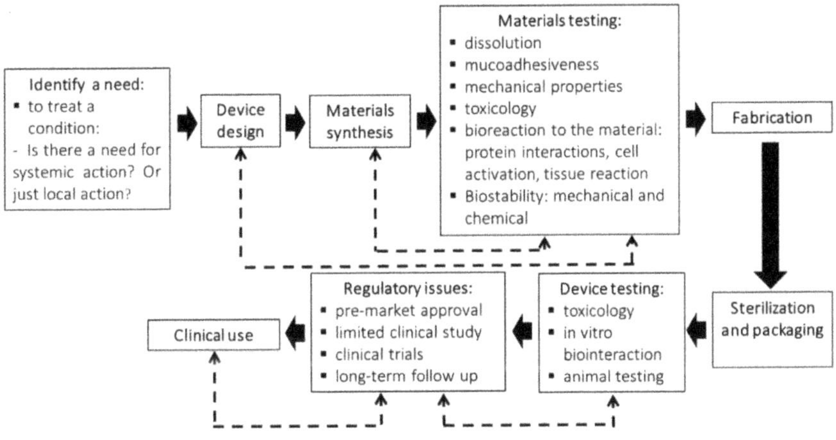

Figure 6. Illustrative diagram of the steps for the development of a new biomaterial designed for drug delivery to the oral cavity. Dashed lines refer to steps in which retro-feedback may occur with the purpose of improving the biomaterial originally designed (adapted from [127,128]).

4. Polymers Used to Produce Biomaterials for the Use in Oral Cavity

The polymers commonly used for the production of biomaterials used in the oral cavity include natural, synthetic, and smart polymers, which are discussed in detail below.

4.1. Natural Polymers

According to the European Chemical Agency (ECHA) [129], natural polymers are those that result from a polymerization process that has taken place in nature, independently of the extraction process used in their production. Natural polymers have the advantage of having generally low toxicity, being biocompatible, biodegradable, and coming from renewable sources. However, this type of polymer can also present a high degree of variability from lot to lot, can be structurally more complex, and may have complex production, extraction, and purification processes of high cost [130].

Natural polymers can be classified in different types [131]. Polysaccharides of plant cell walls, for instance, originate from the cell wall of plants and include mainly cellulose, hemicelluloses, and pectin. Gums and mucilages are polysaccharides, converted to monosaccharides by hydrolysis [132], including guar gum, arabica gum, and karaya gum. Exudate gums are also produced by plants, but as a result of stress, e.g., physical injury or fungal attack. Some examples of this class include arabic gum, acacia gum, cashew gum, karaya gum, and tragacanth gum. Another class refers to inulin, which belongs to a class of carbohydrates known as fructans. The main sources of inulin used in the food industry are chicory and Jerusalem artichoke. However, some vegetables, such as garlic, asparagus, and Dahlia tubers, also have a considerable amount of inulin [133]. Within the plant-derived category, starch, a polysaccharide present in many green plants, may also be used, having a wide range of applications in the pharmaceutical sector as an excipient. There are also seaweed-derived polysaccharides, which are isolated from algae. Seaweed gums are represented by carrageenans, agar, and alginates.

Microbial polysaccharides correspond to molecules obtained as fermentation products of microorganisms such as *Xanthomonas campestris* bacteria (xanthan gum), *Pseudomonas elodea* (gellan gum), *Sclerotium* (scleroglucan), and the fungus *Aureobasidium pullulans* (pulullan). Dextrans, on the other hand, are glucose-based polymers with molecular weights ranging between 1000–40,000,000 Da, being produced by lactic acid bacteria and also by the dental plaque-forming species *Streptococcus mutans*. Dextran is a family of natural polysaccharides that is widely under investigation for use as polymeric carriers in novel drug delivery systems.

Finally, animal polysaccharides and proteins are obtained from animal sources. The most common polysaccharide extracted from animal sources is chitin, from which chitosan can be produced, while the main examples of this class of protein include collagen, gelatin, fibrinogen, silk, elastin, and keratin.

Table 2 presents the main natural polymers used to manufacture oral drug delivery biomaterials and examples of their applications.

Table 2. Main natural polymers used to manufacture oral drug delivery biomaterials and examples of their applications.

Natural Polymer	Main Characteristic	Examples of Use	Tested Model	References
Alginate	Ability to form reversible hydrogels through interaction of carboxylic acid functional groups with metal cations	Film for the delivery of cetirizine dihydrochloride	In vitro release test	[134]
Cellulose	Tunable mechanical properties	Biomaterials for buccal delivery of non-steroidal anti-inflammatory drugs	In vitro release test	[135]
Chitosan	Cationic nature, a property that allows the formation of electrostatic complexes with negatively charged polymers	Gel for the delivery of the antimicrobial *Schinus molle* L essential oil	No models have been tested	[136]
Collagen	Excellent biocompatibility, flexibility and ability to absorb body fluids for delivery of nutrients	Buccal patch for the delivery of lorazepan	In vitro release test and ex vivo drug permeation through bovine buccal mucosa	[137]
Gellan Gum	Excellent gelation capability	Film for the delivery of triamcinolone acetonide	In vitro release test	[138]
Guar Gum	Excellent ability to hydrate rapidly, generating highly viscous solutions	Film-nanoparticle composite for the delivery of alpha-casozepine	In vitro release test	[57]
Gelatin	Thermoresponsiveness	Film for the delivery of propranolol hydrochloride	In vitro and in silico release test	[48]
Hyaluronic acid	Biocompatibility, hydrophilicity, low immunogenicity and excellent viscoelasticity	Mucoadhesive microneedles for the delivery of lidocaine	In vivo drug permeation through rat buccal mucosa	[104]
Pectin	Excellent biodegradability, biocompatibility and possibility of ionic crosslinking	Film for the delivery of triamcinolone acetonide	In vitro release test	[139]

4.2. Synthetic Polymers

Synthetic polymers are artificially produced through chemical reactions using well-defined conditions, e.g., heat, pressure, and catalyst type [139,140]. Most synthetic polymers have good mechanical properties, unlike natural polymers [140]. Compared to natural polymers, synthetic polymers are more biologically inert, the mechanical properties and degradation rates of synthetic polymers can be tailored, and, frequently, these polymers can be processed into various shapes. Since the synthesis conditions are well known and controlled, synthetic polymers offer higher batch-to-batch uniformity and show more reproducible physicochemical properties.

Synthetic polymers of interest in drug delivery for the oral cavity can be divided into different categories, e.g., polyethers, polyesters, poly (N-isopropylacrylamide-co-propylacrylic acid) copolymers, and poloxamers.

Among the polyethers, one of the most common is polyethylene glycol (PEG), which has gained great visibility due to its wide range of applications in the field of pharmacy. This compound is hydrophilic and can be used favorably as a penetration enhancer, especially in topical dermatological preparations. It is also widely used in cosmetics as cleansing surfactants, emulsifiers, skin conditioners, and humectants [141]. The two most important polyesters under intense investigation as drug carriers for active agent delivery are poly (lactic-co-glycolic acid) (PLGA), which is a copolymer of poly lactic acid (PLA) and poly glycolic acid (PGA), and polycaprolactone (PCL). Both can be biodegraded by hydrolysis of the ester linkages.

Poly (N-isopropylacrylamide-co-propylacrylic acid) copolymers (PNIPAAM) and their derivatives are also widely investigated for drug delivery purposes. They have great potential for the delivery of therapeutic proteins and peptides since they are thermoresponsive. However, the clinical use of PNIPAAM and its derivatives is limited, as they are non-biodegradable and, upon contact with blood, activate platelets.

Poloxamers, traded as Pluronic® (BASF), Kolliphor® (BASF), Lutrol® (BASF), Synperonic® (Croda), and Antarox® (Rhodia), are nonionic triblock copolymers. The central part of poloxamers is hydrophobic, composed of polypropylene oxide (PPO), and this region is flanked on both sides by polyethylene oxide (PEO) chains, which are hydrophilic. This class of polymers is thermosensitive and has been intensively studied for the sustained delivery of therapeutic proteins. These polymers are inert and known to maintain the stability of incorporated therapeutic proteins and peptides, with an increased in vivo half-life when compared to other drug delivery systems.

4.3. Smart Polymers

Smart polymers or stimuli-responsive polymers undergo reversible physico-chemical modifications with small alterations in their surrounding environment [142]. Frequently, these polymers exhibit a non-linear response to a small stimulus, leading to a macroscopic alteration in their structure or properties. Several changes can be observed in this sense, from swelling and contraction to complete disintegration. These variations can be caused by chemical events, including simple reactions, such as oxidation, acid-base reaction, reduction, and hydrolysis of fractions linked to the polymer chain. More severe changes can also occur, resulting in irreversible bond breakage in response to external stimuli, causing polymer structural degradation [143]. Physical-chemical changes can occur due to variation in pH, temperature, as well as due to the presence of enzymes and exposure to some types of radiation. Polymers that are stimulated via electric [144] and magnetic [145] fields are not as frequently reported in the literature for drug delivery, and therefore they will not be covered in this work.

4.3.1. Temperature-Responsive Polymers

Thermoresponsive polymers are the most widely employed smart polymers, whose main characteristic is the reversible phase (or volume) transition that occurs in response to temperature changes. This property allows the manipulation of polymers in a remote and switchable way by controlling the temperature [30,146].

According to Hoffman (2013), different working mechanisms can be exploited for the development of temperature-responsive polymers. The three main classes of temperature-responsive polymers are: shape-memory materials, liquid crystalline materials, and responsive polymer solutions [146].

Shape-memory polymers (SMPs) belong to a class of smart materials that are mechanically modified by external stimuli. More directly, these materials can "remember" shapes. As a simple example, a complex three-dimensional SMP shape can be compressed into a smaller size shape (suitable for delivery of a catheter to the body or to fit in a compact space) by applying some source of heating or cooling, for instance. In a second moment, the application of heat, for example, can trigger a return to equilibrium, causing the object to go back to a complex shape through the mobilization of the network chain [147].

Liquid crystalline polymers (LCPs) show properties of both solids and liquids. These polymers can be of many types, depending on the position and type of the mesogenic units in the molecular architecture. LCPs are mainly classified as main-chain, side-chain, and crosslinked polymers, among others [148]. Main chain-type polymers elongate in the liquid crystalline phase and contract when heated in the isotropic phase, with full reversibility to the initial stage.

The most common types of thermoresponsive polymers are those that undergo a solution liquid-liquid phase transition in response to variation of the temperature, in which phase separation occurs from a homogeneous solution to a concentrated polymer phase and a dilute polymer phase. This effect is associated with the transition from a clear phase to a phase with clouds. Clouds are observed due to the formation of polymer rich droplets, causing phase separation. When separation occurs at a high temperature, this is referred to as lower critical solution temperature (LCST) transition, while the reversed-phase behavior is known as upper critical solution temperature (UCST) transition [149].

The literature reports several examples of the use of temperature-responsive polymers for drug delivery in the oral cavity. Gao et al. (2010) developed a thermosensitive PLGA-PEG-PLGA hydrogel for the sustained release of docetaxel to treat lung cancer [150]. Choi et al. (2014) produced hydrogels with combined properties using the thermosensitive polymer Pluronic F127 (PF127) and the mucoadhesive polymer polyethylene oxide (PEO) to deliver the anticancer drug paclitaxel incorporated in dimethyl-β-cyclodextrin in the oral mucosa [151]. Shin et al. (2013) prepared a thermoreversible hydrogel of the polymer Pluronic F127 (PF127) with mucoadhesive properties due to the inclusion of the polymer Carbopol 934P (C934P) in the formulation to deliver the anti-inflammatory and anti-pain drug naproxen [152].

4.3.2. pH-Responsive Polymers

pH-responsive polymers represent a group of stimuli-responsive polymers that have in their chemical structure weak acidic or basic groups that either accept or release protons in response to a change in the environmental pH. They show different architectures, having linear (as in homopolymers, di, ter, multiblock copolymers, and organic/inorganic hybrid polymers), nonlinear (branched or hyperbranched polymers), and lightly or highly cross-linked network polymers.

These polymers respond to changes in the pH of the environment undergoing structural changes, e.g., in chain conformation alteration and configuration, and with property changes, such as surface activity and solubility [153,154]. The protonation or deprotonation of functional groups they experience may result in chain flocculation and/or collapse. These molecules can also self-assemble in larger structures or swell [154], as illustrated in Figure 7.

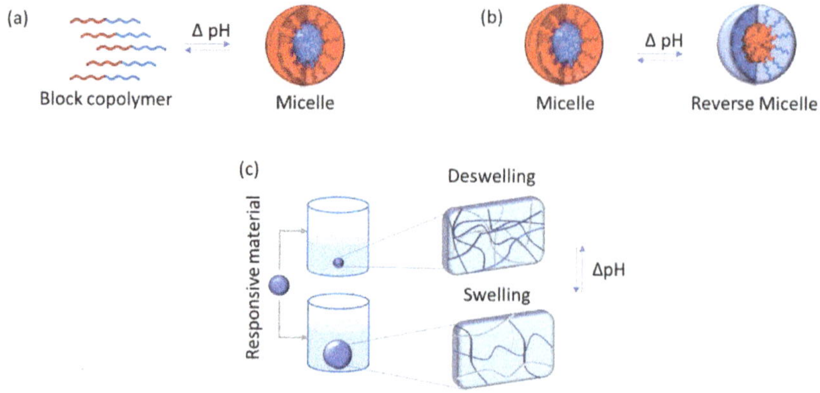

Figure 7. Possible transformations caused by pH modification in the surroundings of the polymeric environment.

The most commonly used pH-sensitive polymers are linear homopolymers amphiphilic in nature and double hydrophilic block copolymers which form micelles, vesicles, stars, branched and hyperbranched polymers, polymer brushes, dendrimers, nanogels, micro-gels, and hydrogels (macrogels) [154,155].

There are many studies in the literature that report the use of this type of material for drug delivery to the oral cavity. Hu et al. (2019) synthesized a series of amphiphilic pH-sensitive block copolymer poly(methyl methacrylate-co-methacrylic acid)-b-poly(2-amino ethyl methacrylate) [P(MMA-co-MAA)-b-PAEMA] via activators regenerated by electron transfer atom transfer radical polymerization (ARGET ATRP) and further self-assembled into pH-responsive cationic polymeric micelles (PCPMs) for oral insulin delivery [156]. Zamani et al. (2019) carried out an in vivo study of the use of poly (ethylene glycol)-poly(caprolactone)-modified folic acid nanocarriers as a pH-responsive system for tumor-targeted co-delivery of tamoxifen and quercetin [157]. Arjama et al. (2018) developed a sericin/rice bran albumin embedded gellan gum-based smart nanosystem for pH-responsive delivery of doxorubicin [158]. Sherje and Londhe (2017) developed and evaluated a pH-responsive cyclodextrin-based in situ gel of paliperidone for intranasal delivery [159]. Ishak et al. (2020) developed pH-responsive gamma-irradiated poly(acrylic acid)-cellulose-nanocrystal-reinforced hydrogels and concluded that they show a good response toward pH, thus suggesting their potential as a drug-delivery system [160]. Jamshidzadeh et al. (2020) investigated the modification of halloysite nanotubes by chitosan (CTS) and pectin (PCN) to produce new pH-sensitive bionanocomposites via a layer-by-layer method. The main objective of this study was to improve loading efficiency and control the release of phenytoin sodium (PHT) [161].

Although many of the above-mentioned works do not necessarily focus on drug delivery to oral mucosa lesions, pH-responsive polymers have a great potential to be further explored in this field.

4.3.3. Bioresponsive Polymers

Bioresponsive polymers are those that change their properties due to stimuli provided by biologically relevant molecules such as glucose, ATP, enzymes, antibodies, etc. These stimuli can cause changes in molecular interactions of the polymer, which eventually translate into macroscopic responses, such as solution-to-gel transitions, polymer swelling, or collapse [162,163].

Ulijn et al. (2007) categorized bioresponsive hydrogels based on three types of stimuli [164]. The first includes hydrogels modified with small biomolecules able to selectively bind to protein receptors, antibodies, or other biomacromolecules. This interaction can trigger a macroscopic transition in the hydrogel. The second type refers to hydrogels modified with sensitive enzymes. In this case, molecular recognition can cause a chemical change that involves breaking the bonds of a substrate molecule. Hydrogels having biomacromolecules incorporated into their structure comprise the third case. The biomacromolecules, such as enzymes, can recognize small biomolecules as substrates and convert them into molecules with different physical properties. This modification can trigger swelling or collapse of the hydrogel, and the signal can be utilized, e.g., to produce a biosensor.

Maitz et al. (2013) developed bio-responsive polymer hydrogels capable to homeostatically regulate blood coagulation [165]. The authors synthesized unmodified amino-terminated starPEG or the peptide-functionalized starPEG (PEG-FXRS), which were used to form non-responsive or thrombin-responsive hydrogels, respectively, useful in case of bleeding lesions of the oral cavity. The reported system was shown to be a simple yet powerful and clinically relevant example, illustrating the enormous potential of reciprocally triggered bio-responsive polymers.

5. Methods for the Production of Drug Delivery Systems

The manufacture of biomaterials for the oral cavity may be carried out by various methods, some as simple as mixing the excipients and the active agents (e.g., in the case of a number of gels and ointments), and some more elaborated, such as solvent casting, electrospinning, hot-melt extrusion, and 3D printing, among many others. The most commonly used methods will be discussed below.

5.1. Solvent Casting Method

The solvent casting method is the most common technique applied for the preparation of single or multilayer films on a lab scale as well as in continuous industrial production processes [18,166,167]. The solvent casting method is based on three steps: preparing a homogeneous mixture of components, obtaining a dry laminate by evaporating the solvent, and, lastly, cutting the laminate into the desired shape and dimensions [167].

In the first step, the drug and excipients (stabilizers, plasticizers, and other products necessary for adequate film formation) are dissolved or dispersed in an appropriate solvent or solution using, for instance, a stirring tank at an industrial level or a beaker in a laboratory/pharmacy setting. Then, the formed mass is transferred to a cast and dried to produce a film with constant thickness and uniformity of drug content.

The solvent casting method can be used industrially for pharmaceutical applications, in which the drug is suspended or dissolved in a solution of polymers in a volatile solvent. The film is deposited on a continuous-release roller impregnated on a plastic base and then passed through a drying apparatus to remove the solvents (Figure 8). The dry film is then cut to sizes suitable for proper use [167].

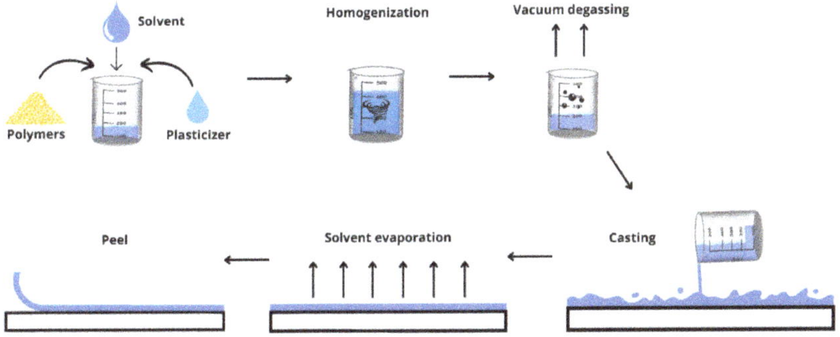

Figure 8. Scheme of the production, in a small scale, of a polymeric film using the solvent casting technique.

Figures 8 and 9 illustrate the steps for producing a polymeric film using the solvent casting technique on laboratory and industrial scales, respectively.

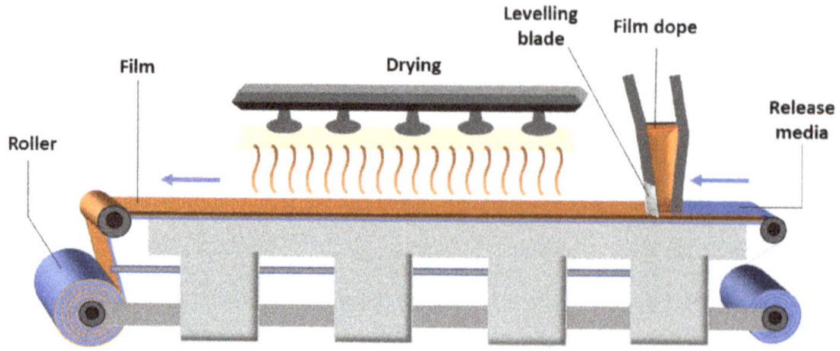

Figure 9. Scheme of industrial production of a polymeric film using the solvent casting technique (adapted from [168]).

5.2. Electrospinning

Electrospinning is a versatile technique useful to produce ultrathin fibers. This technique involves an electrohydrodynamic process, during which a liquid droplet is electrified, stretched, and elongated to generate fibers, as a result of exposure to a high-voltage electric field. The solvent in the fibers evaporates, and the fibers are deposited onto a collector [169].

The electrospinning equipment is basically composed of four elements: (1) a positive displacement pump (or a syringe pump at laboratorial scale), to ensure constant flow of the polymeric solution; (2) a high voltage power supply, which is responsible for yielding the driving force necessary to attract the fibers from the polymeric solution to the collector; (3) a metal tip or needle that directs the solution into the high voltage electric field; and (4) a grounded collector used in static or rotating mode, over which the fibers are deposited and form a non-woven film [169,170].

To produce the fibers, a high voltage (normally between 15 and 20 kV) is applied to the chamber, between the tip of the needle and the collector. The polymeric solution is pumped and tends to assume a conical shape on the tip of the needle (forming a Taylor cone), where the liquid voltage is in equilibrium with the electric field. When the electric field intensity exceeds the surface tension, the polymer jet is ejected, and the solution is electrospun and deposited on the collector. When jetting occurs, evaporation of the solvent also occurs, followed by weaving of the fibers, which produces a mat with high surface area to volume ratio. The basic setup for electrospinning is shown in Figure 10.

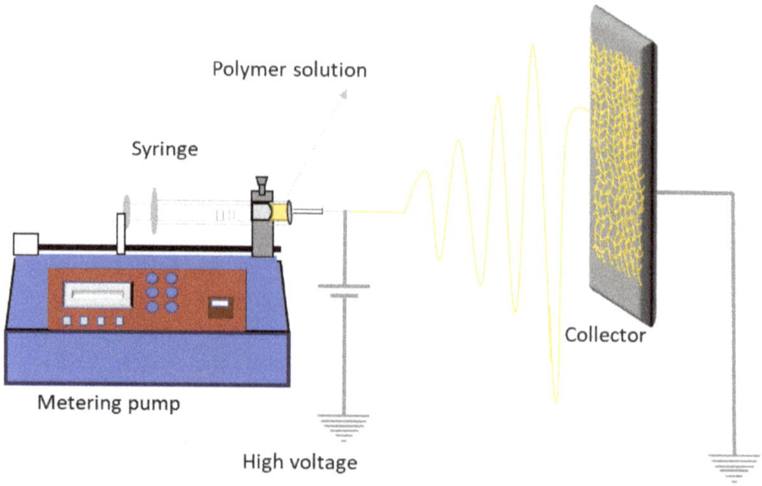

Figure 10. Basic configuration of electrospinning equipment (adapted from [171]).

Several factors affect the characteristics of the fibers formed, e.g., electrospinning conditions, solution formulation, and environmental parameters. Electrospinning parameters comprise the intensity of the electric field, the separation distance between the needle and the collector, solution flow rate, and needle diameter. The solution-related parameters are polymer properties and concentration, viscosity, solvent type, and solution conductivity. The environmental parameters include relative humidity and temperature. All these parameters must be taken into account so that the required design of the material can be achieved [172].

Similar to the solvent casting method, the electrospinning technique is also widely investigated for the production of biomaterials that can be used for oral drug delivery. Table 4 illustrates some of the works found in the literature in this sense.

Although widely used on laboratory scale, electrospinning has a few limitations regarding its scale-up for the production of biomaterials by the pharmaceutical industry, such as the low productivity and the use of organic solvents, in which case the manufacturer must guarantee that the final product does not contain harmful levels of residual solvent. Furthermore, strategies to increase productivity have been studied, such as multi-jet electrospinning, multi-needle electrospinning, and free-surface electrospinning, in which the needle is replaced by an open surface. Although these technologies also have drawbacks, studies have been carried out to address these issues and some industrial scale equipment is already available [173].

5.3. Hot Melt Extrusion

The hot melt extrusion (HME) technique makes use of heat and pressure in mixing a set of materials using a screw in a barrel, which transports the molten material through a matrix to mold it into the desired shape [174]. The pharmaceutical process involves pumping polymeric materials with a rotating screw at temperatures above their glass transition temperature (T_g) or above the melting temperature (T_m) to promote their effective mixing with the active compounds at a molecular level. This molecular mixing process converts the components into an amorphous product with uniform shape and density, increasing the dissolution profile of poorly water-soluble drugs [175].

During the design and optimization of an HME process, two main classes of factors should be evaluated, one related to the equipment design itself and its operating conditions, and the other referring to the characteristics of the chemical compounds used in the production of the biomaterial. Process variables include the selection of the appropriate equipment, mass feeding rate, process temperature, and tolerable shear stress, while physicochemical factors must also be considered, such as drug and excipient properties, possible interactions among the components of the formulation, mixture physical state, and stability.

The design and optimization of an HME process can be described in terms of preformulation, formulation, and post-formulation phases and should be investigated. Factors, such as the chemical and thermal stability of extrudates, the solid physical state of extrudates, drug–polymer interaction, miscibility or solubility of the drug–polymer system, rheological properties of extrudates, physicomechanical properties of films produced by hot melt extrusion, and drug particle dissolution from extrudates [176], are frequently analyzed.

This technology, which follows basically the same procedures and steps at small and large scale, may offer advantages over conventional pharmaceutical manufacturing processes, such as shorter and more efficient time to achieve the final product, environmental advantages due to the elimination of solvent use, increased solubility and bioavailability of the drugs, taste masking, and production of amorphous materials. Thus, HME has emerged as an alternative platform technology to other traditional techniques for manufacturing pharmaceutical dosage forms, such as tablets, capsules, films, and implant, for drug delivery not only via oral routes, but also for transdermal and transmucosal administration [175–177]. Figure 11 shows a schematic of hot-melt extrusion technique used to produce films, tablets, capsules, and pellets.

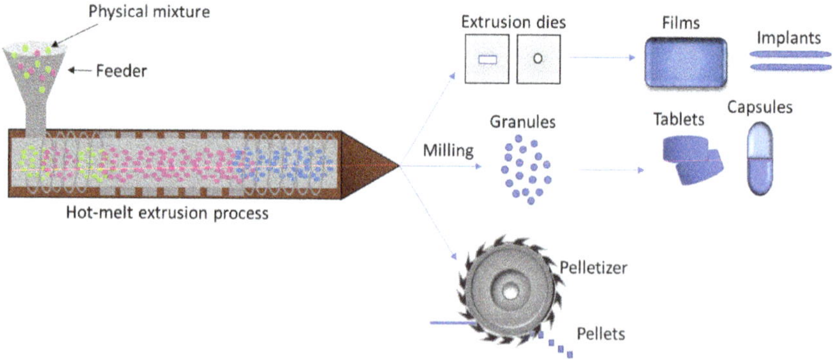

Figure 11. Schematic of hot-melt extrusion of films, tablets, capsules, and pellets (adapted from [178]).

5.4. 3D Printing Method

Three-dimensional (3D) printing is an additive manufacturing (AM) technique for fabricating a wide range of structures with simple or complex geometries from three-dimensional model data. The process consists of printing successive layers one above the other until the desired shape and dimensions are obtained [177]. According to the ASTM F2792-12A [179], the manufacturing process techniques of AM are divided into seven main categories, as shown in Table 3 [180,181].

Table 3. ASTM F2792-12A standard terminology for additive manufacturing processes.

Type of Process	Basic Description
Material extrusion	Material is extruded through a nozzle or orifice and deposited on the surface
Material jetting	Drops of material are deposited on the surface until the desired layers are formed
Binder jetting	A liquid bonding agent is deposited to join powder materials
Sheet lamination	Material sheets are deposited to form the final desired object
Vat photopolymerization	Liquid photopolymer in a vat is selectively cured by light-activated polymerization
Powder bed fusion	Thermal energy selectively fuses regions of a powder bed
Directed energy deposition	Focused thermal energy is used to fuse materials by melting as the material is deposited

With the 3D printing technique, it is possible to produce highly customized, functional parts of varied composition. These materials can be metals, ceramics, polymers, or even their combinations. To produce polymeric biomaterials, extrusion techniques and those based on resin and powder processes are commonly used. Each type of process allows the manufacture of pieces in different ways, using exclusive steps [181,182]. In Figure 12, three different 3D printing techniques are illustrated.

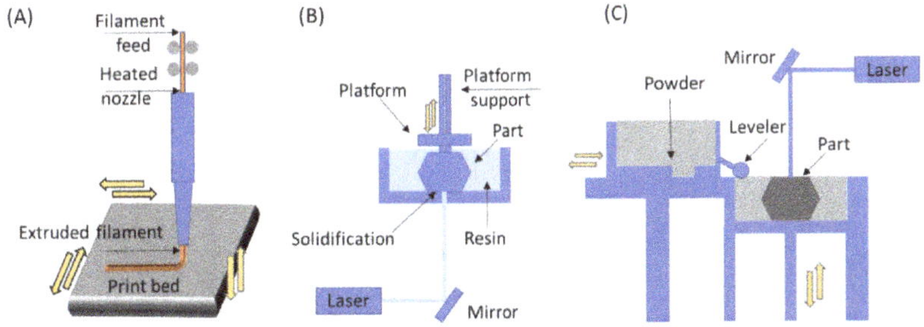

Figure 12. Representation of different 3D printing techniques: (**A**) fused deposition modeling, (**B**) stereolithography, and (**C**) selective laser sintering (adapted from [183]).

Fused deposition modeling (FDM) is an extrusion method used to deposit filaments of thermoplastics. The layout for FDM consists of a printhead able to move along X and Y directions above a built platform. The polymer is extruded through the heated nozzle and laid down as filaments according to the CAD design. The build platform is then moved along Z direction, and layers can be built until the object has the desired shape completed [184].

By the stereolithography (SLA) printing process, it is possible to fabricate 3D objects using light to selectively solidify a liquid resin through a photopolymerization reaction, achieving a resolution of about 20–40 µm [185,186]. This technique can be used to directly produce biomaterials from photopolymers and photoinitiators, as well as to create a negative replica of the desired structure. In the latter, the structure can then be filled with ceramic or metallic slurries, followed by sintering. In this way, a broader range of materials may be used [186].

In direct laser writing, ultraviolet light is directed towards a vat of photosensitive resin to form solid layers with a moving build platform. Therefore, this technique is based on the non-linear absorption of photons by the photopolymers, for which a laser beam is focused on the volume of a transparent material, leading to the absorption of two or more photons and polymerizing locally. This laser can be moved according to a path previously established in a computer, being able to reproduce a CAD model, for instance [187].

Products designed for efficient and controlled oral drug delivery obtained by this technique are exemplified in Table 4.

Table 4. *Cont.*

Biomaterial Production Technique	Biomaterial Application	Matrix Composition	Active Agent	Main Conclusion	Reference
3D printing	Mucoadhesive oral film	HPMC	Catechin hydrate	Flexible application of 3D bioprinters (semi-solid extrusion-type 3D printers) to prepare film formulations	[204]
	Oral disintegrating tablets	PVP/Starch/ Microcrystalline Cellulose	Warfarin sodium	Tablets prepared by the 3D technique showed uniform drug content, good mechanical properties, and presented fast disintegration and fast dissolution	[205]
	Oral fast-dissolving films	PEO/PVP/ Poloxamer (P 407 and P188)	Olanzapine	The films showed increased dissolution rates of the poorly water-soluble drug, consisting in a suitable formulation for fast drug absorption	[206]

6. Market Perspectives of Drug Delivery Biomaterials for the Oral Cavity

Several companies that provide market research services monitor the market and global tendencies of investment in the areas of drug delivery systems, materials for drug delivery in the oral cavity, and mucoadhesive drug delivery systems, e.g., Databridge [207], Future Marketing Insights [208], Grand View Research [209], Transparency Market Research [210], and Biospace [211], among others.

The data provided by these companies, despite being quite divergent in many points, show that these fields of investment are very attractive, given their market value and growth projections for the next years.

The Future Marketing Insights company, for instance, estimates the market value of oral controlled release drug delivery in US$ 34.1 billion for 2022, projecting a market value of US$ 68.4 billion by 2032. However, these data are focused on oral systems designed to deliver drugs in the gastrointestinal tract [208].

In the case of evaluation of oral transmucosal drugs market, most of the analyses cover different routes of penetration, such as buccal and sublingual mucosa, gingival and palatal tissues, as well as the evaluation of preferred distribution channels (e.g., retail, hospital and online pharmacies). In 2020, according to Transparency Market Research, the buccal mucosa segment dominated the global market and retail pharmacies were the favorite distribution channel (around 50% of the total), with significant growth being expected in the online pharmacy distribution channel segment [210].

The company Biospace pointed that the global oral transmucosal drugs market was of US$ 14 billion in 2020 [211]. Transparency Market Research predicts an increase to US$ 27 billion in 2031 for the same market, with a compound annual growth rate (CAGR) of around 6% [211]. Other companies estimate CAGR values for oral transmucosal drug market in the same neighborhoods: 6% (Transparent Market and Biospace) [210,211]; 7.2% (Future Marketing Insights) [208]; 7.6% (Databridge Company) [207], 9.2% (Grand View Research) [205].

There are some key factors that are driving the market growth, including the increasing prevalence of chronic disorders, growing importance to develop drug delivery systems tailored to the patient needs, and the increasing R&D expenditure related to drug delivery systems by pharma companies. The use of drug delivery systems, moreover, reduces the chances of side effects by optimizing the drug presence at the target site. This helps in minimizing the dose and undesired effects. In addition, this helps in reducing the overall product price by decreasing the amount of active pharmaceutical ingredient required.

According to Grand View Research [209], North America dominated the buccal drug delivery systems market with a share of 31.6% in 2020. Delivery devices in the oral cavity have gained prominence due to constant research and development in this field. The delivery of drugs via the oral mucosa is considered a convenient and highly appreciated choice, especially among elderly patients. With this delivery pathway, it is possible to eliminate first-pass metabolism, allowing pH-sensitive and easily degradable drugs to be administered more effectively. This route also allows rapid administration, making it a field of many opportunities.

According to Future Market Insights, the key companies profiled in the oral controlled release drug delivery technology segment include AbbVie (North Chicago, IL, USA), Amgen (Thousand Oaks, CA, USA), AstraZeneca Plc. (Cambridge, UK), Bayer (Leverkusen, Germany), BioNTech (Mainz, Germany), Boehringer Ingelheim (Ingelheim am Rhein, Germany), Bristol-Myers Squibb Company (New York, NY, USA), Eli Lilly (Indianapolis, IN, USA), F. Hoffmann-La Roche Ltd. (Basel, Switzerland), Gilead Sciences (Foster City, CA, USA), GlaxoSmithKline Plc. (London, UK), Johnson & Johnson (New Brunswick, NJ, USA), Merck & Co. Inc. (Rahway, NJ, USA), Moderna (Cambridge, MA, USA), Novartis AG (Cambridge, MA, USA), Pfizer Inc. (New York, NY, USA), Sanofi S.A. (Paris, France), Sun Pharmaceuticals (Mumbai, India), Takeda (Tokyo, Japan), and Viatris (Canonsburg, PA, USA) [208].

Specifically in the oral transmucosal drugs market, the company Transparency Market Research pointed as main players, in addition to AstraZeneca (Cambridge, UK), Bristol-Myers Squibb Company (New York, NY, USA) and GlaxoSmithKline Plc (Brentford, UK), Access Pharmaceutical Inc. (Tokyo, Japan), Aquestive Therapeutics Inc. (Warren, NJ, USA), C.L Pharm (Seoul, Republic of Korea), Cure Pharmaceutical, Eisai Co. Ltd. Inc. (Tokyo, Japan), Solvay S.A. (Brussels, Belgium), IntelGenx Corp. (Montreal, QC, Canada), Izun Pharmaceuticals, Ltd. (New York, NY, USA), Pfizer Inc. (New York, NY, USA), LTS Lohmann Therapie-Systeme AG (Andernach, Germany), Mylan N.V. (Canonsburg, PA, USA), NAL Pharma (Hong Kong, China), Otsuka Pharmaceutical Co. (Tokyo, Japan), Seoul Pharmaceuticals (Seoul, Republic of Korea), Soligenix (Princeton, NJ, USA), Teva Pharmaceutical Industries Ltd. (Tel Aviv, Israel), and ZIM Laboratories Limited (Nagpur, India) [210].

The high number of companies operating in this market reinforces the perception of a trend towards increasing investments and technology development in this sector.

7. Conclusions

Although numerous formulations and biomaterials for drug delivery are described in the literature, drug delivery in the oral cavity is still relatively unexplored, especially when compared to drug delivery to the skin. There is a therapeutic need for drug delivery in this complex region which still presents many challenges for treating local oral diseases. This approach is also interesting because it is easy to be used by children, the elderly, and other people with difficulties, e.g., swallowing medication. In the literature, several studies describe the use of fast dissolving oral films, mucoadhesive formulations, such as tablets, films, and gels; and microneedles, which is a more recent category with a wide range of applications and is highly promising. Biomaterials for application in the oral cavity can be composed of synthetic and natural polymers. Natural polymers represent a highly promising class, as they are more biocompatible than synthetic ones. Finally, these devices can be produced by solvent casting, electrospinning, hot-melt extrusion, and 3D printing, among other techniques. The 3D printing technique is one of the most applied due to its versatility and for being customizable. While many challenges are still present in this field, there are several opportunities to produce oral cavity systems to promote drug delivery, which is well illustrated by the high number of companies focusing on this particular area.

Author Contributions: Conceptualization, P.d.F.R.R., Â.M.M., M.H.T.N. and R.A.A.; methodology P.d.F.R.R., Â.M.M., M.H.T.N. and R.A.A., resources, P.d.F.R.R., M.H.T.N. and R.A.A.; data curation, P.d.F.R.R. and Â.M.M.; writing—original draft preparation, P.d.F.R.R., Â.M.M., M.H.T.N., R.A.A., M.A.d. and M.F.-M.; writing—review and editing, P.d.F.R.R. and Â.M.M.; supervision, Â.M.M., M.A.d. and M.F.-M.; project administration, P.d.F.R.R. and Â.M.M.; funding acquisition, P.d.F.R.R. and Â.M.M. All authors have read and agreed to the published version of the manuscript.

Funding: The authors acknowledge the support from the São Paulo Research Foundation (FAPESP, process # 2018/05237-2), the National Council for Scientific and Technological Development (CNPq, grant # 314724/2021-4) and the Coordination for the Improvement of Higher Education Personnel (CAPES, finance code 001), all from Brazil.

Institutional Review Board Statement: Not applicable.

Informed Consent Statement: Not applicable.

Data Availability Statement: Not applicable.

Acknowledgments: The authors are grateful to the São Paulo Research Foundation (FAPESP), the National Council for Scientific and Technological Development (CNPq) and the Coordination for the Improvement of Higher Education Personnel (CAPES) for the support of this investigation.

Conflicts of Interest: The authors declare no conflict of interest.

References

1. Kwatra, S.; Taneja, G.; Nasa, N. Alternative routes of drug administration—Transdermal, pulmonary & parenteral. *Indo Glob. J. Pharm. Sci.* **2012**, *2*, 409–426. [CrossRef]
2. Alqahtani, M.S.; Kazi, M.; Alsenaidy, M.A.; Ahmad, M.Z. Advances in oral drug delivery. *Front. Pharmacol.* **2021**, *12*, 618411. [CrossRef] [PubMed]
3. Alves, L.M.T.; Cassiani, R.D.A.; Dos Santos, C.M.; Dantas, R.O. Gender effect on the clinical measurement of swallowing. *Arq. Gastroenterol.* **2007**, *44*, 227–229. [CrossRef] [PubMed]
4. Clavé, P.; Shaker, R. Dysphagia: Current reality and scope of the problem. *Nat. Rev. Gastroenterol. Hepatol.* **2015**, *12*, 259–270. [CrossRef]
5. Camargo, L.G.; Remiro, P.D.F.R.; Rezende, G.S.; Santos, S.D.C.; Franz-Montan, M.; Moraes, Â.M. Development of bioadhesive polysaccharide-based films for topical release of the immunomodulatory agent imiquimod on oral mucosa lesions. *Eur. Polym. J.* **2021**, *151*, 110422. [CrossRef]
6. Donnelly, R.F.; Shaikh, R.; Singh, R.R.T.; Garland, M.J.; Woolfson, A.D. Mucoadhesive drug delivery systems. *J. Pharm. Bioallied Sci.* **2011**, *3*, 89–100. [CrossRef]
7. Franz-Montan, M.; de Araújo, D.R.; de Morais Ribeiro, L.N.; de Melo, N.F.S.; de Paula, E. Nanostructured systems for transbuccal drug delivery. In *Nanostructures for Oral Medicine*; Elsevier: Amsterdam, The Netherlands, 2017; pp. 87–121.
8. Kulkarni, U.; Mahalingam, R.; Pather, S.I.; Li, X.; Jasti, B. Porcine buccal mucosa as an in vitro model: Relative contribution of epithelium and connective tissue as permeability barriers. *J. Pharm. Sci.* **2009**, *98*, 471–483. [CrossRef]
9. Hassan, N.; Ahad, A.; Ali, M.; Ali, J. Chemical permeation enhancers for transbuccal drug delivery. *Expert Opin. Drug Deliv.* **2010**, *7*, 97–112. [CrossRef]
10. Paderni, C.; Compilato, D.; Giannola, L.I.; Campisi, G. Oral local drug delivery and new perspectives in oral drug formulation. *Oral Surg. Oral Med. Oral Pathol. Oral Radiol.* **2012**, *114*, e25–e34. [CrossRef]
11. Sanz, R.; Calpena, A.C.; Mallandrich, M.; Clares, B. Enhancing topical analgesic administration: Review and prospect for transdermal and transbuccal drug delivery systems. *Curr. Pharm. Des.* **2015**, *21*, 2867–2882. [CrossRef]
12. Franz-Montan, M.; Ribeiro, L.N.D.M.; Volpato, M.C.; Cereda, C.M.S.; Groppo, F.C.; Tofoli, G.R.; Araújo, D.R.; Santi, P.; Padula, C.; de Paula, E. Recent advances and perspectives in topical oral anesthesia. *Expert Opin. Drug Deliv.* **2017**, *14*, 673–684. [CrossRef] [PubMed]
13. Triggle David, J.; Taylor John, B. (Eds.) *Comprehensive Medicinal Chemistry II*; Elsevier: Amsterdam, The Netherlands, 2006.
14. Guilherme, V.A.; Ribeiro, L.N.; Tofoli, G.R.; Franz-Montan, M.; de Paula, E.; de Jesus, M.B. Current challenges and future of lipid nanoparticles formulations for topical drug application to oral mucosa, skin, and eye. *Curr. Pharm. Des.* **2017**, *23*, 6659–6675. [CrossRef] [PubMed]
15. Dixit, R.P.; Puthli, S.P. Oral strip technology: Overview and future potential. *J. Control. Release* **2009**, *139*, 94–107. [CrossRef] [PubMed]
16. Bala, R.; Khanna, S.; Pawar, P.; Arora, S. Orally dissolving strips: A new approach to oral drug delivery system. *Int. J. Pharm. Investig.* **2013**, *3*, 67–76. [CrossRef] [PubMed]
17. Bhatia, S. Natural Polymers vs. Synthetic Polymer. In *Natural Polymer Drug Delivery Systems*; Springer: Berlin/Heidelberg, Germany, 2016.
18. Łyszczarz, E.; Brniak, W.; Szafraniec-Szczęsny, J.; Majka, T.; Majda, D.; Zych, M.; Pielichowski, K.; Jachowicz, R. The Impact of the Preparation Method on the Properties of Orodispersible Films with Aripiprazole: Electrospinning vs. Casting and 3D Printing Methods. *Pharmaceutics* **2021**, *13*, 1122. [CrossRef]
19. Salawi, A. An Insight into Preparatory Methods and Characterization of Orodispersible Film—A Review. *Pharmaceuticals* **2022**, *15*, 844. [CrossRef]
20. Kalyan, S.; Bansal, M. Recent Trends in the Development of Oral dissolving Film. *Int. J. Pharm. Tech. Res.* **2012**, *4*, 725–733.
21. Mostafa, D.A.E. Fast dissolving oral film: Overview. *Eur. J. Biomed. Pharm. Sci.* **2018**, *5*, 86–101.
22. Bhyan, B.; Jangra, S.; Kaur, M.; Singh, H. Orally fast dissolving films. Innovations in formulation and technology. *Int. J. Pharm. Sci. Rev. Res.* **2011**, *9*, 009.
23. Gavaskar, B.; Vijaya Kumar, S.; Sharan, G.; Madhusudan Rao, Y. Overview on fast dissolving films. *Int. J. Pharm. Pharm Sci.* **2010**, *2*, 0975–1491.
24. Prabhu, P.; Malli, R.; Koland, M.; Vijaynarayana, K.; D'souza, U.; Harish, N.; Shastry, C.; Charyulu, R. Formulation and evaluation of fast dissolving films of levocitirizine di hydrochloride. *Int. J. Pharm. Investig.* **2011**, *1*, 99–104. [CrossRef] [PubMed]
25. Londhe, V.Y.; Umalkar, K.B. Formulation Development and Evaluation of Fast Dissolving Film of Telmisartan. *Indian J. Pharm. Sci.* **2012**, *74*, 122–126. [CrossRef] [PubMed]
26. Illangakoon, U.E.; Gill, H.; Shearman, G.C.; Parhizkar, M.; Mahalingam, S.; Chatterton, N.P.; Williams, G.R. Fast dissolving paracetamol/caffeine nanofibers prepared by electrospinning. *Int. J. Pharm.* **2014**, *477*, 369–379. [CrossRef] [PubMed]
27. Sheela, A.; Haque, S.E. Development of polymer-bound fast-dissolving metformin buccal film with disintegrants. *Int. J. Nanomed.* **2015**, *10*, 199–205. [CrossRef]
28. Chonkar, A.D.; Rao, J.V.; Managuli, R.S.; Mutalik, S.; Dengale, S.; Jain, P.; Udupa, N. Development of fast dissolving oral films containing lercanidipine HCl nanoparticles in semicrystalline polymeric matrix for enhanced dissolution and ex vivo permeation. *Eur. J. Pharm. Biopharm.* **2016**, *103*, 179–191. [CrossRef]
29. Nippani, A.; Vijendar, C.; Dindigala, A.; Kandhula, A.G.; Chandra, S.K.; Alabadri, A. Preparation and In-Vitro Evaluation of Mirtazapine Oral Films. *Res. Rev. J. Pharm. Sci.* **2016**, *5*, 96–103.

30. Kim, B.-S.; Park, G.-T.; Park, M.-H.; Shin, Y.G.; Cho, C.-W. Preparation and evaluation of oral dissolving film containing local anesthetic agent, lidocaine. *J. Pharm. Investig.* **2016**, *47*, 575–581. [CrossRef]
31. Sallam, N.M.; Sanad, R.A.; Kharshoom, R.M.; Zeneldin, M.A. Development of Salbutamol Sulphate Sublingual Films in Pullulan Matrix for Enhanced Bioavailability & Clinical Efficacy. *Curr. Drug. Deliv.* **2017**, *14*, 503–515. [CrossRef]
32. Garcia, V.A.D.S.; Borges, J.G.; Maciel, V.B.V.; Mazalli, M.R.; Lapa-Guimaraes, J.D.G.; Vanin, F.M.; de Carvalho, R.A. Gelatin/starch orally disintegrating films as a promising system for vitamin C delivery. *Food Hydrocoll.* **2018**, *79*, 127–135. [CrossRef]
33. Qin, Z.Y.; Jia, X.W.; Liu, Q.; Kong, B.H.; Wang, H. Fast dissolving oral films for drug delivery prepared from chitosan/pullulan electrospinning nanofibers. *Int. J. Biol. Macromol.* **2019**, *137*, 224–231. [CrossRef]
34. Wannaphatchaiyong, S.; Heng, P.W.S.; Suksaeree, J.; Boonme, P.; Pichayakorn, W. Lidocaine loaded gelatin/gelatinized tapioca starch films for buccal delivery and the irritancy evaluation using chick chorioallantoic membrane. *Saudi Pharm. J.* **2019**, *27*, 1085–1095. [CrossRef] [PubMed]
35. Al-Mogherah, A.I.; Ibrahim, M.A.; Hassan, M.A. Optimization and evaluation of venlafaxine hydrochloride fast dissolving oral films. *Saudi Pharm. J.* **2020**, *28*, 1374–1382. [CrossRef] [PubMed]
36. Elshafeey, A.H.; El-Dahmy, R.M. Formulation and Development of Oral Fast-Dissolving Films Loaded with Nanosuspension to Augment Paroxetine Bioavailability: In Vitro Characterization, Ex Vivo Permeation, and Pharmacokinetic Evaluation in Healthy Human Volunteers. *Pharmaceutics* **2021**, *13*, 1869. [CrossRef]
37. Celik, B. Risperidone mucoadhesive buccal tablets: Formulation design, optimization and evaluation. *Drug Des. Dev. Ther.* **2017**, *11*, 3355–3365. [CrossRef]
38. Gajdziok, J.; Bajerová, M.; Chalupová, Z.; Rabišková, M. Oxycellulose as mucoadhesive polymer in buccal tablets. *Drug Dev. Ind. Pharm.* **2010**, *36*, 1115–1130. [CrossRef] [PubMed]
39. Abruzzo, A.; Cerchiara, T.; Bigucci, F.; Gallucci, M.C.; Luppi, B. Mucoadhesive Buccal Tablets Based on Chitosan/Gelatin Microparticles for Delivery of Propranolol Hydrochloride. *J. Pharm. Sci.* **2015**, *104*, 4365–4372. [CrossRef]
40. Carvalho, F.C.; Bruschi, M.L.; Evangelista, R.C.; Gremião, M.P.D. Mucoadhesive drug delivery systems. *Braz. J. Pharm. Sci.* **2010**, *46*, 1–17. [CrossRef]
41. Pham, Q.D.; Nöjd, S.; Edman, M.; Lindell, K.; Topgaard, D.; Wahlgren, M. Mucoadhesion: Mucin-polymer molecular interactions. *Int. J. Pharm.* **2021**, *610*, 121245. [CrossRef]
42. Laffleur, F. Mucoadhesive polymers for buccal drug delivery. *Drug Dev. Ind. Pharm.* **2014**, *40*, 591–598. [CrossRef]
43. Boyapally, H.; Nukala, R.K.; Bhujbal, P.; Douroumis, D. Controlled release from directly compressible theophylline buccal tablets. *Colloids Surf. B Biointerfaces* **2010**, *77*, 227–233. [CrossRef]
44. Koirala, S.; Nepal, P.; Ghimire, G.; Basnet, R.; Rawat, I.; Dahal, A.; Pandey, J.; Parajuli-Baral, K. Formulation and evaluation of mucoadhesive buccal tablets of aceclofenac. *Heliyon* **2021**, *7*, e06439. [CrossRef] [PubMed]
45. Chandira, M.; Sachin, D.B.; Jayakar, B. Formulation and evaluation of mucoadhesive oral tablet of Clarithromycin. *Pharm. Res.* **2009**, *2*, 30–42.
46. Salehi, S.; Boddohi, S. Design and optimization of kollicoat ®IR based mucoadhesive buccal film for co-delivery of rizatriptan benzoate and propranolol hydrochloride. *Mater. Sci. Eng. C Mater. Biol. Appl.* **2018**, *97*, 230–244. [CrossRef]
47. Morales, J.O.; McConville, J.T. Manufacture and characterization of mucoadhesive buccal films. *Eur. J. Pharm. Biopharm.* **2011**, *77*, 187–199. [CrossRef] [PubMed]
48. Jovanović, M.; Tomić, N.; Cvijić, S.; Stojanović, D.; Ibrić, S.; Uskoković, P. Mucoadhesive gelatin buccal films with propranolol hydrochloride: Evaluation of mechanical, mucoadhesive, and biopharmaceutical properties. *Pharmaceutics* **2021**, *13*, 273. [CrossRef]
49. Fini, A.; Bergamante, V.; Ceschel, G.C. Mucoadhesive Gels Designed for the Controlled Release of Chlorhexidine in the Oral Cavity. *Pharmaceutics* **2011**, *3*, 665–679. [CrossRef] [PubMed]
50. Tangsuksan, P.; Chuerduangphui, J.; Yupanqui, C.T.; Srichana, T.; Hitakomate, E.; Pientong, C.; Ekalaksananan, T.; Nittayananta, W. Mucoadhesive film containing α-mangostin shows potential role in oral cancer treatment. *BMC Oral Health* **2021**, *21*, 512. [CrossRef]
51. del Consuelo, I.D.; Falson, F.; Guy, R.H.; Jacques, Y. Ex vivo evaluation of bioadhesive films for buccal delivery of fentanyl. *J. Control. Release* **2007**, *122*, 135–140. [CrossRef]
52. Nair, A.; Shah, J.; Jacob, S.; Al-Dhubiab, B.; Patel, V.; Sreeharsha, N.; Shinu, P. Development of Mucoadhesive Buccal Film for Rizatriptan: In Vitro and In Vivo Evaluation. *Pharmaceutics* **2021**, *13*, 728. [CrossRef]
53. Batista, P.; Castro, P.; Madureira, A.R.; Sarmento, B.; Pintado, M. Development and Characterization of Chitosan Microparticles-in-Films for Buccal Delivery of Bioactive Peptides. *Pharmaceuticals* **2019**, *12*, 32. [CrossRef]
54. Shao, Y.; Zhou, H. Clinical evaluation of an oral mucoadhesive film containing chitosan for the treatment of recurrent aphthous stomatitis: A randomized, double-blind study. *J. Dermatol. Treat.* **2020**, *31*, 739–743. [CrossRef] [PubMed]
55. Muniz, B.V.; Baratelli, D.; Di Carla, S.; Serpe, L.; da Silva, C.B.; Guilherme, V.A.; Ribeiro, L.N.M.; Cereda, C.M.S.; de Paula, E.; Volpato, M.C.; et al. Hybrid hydrogel composed of polymeric nanocapsules co-loading lidocaine and prilocaine for topical intraoral anesthesia. *Sci. Rep.* **2018**, *8*, 17972. [CrossRef] [PubMed]
56. Ribeiro, L.N.M.; Franz-Montan, M.; Alcântara, A.C.S.; Breitkreitz, M.C.; Castro, S.R.; Guilherme, V.A.; Muniz, B.V.; da Silva, G.H.R.; De Paula, E. Hybrid nanofilms as topical anesthetics for pain-free procedures in dentistry. *Sci. Rep.* **2020**, *10*, 11341. [CrossRef] [PubMed]
57. Castro, P.M.; Baptista, P.; Zuccheri, G.; Madureira, A.R.; Sarmento, B.; Pintado, M.E. Film-nanoparticle composite for enhanced oral delivery of alpha-casozepine. *Colloids Surf. B Biointerfaces* **2019**, *181*, 149–157. [CrossRef] [PubMed]

58. Wang, M.; Hu, L.; Xu, C. Recent advances in the design of polymeric microneedles for transdermal drug delivery and biosensing. *Lab Chip* **2017**, *17*, 1373–1387. [CrossRef]
59. Adams, D. Penetration of Water through Human and Rabbi Oral Mucosa in vitro. *Arch. Oral Biol.* **1974**, *19*, 865–872. [CrossRef]
60. Santos, S.D.C.; Fávaro-Moreira, N.C.; Abdalla, H.B.; Augusto, G.G.X.; Costa, Y.M.; Volpato, M.C.; Groppo, F.C.; Gill, H.S.; Franz-Montan, M. A crossover clinical study to evaluate pain intensity from microneedle insertion in different parts of the oral cavity. *Int. J. Pharm.* **2020**, *592*, 120050. [CrossRef]
61. Ma, Y.; Boese, S.E.; Luo, Z.; Nitin, N.; Gill, H.S. Drug coated microneedles for minimally-invasive treatment of oral carcinomas: Development and in vitro evaluation. *Biomed. Microdevices* **2015**, *17*, 44. [CrossRef]
62. Zhao, X.; Coulman, S.A.; Hanna, S.; Wong, F.S.; Dayan, C.M.; Birchall, J.C. Formulation of hydrophobic peptides for skin delivery via coated microneedles. *J. Control. Release* **2017**, *265*, 2–13. [CrossRef]
63. Kim, Y.-C.; Quan, F.-S.; Yoo, D.-G.; Compans, R.W.; Kang, S.-M.; Prausnitz, M.R. Improved influenza vaccination in the skin using vaccine coated microneedles. *Vaccine* **2009**, *27*, 6932–6938. [CrossRef]
64. Jiang, J.; Gill, H.S.; Ghate, D.; McCarey, B.E.; Patel, S.R.; Edelhauser, H.F.; Prausnitz, M.R. Coated Microneedles for Drug Delivery to the Eye. *Investig. Opthalmol. Vis. Sci.* **2007**, *48*, 4038–4043. [CrossRef] [PubMed]
65. Kim, Y.C.; Grossniklaus, H.E.; Edelhauser, H.F.; Prausnitz, M.R. Intrastromal Delivery of Bevacizumab Using Microneedles to Treat Corneal Neovascularization. *Investig. Opthalmol. Vis. Sci.* **2014**, *55*, 7376–7386. [CrossRef] [PubMed]
66. Bhatnagar, S.; Gadeela, P.R.; Thathireddy, P.; Venuganti, V.V.K. Microneedle-based drug delivery: Materials of construction. *J. Chem. Sci.* **2019**, *131*, 90. [CrossRef]
67. Ashraf, M.W.; Tayyaba, S.; Afzulpurkar, N.; Nisar, A. Fabrication and analysis of tapered tip silicon microneedles for mems based drug delivery system. *Sens. Transducer* **2010**, *122*, 158–173.
68. Ashraf, M.W.; Tayyaba, S.; Afzulpurkar, N. Micro Electromechanical Systems (MEMS) Based Microfluidic Devices for Biomedical Applications. *Int. J. Mol. Sci.* **2011**, *12*, 3648–3704. [CrossRef]
69. Cahill, E.M.; Keaveney, S.; Stuettgen, V.; Eberts, P.; Ramos-Luna, P.; Zhang, N.; Dangol, M.; O'Cearbhaill, E.D. Corrigendum to "Metallic microneedles with interconnected porosity: A scalable platform for biosensing and drug delivery". *Acta Biomater.* **2018**, *86*, 493–495. [CrossRef]
70. Davis, S.P.; Martanto, W.; Allen, M.G.; Prausnitz, M.R. Hollow metal microneedles for insulin delivery to diabetic rats. *IEEE Trans. Biomed. Eng.* **2005**, *52*, 909–915. [CrossRef]
71. Ullah, A.; Kim, C.M.; Kim, G.M. Porous polymer coatings on metal microneedles for enhanced drug delivery. *R. Soc. Open Sci.* **2018**, *5*, 171609. [CrossRef]
72. Ma, Y.; Tao, W.; Krebs, S.J.; Sutton, W.F.; Haigwood, N.L.; Gill, H.S. Vaccine Delivery to the Oral Cavity Using Coated Microneedles Induces Systemic and Mucosal Immunity. *Pharm. Res.* **2014**, *31*, 2393–2403. [CrossRef]
73. Wang, P.M.; Cornwell, M.; Hill, J.; Prausnitz, M.R. Precise Microinjection into Skin Using Hollow Microneedles. *J. Investig. Dermatol.* **2006**, *126*, 1080–1087. [CrossRef]
74. Bolton, C.J.W.; Howells, O.; Blayney, G.J.; Eng, P.F.; Birchall, J.C.; Gualeni, B.K.; Roberts, K.; Ashraf, H.; Guy, O.J. Hollow silicon microneedle fabrication using advanced plasma etch technologies for applications in transdermal drug delivery. *Lab Chip* **2020**, *20*, 2788–2795. [CrossRef] [PubMed]
75. Narayanan, S.P.; Raghavan, S. Solid silicon microneedles for drug delivery applications. *Int. J. Adv. Manuf. Technol.* **2016**, *93*, 407–422. [CrossRef]
76. Li, Y.; Zhang, H.; Yang, R.; Laffitte, Y.; Schmill, U.; Hu, W.; Kaddoura, M.; Blondeel, E.J.M.; Cui, B. Fabrication of sharp silicon hollow microneedles by deep-reactive ion etching towards minimally invasive diagnostics. *Microsyst. Nanoeng.* **2019**, *5*, 41. [CrossRef] [PubMed]
77. Battisti, M.; Vecchione, R.; Casale, C.; Pennacchio, F.A.; Lettera, V.; Jamaledin, R.; Profeta, M.; Di Natale, C.; Imparato, G.; Urciuolo, F.; et al. Non-invasive Production of Multi-Compartmental Biodegradable Polymer Microneedles for Controlled Intradermal Drug Release of Labile Molecules. *Front. Bioeng. Biotechnol.* **2019**, *7*, 296. [CrossRef]
78. Luzuriaga, M.A.; Berry, D.R.; Reagan, J.C.; Smaldone, R.A.; Gassensmith, J.J. Biodegradable 3D printed polymer microneedles for transdermal drug delivery. *Lab Chip* **2018**, *18*, 1223–1230. [CrossRef] [PubMed]
79. Yu, W.; Jiang, G.; Zhang, Y.; Liu, D.; Xu, B.; Zhou, J. Polymer microneedles fabricated from alginate and hyaluronate for transdermal delivery of insulin. *Mater. Sci. Eng. C Mater. Biol. Appl.* **2017**, *80*, 187–196. [CrossRef] [PubMed]
80. Bonfante, G.; Lee, H.; Bao, L.; Park, J.; Takama, N.; Kim, B. Comparison of polymers to enhance mechanical properties of microneedles for bio-medical applications. *Micro Nano Syst. Lett.* **2020**, *8*, 13. [CrossRef]
81. Demir, Y.K.; Kerimoglu, O. Novel use of pectin as a microneedle base. *Chem. Pharm. Bull.* **2015**, *63*, 300–304. [CrossRef]
82. Vallhov, H.; Xia, W.; Engqvist, H.; Scheynius, A. Bioceramic microneedle arrays are able to deliver OVA to dendritic cells in human skin. *J. Mater. Chem. B* **2018**, *6*, 6808–6816. [CrossRef]
83. Jiang, J.; Moore, J.; Edelhauser, H.F.; Prausnitz, M.R. Intrascleral Drug Delivery to the Eye Using Hollow Microneedles. *Pharm. Res.* **2008**, *26*, 395–403. [CrossRef]
84. Lyon, B.J.; Aria, A.I.; Gharib, M. Fabrication of carbon nanotube-polyimide composite hollow microneedles for transdermal drug delivery. *Biomed. Microdevices* **2014**, *16*, 879–886. [CrossRef] [PubMed]
85. Rzhevskiy, A.S.; Singh, T.R.R.; Donnelly, R.F.; Anissimov, Y.G. Microneedles as the technique of drug delivery enhancement in diverse organs and tissues. *J. Control. Release* **2018**, *270*, 184–202. [CrossRef]

86. Donnelly, R.F.; Singh, R.R.T.; Woolfson, A.D. Microneedle-based drug delivery systems: Microfabrication, drug delivery, and safety. *Drug Deliv.* **2010**, *17*, 187–207. [CrossRef] [PubMed]
87. Tucak, A.; Sirbubalo, M.; Hindija, L.; Rahić, O.; Hadžiabdić, J.; Muhamedagić, K.; Čekić, A.; Vranić, E. Microneedles: Characteristics, Materials, Production Methods and Commercial Development. *Micromachines* **2020**, *11*, 961. [CrossRef] [PubMed]
88. Nair, K.J. *Micro-Injection Moulded Microneedles for Drug Delivery*; University of Bradford: Bradford, UK, 2014.
89. Ingrole, R.S.J.; Gill, H.S. Microneedle Coating Methods: A Review with a Perspective. *J. Pharmacol. Exp. Ther.* **2019**, *370*, 555–569. [CrossRef]
90. Gill, H.S.; Prausnitz, M.R. Coated microneedles for transdermal delivery. *J. Control. Release* **2007**, *117*, 227–237. [CrossRef]
91. Ita, K. Dissolving microneedles for transdermal drug delivery: Advances and challenges. *Biomed. Pharm.* **2017**, *93*, 1116–1127. [CrossRef]
92. Qiu, Y.; Li, C.; Zhang, S.; Yang, G.; He, M.; Gao, Y. Systemic delivery of artemether by dissolving microneedles. *Int. J. Pharm.* **2016**, *508*, 1–9. [CrossRef]
93. Yan, Q.; Wang, W.; Weng, J.; Zhang, Z.; Yin, L.; Yang, Q.; Guo, F.; Wang, X.; Chen, F.; Yang, G. Dissolving microneedles for transdermal delivery of huperzine A for the treatment of Alzheimer's disease. *Drug Deliv.* **2020**, *27*, 1147–1155. [CrossRef]
94. McCrudden, M.T.; Alkilani, A.Z.; McCrudden, C.M.; McAlister, E.; McCarthy, H.O.; Woolfson, A.D.; Donnelly, R.F. Design and physicochemical characterisation of novel dissolving polymeric microneedle arrays for transdermal delivery of high dose, low molecular weight drugs. *J. Control. Release* **2014**, *180*, 71–80. [CrossRef]
95. Lee, J.W.; Choi, S.-O.; Felner, E.I.; Prausnitz, M.R. Dissolving Microneedle Patch for Transdermal Delivery of Human Growth Hormone. *Small* **2011**, *7*, 531–539. [CrossRef] [PubMed]
96. Seon-Woo, H.-S.; Kim, H.J.; Roh, J.Y.; Park, J.-H. Dissolving Microneedle Systems for the Oral Mucosal Delivery of Triamcinolone Acetonide to Treat Aphthous Stomatitis. *Macromol. Res.* **2018**, *27*, 282–289. [CrossRef]
97. Caffarel-Salvador, E.; Kim, S.; Soares, V.; Tian, R.Y.; Stern, S.R.; Minahan, D.; Yona, R.; Lu, X.; Zakaria, F.R.; Collins, J.; et al. A microneedle platform for buccal macromolecule delivery. *Sci. Adv.* **2021**, *7*, eabe2620. [CrossRef] [PubMed]
98. Kim, Y.-C.; Park, J.-H.; Prausnitz, M.R. Microneedles for drug and vaccine delivery. *Adv. Drug Deliv. Rev.* **2012**, *64*, 1547–1568. [CrossRef] [PubMed]
99. Chai, Q.; Jiao, Y.; Yu, X. Hydrogels for Biomedical Applications: Their Characteristics and the Mechanisms behind Them. *Gels* **2017**, *3*, 6. [CrossRef]
100. Turner, J.G.; White, L.R.; Estrela, P.; Leese, H.S. Hydrogel-Forming Microneedles: Current Advancements and Future Trends. *Macromol. Biosci.* **2020**, *21*, e2000307. [CrossRef] [PubMed]
101. Hong, X.; Wu, Z.; Chen, L.; Wu, F.; Wei, L.; Yuan, W. Hydrogel Microneedle Arrays for Transdermal Drug Delivery. *Nano-Micro Lett.* **2014**, *6*, 191–199. [CrossRef]
102. Chu, L.Y.; Choi, S.-O.; Prausnitz, M.R. Fabrication of Dissolving Polymer Microneedles for Controlled Drug Encapsulation and Delivery: Bubble and Pedestal Microneedle Designs. *J. Pharm. Sci.* **2010**, *99*, 4228–4238. [CrossRef]
103. Hong, X.; Wei, L.; Wu, F.; Wu, Z.; Chen, L.; Liu, Z.; Yuan, W. Dissolving and biodegradable microneedle technologies for transdermal sustained delivery of drug and vaccine. *Drug Des. Dev. Ther.* **2013**, *7*, 945–952. [CrossRef]
104. Zhu, T.; Yu, X.; Yi, X.; Guo, X.; Li, L.; Hao, Y.; Wang, W. Lidocaine-Loaded Hyaluronic Acid Adhesive Microneedle Patch for Oral Mucosal Topical Anesthesia. *Pharmaceutics* **2022**, *14*, 686. [CrossRef]
105. Ye, Y.; Yu, J.; Wang, C.; Nguyen, N.Y.; Walker, G.M.; Buse, J.B.; Gu, Z. Microneedles integrated with pancreatic cells and synthetic glucose-signal amplifiers for smart insulin delivery. *Adv. Mater.* **2016**, *28*, 3115–3121. [CrossRef] [PubMed]
106. Singh, P.; Carrier, A.; Chen, Y.; Lin, S.; Wang, J.; Cui, S.; Zhang, X. Polymeric microneedles for controlled transdermal drug delivery. *J. Control. Release* **2019**, *315*, 97–113. [CrossRef] [PubMed]
107. Ye, Y.; Yu, J.; Wen, D.; Kahkoska, A.R.; Gu, Z. Polymeric microneedles for transdermal protein delivery. *Adv. Drug Deliv. Rev.* **2018**, *127*, 106–118. [CrossRef] [PubMed]
108. Donnelly, R.F.; Morrow, D.I.J.; Singh, T.R.R.; Migalska, K.; McCarron, P.A.; O'Mahony, C.; Woolfson, A.D. Processing difficulties and instability of carbohydrate microneedle arrays. *Drug Dev. Ind. Pharm.* **2009**, *35*, 1242–1254. [CrossRef]
109. Vora, L.K.; Courtenay, A.J.; Tekko, I.A.; Larrañeta, E.; Donnelly, R.F. Pullulan-based dissolving microneedle arrays for enhanced transdermal delivery of small and large biomolecules. *Int. J. Biol. Macromol.* **2020**, *146*, 290–298. [CrossRef]
110. Zhang, N.; Zhou, X.; Liu, L.; Zhao, L.; Xie, H.; Yang, Z. Dissolving Polymer Microneedles for Transdermal Delivery of Insulin. *Front. Pharmacol.* **2021**, *12*, 719905. [CrossRef]
111. Nguyen, H.X.; Bozorg, B.D.; Kim, Y.; Wieber, A.; Birk, G.; Lubda, D.; Banga, A.K. Poly (vinyl alcohol) microneedles: Fabrication, characterization, and application for transdermal drug delivery of doxorubicin. *Eur. J. Pharm. Biopharm.* **2018**, *129*, 88–103. [CrossRef]
112. Cole, G.; McCaffrey, J.; Ali, A.A.; McBride, J.W.; McCrudden, C.M.; Vincente-Perez, E.M.; Donnelly, R.; McCarthy, H.O. Dissolving microneedles for DNA vaccination: Improving functionality via polymer characterization and RALA complexation. *Hum. Vaccines Immunother.* **2017**, *13*, 50–62. [CrossRef]
113. Zhao, X.; Li, X.; Zhang, P.; Du, J.; Wang, Y. Tip-loaded fast-dissolving microneedle patches for photodynamic therapy of subcutaneous tumor. *J. Control. Release* **2018**, *286*, 201–209. [CrossRef]
114. Du, H.; Liu, P.; Zhu, J.; Lan, J.; Li, Y.; Zhang, L.; Zhu, J.; Tao, J. Hyaluronic acid-based dissolving microneedle patch loaded with methotrexate for improved treatment of psoriasis. *ACS Appl. Mater. Interfaces* **2019**, *11*, 43588–43598. [CrossRef]
115. Aung, N.N.; Ngawhirunpat, T.; Rojanarata, T.; Patrojanasophon, P.; Opanasopit, P.; Pamornpathomkul, B. HPMC/PVP Dissolving Microneedles: A Promising Delivery Platform to Promote Trans-Epidermal Delivery of Alpha-Arbutin for Skin Lightening. *AAPS PharmSciTech* **2019**, *21*, 25. [CrossRef] [PubMed]

116. Kim, J.-Y.; Han, M.-R.; Kim, Y.-H.; Shin, S.-W.; Nam, S.-Y.; Park, J.-H. Tip-loaded dissolving microneedles for transdermal delivery of donepezil hydrochloride for treatment of Alzheimer's disease. *Eur. J. Pharm. Biopharm.* **2016**, *105*, 148–155. [CrossRef] [PubMed]
117. Fukushima, K.; Ise, A.; Morita, H.; Hasegawa, R.; Ito, Y.; Sugioka, N.; Takada, K. Two-layered dissolving microneedles for percutaneous delivery of peptide/protein drugs in rats. *Pharm. Res.* **2011**, *28*, 7–21. [CrossRef] [PubMed]
118. Ling, M.-H.; Chen, M.-C. Dissolving polymer microneedle patches for rapid and efficient transdermal delivery of insulin to diabetic rats. *Acta Biomater.* **2013**, *9*, 8952–8961. [CrossRef] [PubMed]
119. Arshad, M.S.; Hassan, S.; Hussain, A.; Abbas, N.; Kucuk, I.; Nazari, K.; Ali, R.; Ramzan, S.; Alqahtani, A.; Andriotis, E.G.; et al. Improved transdermal delivery of cetirizine hydrochloride using polymeric microneedles. *DARU J. Pharm. Sci.* **2019**, *27*, 673–681. [CrossRef]
120. Demir, Y.K.; Akan, Z.; Kerimoglu, O. Characterization of Polymeric Microneedle Arrays for Transdermal Drug Delivery. *PLoS ONE* **2013**, *8*, e77289. [CrossRef]
121. Castilla-Casadiego, D.A.; Carlton, H.; Gonzalez-Nino, D.; Miranda-Muñoz, K.A.; Daneshpour, R.; Huitink, D.; Prinz, G.; Powell, J.; Greenlee, L.; Almodovar, J. Design, characterization, and modeling of a chitosan microneedle patch for transdermal delivery of meloxicam as a pain management strategy for use in cattle. *Mater. Sci. Eng. C Mater. Biol. Appl.* **2021**, *118*, 111544. [CrossRef]
122. Na, Y.-G.; Kim, M.; Han, M.; Huh, H.W.; Kim, J.-S.; Kim, J.C.; Park, J.-H.; Lee, H.-K.; Cho, C.-W. Characterization of Hepatitis B Surface Antigen Loaded Polylactic Acid-Based Microneedle and Its Dermal Safety Profile. *Pharmaceutics* **2020**, *12*, 531. [CrossRef]
123. Nguyen, H.X.; Banga, A.K. Delivery of Methotrexate and Characterization of Skin Treated by Fabricated PLGA Microneedles and Fractional Ablative Laser. *Pharm. Res.* **2018**, *35*, 68. [CrossRef]
124. Wang, Q.L.; Zhu, D.D.; Chen, Y.; Guo, X.D. A fabrication method of microneedle molds with controlled microstructures. *Mater. Sci. Eng. C* **2016**, *65*, 135–142. [CrossRef]
125. Elahpour, N.; Pahlevanzadeh, F.; Kharaziha, M.; Bakhsheshi-Rad, H.R.; Ramakrishna, S.; Berto, F. 3D printed microneedles for transdermal drug delivery: A brief review of two decades. *Int. J. Pharm.* **2021**, *597*, 120301. [CrossRef] [PubMed]
126. Dabbagh, S.R.; Sarabi, M.R.; Rahbarghazi, R.; Sokullu, E.; Yetisen, A.K.; Tasoglu, S. 3D-printed microneedles in biomedical applications. *iScience* **2020**, *24*, 102012. [CrossRef] [PubMed]
127. Pires, A.L.R.; Bierhalz, A.C.; Moraes, A.M. Biomaterials: Types, applications, and market. *Química Nova* **2015**, *38*, 957–971. [CrossRef]
128. Ratner, B.D.; Hoffman, A.S.; Shoen, F.J.; Lemons, J.E. *Biomaterials Science: An Introduction to Materials in Medicine*; Elsevier: Amsterdam, The Netherlands, 2013.
129. ECHA. Available online: https://echa.europa.eu/documents/10162/23036412/polymers_en.pdf/9a74545f-05be-4e10-8555-4d7cf051bbed. (accessed on 4 November 2021).
130. Muhamad, I.; Selvakumaran, S.; Lazim, N. Designing polymeric nanoparticles for targeted drug delivery system. In *Nanomedicine*, 1st ed.; Seifalian, A., De Mel, A., Kalaskar, D.M., Eds.; One Central Press: Manchester, UK, 2014; pp. 287–313.
131. Ogaji, I.J.; Nep, E.; Audu-Peter, J.D. Advances in Natural Polymers as Pharmaceutical Excipients. *Pharm. Anal. Acta* **2012**, *3*, 146. [CrossRef]
132. Amiri, M.S.; Mohammadzadeh, V.; Yazdi, M.E.T.; Barani, M.; Rahdar, A.; Kyzas, G.Z. Plant-Based Gums and Mucilages Applications in Pharmacology and Nanomedicine: A Review. *Molecules* **2021**, *26*, 1770. [CrossRef]
133. Kaur, N.; Gupta, A.K. Applications of inulin and oligofructose in health and nutrition. *J. Biosci.* **2002**, *27*, 703–714. [CrossRef]
134. Pamlényi, K.; Kristó, K.; Jójárt-Laczkovich, O.; Regdon, G., Jr. Formulation and Optimization of Sodium Alginate Polymer Film as a Buccal Mucoadhesive Drug Delivery System Containing Cetirizine Dihydrochloride. *Pharmaceutics* **2021**, *13*, 619. [CrossRef]
135. Nazari, K.; Kontogiannidou, E.; Ahmad, R.H.; Gratsani, A.; Rasekh, M.; Arshad, M.S.; Sunar, B.S.; Armitage, D.; Bouropoulos, N.; Chang, M.-W.; et al. Development and characterisation of cellulose based electrospun mats for buccal delivery of non-steroidal anti-inflammatory drug (NSAID). *Eur. J. Pharm. Sci.* **2017**, *102*, 147–155. [CrossRef]
136. Alves, M.C.; Chaves, D.S.; Benevenuto, B.R.; De Farias, B.O.; Coelho, S.M.; Ferreira, T.P.; Pereira, G.A.; Dos Santos, G.C.; Moreira, L.O.; De Freitas, J.P.; et al. Chitosan gels for buccal delivery of *Schinus molle* L. essential oil in dogs: Characterization and antimicrobial activity in vitro. *An. Da Acad. Bras. De Ciências* **2020**, *92*, e20200562. [CrossRef]
137. Lakshmi, V.S.; Menon, R.B.; Raju, K.; Aiswaryam, M.U.; Nair, S.C. Formulation and evaluation of lorazepam encapsulated collagen/pectin buccal patch. *Int. J. Appl. Pharm.* **2019**, *11*, 200–209. [CrossRef]
138. Fernandes, F.P.; Fortes, A.C.; Fonseca, S.G.D.C.; Breitkreutz, J.; Ferraz, H.G. Manufacture and Characterization of Mucoadhesive Buccal Films Based on Pectin and Gellan Gum Containing Triamcinolone Acetonide. *Int. J. Polym. Sci.* **2018**, 1–10. [CrossRef]
139. Shrivastava, A. *Introduction to Plastics Engineering*; William Andrew: Norwich, NY, USA, 2018; pp. 1–16.
140. Liu, F.; Wang, X. Synthetic Polymers for Organ 3D Printing. *Polymers* **2020**, *12*, 1765. [CrossRef] [PubMed]
141. Jang, H.-J.; Shin, C.Y.; Kim, K.-B. Safety Evaluation of Polyethylene Glycol (PEG) Compounds for Cosmetic Use. *Toxicol. Res.* **2015**, *31*, 105–136. [CrossRef] [PubMed]
142. Aguilar, M.R.; Elvira, C.; Gallardo, A.; Vázquez, B.; San Román, J. Smart polymers and their applications as biomaterials. In *Topics in Tissue Engineering e-Book*; Ashammakhi, N., Reis, R.L., Chiellini, E., Eds.; Expertissues: Alicante, Spain, 2007.
143. James, H.P.; John, R.; Alex, A.; Anoop, K. Smart polymers for the controlled delivery of drugs – a concise overview. *Acta Pharm. Sin. B* **2014**, *4*, 120–127. [CrossRef]
144. Miar, S.; Ong, J.L.; Bizios, R.; Guda, T. Electrically stimulated tunable drug delivery from polypyrrole-coated polyvinylidene fluoride. *Front. Chem.* **2021**, *9*, 599631. [CrossRef] [PubMed]
145. Frachini, E.; Petri, D. Magneto-Responsive Hydrogels: Preparation, Characterization, Biotechnological and Environmental Applications. *J. Braz. Chem. Soc.* **2019**, *30*, 2010–2028. [CrossRef]

146. Hoffman, A.S. Stimuli-responsive polymers: Biomedical applications and challenges for clinical translation. *Adv. Drug Deliv. Rev.* **2013**, *65*, 10–16. [CrossRef]
147. Mather, P.T.; Luo, X.; Rousseau, I.A. Shape memory polymer research. *Annu. Rev. Mater. Res.* **2009**, *39*, 445–471. [CrossRef]
148. Banerjee, S.; Kar, K.K. Introduction to Liquid Crystalline Polymers. In *Polymers and Polymeric Composites: A Reference Series*; Kar, K., Ed.; Springer: Berlin/Heidelberg, Germany, 2020. [CrossRef]
149. Hoogenboom, R. Temperature-responsive polymers: Properties, synthesis, and applications. In *Smart Polymers and Their Applications*; Woodhead Publishing: Cambridge, UK, 2019; pp. 13–44.
150. Gao, Y.; Ren, F.; Ding, B.; Sun, N.; Liu, X.; Ding, X.; Gao, S. A thermo-sensitive PLGA-PEG-PLGA hydrogel for sustained release of docetaxel. *J. Drug Target.* **2010**, *19*, 516–527. [CrossRef]
151. Gil Choi, S.; Lee, S.-E.; Kang, B.-S.; Ng, C.L.; Davaa, E.; Park, J.-S. Thermosensitive and Mucoadhesive Sol-Gel Composites of Paclitaxel/Dimethyl-β-Cyclodextrin for Buccal Delivery. *PLoS ONE* **2014**, *9*, e109090. [CrossRef]
152. Shin, B.K.; Baek, E.J.; Choi, S.G.; Davaa, E.; Nho, Y.C.; Lim, Y.M.; Park, J.S.; Huh, K.M.; Park, J.S. Preparation and irradiation of Pluronic F127-based thermoreversible and mucoadhesive hydrogel for local delivery of naproxen. *Drug Dev. Ind. Pharm.* **2013**, *39*, 1874–1880. [CrossRef] [PubMed]
153. Reyes-Ortega, F. pH-responsive polymers: Properties, synthesis and applications. In *Smart Polymers and Their Applications*; Woodhead Publishing: Cambridge, UK, 2014; pp. 45–92.
154. Kocak, G.; Tuncer, C.A.N.S.E.L.; Bütün, V.J.P.C. pH-Responsive polymers. *Polym. Chem.* **2017**, *8*, 144–176. [CrossRef]
155. Ofridam, F.; Tarhini, M.; Lebaz, N.; Gagniere, E.; Mangin, D.; Elaïssari, A. pH-sensitive polymers: Classification and some fine potential applications. *Polym. Adv. Technol.* **2021**, *32*, 1455–1484. [CrossRef]
156. Hu, W.Y.; Wu, Z.M.; Yang, Q.Q.; Liu, Y.J.; Li, J.; Zhang, C.Y. Smart pH-responsive polymeric micelles for programmed oral delivery of insulin. *Colloids Surf. B Biointerfaces* **2019**, *183*, 110443. [CrossRef] [PubMed]
157. Zamani, M.; Aghajanzadeh, M.; Rostamizadeh, K.; Manjili, H.K.; Fridoni, M.; Danafar, H. In vivo study of poly (ethylene glycol)-poly (caprolactone)-modified folic acid nanocarriers as a pH responsive system for tumor-targeted co-delivery of tamoxifen and quercetin. *J. Drug Deliv. Sci. Technol.* **2019**, *54*, 101283. [CrossRef]
158. Arjama, M.; Mehnath, S.; Rajan, M.; Jeyaraj, M. Sericin/RBA embedded gellan gum based smart nanosystem for pH responsive drug delivery. *Int. J. Biol. Macromol.* **2018**, *120 Pt B*, 1561–1571. [CrossRef]
159. Sherje, A.P.; Londhe, V. Development and Evaluation of pH-Responsive Cyclodextrin-Based in situ Gel of Paliperidone for Intranasal Delivery. *AAPS PharmSciTech* **2017**, *19*, 384–394. [CrossRef]
160. Ishak, W.H.W.; Jia, O.Y.; Ahmad, I. pH-Responsive Gamma-Irradiated Poly(Acrylic Acid)-Cellulose-Nanocrystal-Reinforced Hydrogels. *Polymers* **2020**, *12*, 1932. [CrossRef]
161. Jamshidzadeh, F.; Mohebali, A.; Abdouss, M. Three-ply biocompatible pH-responsive nanocarriers based on HNT sandwiched by chitosan/pectin layers for controlled release of phenytoin sodium. *Int. J. Biol. Macromol.* **2020**, *150*, 336–343. [CrossRef]
162. Sawant, S.N. Development of biosensors from biopolymer composites. In *Biopolymer Composites in Electronics*; Elsevier: Amsterdam, The Netherlands, 2017; pp. 353–383.
163. Radhakrishnan, K.; Tripathy, J.; Datey, A.; Chakravortty, D.; Raichur, A.M. Mesoporous silica–chondroitin sulphate hybrid nanoparticles for targeted and bio-responsive drug delivery. *New J. Chem.* **2015**, *39*, 1754–1760. [CrossRef]
164. Ulijn, R.V.; Bibi, N.; Jayawarna, V.; Thornton, P.; Todd, S.J.; Mart, R.; Smith, A.; Gough, J.E. Bioresponsive hydrogels. *Mater. Today* **2007**, *10*, 40–48. [CrossRef]
165. Maitz, M.F.; Freudenberg, U.; Tsurkan, M.; Fischer, M.; Beyrich, T.; Werner, C. Bio-responsive polymer hydrogels homeostatically regulate blood coagulation. *Nat. Commun.* **2013**, *4*, 2168. [CrossRef] [PubMed]
166. Thabet, Y.; Lunter, D.; Breitkreutz, J. Continuous manufacturing and analytical characterization of fixed-dose, multilayer orodispersible films. *Eur. J. Pharm. Sci.* **2018**, *117*, 236–244. [CrossRef] [PubMed]
167. Musazzi, U.M.; Khalid, G.M.; Selmin, F.; Minghetti, P.; Cilurzo, F. Trends in the production methods of orodispersible films. *Int. J. Pharm.* **2019**, *576*, 118963. [CrossRef]
168. Karki, S.; Kim, H.; Na, S.-J.; Shin, D.; Jo, K.; Lee, J. Thin films as an emerging platform for drug delivery. *Asian J. Pharm. Sci.* **2016**, *11*, 559–574. [CrossRef]
169. Xue, J.; Wu, T.; Dai, Y.; Xia, Y. Electrospinning and Electrospun Nanofibers: Methods, Materials, and Applications. *Chem. Rev.* **2019**, *119*, 5298–5415. [CrossRef]
170. Haider, A.; Haider, S.; Kang, I.-K. A comprehensive review summarizing the effect of electrospinning parameters and potential applications of nanofibers in biomedical and biotechnology. *Arab. J. Chem.* **2018**, *11*, 1165–1188. [CrossRef]
171. Chinnappan, B.A.; Krishnaswamy, M.; Xu, H.; Hoque, E. Electrospinning of Biomedical Nanofibers/Nanomembranes: Effects of Process Parameters. *Polymers* **2022**, *14*, 3719. [CrossRef]
172. Omer, S.; Forgách, L.; Zelkó, R.; Sebe, I. Scale-up of electrospinning: Market overview of products and devices for pharmaceutical and biomedical purposes. *Pharmaceutics* **2021**, *13*, 286. [CrossRef]
173. Wilson, M.; A Williams, M.; Jones, D.S.; Andrews, G.P. Hot-melt extrusion technology and pharmaceutical application. *Ther. Deliv.* **2012**, *3*, 787–797. [CrossRef]
174. Patil, H.; Tiwari, R.V.; Repka, M.A. Hot-Melt Extrusion: From Theory to Application in Pharmaceutical Formulation. *AAPS PharmSciTech* **2016**, *17*, 20–42. [CrossRef] [PubMed]

175. Censi, R.; Gigliobianco, M.R.; Casadidio, C.; Di Martino, P. Hot Melt Extrusion: Highlighting Physicochemical Factors to Be Investigated While Designing and Optimizing a Hot Melt Extrusion Process. *Pharmaceutics* **2018**, *10*, 89. [CrossRef] [PubMed]
176. Madan, S.; Madan, S. Hot melt extrusion and its pharmaceutical applications. *Asian J. Pharm.* **2012**, *7*, 123–133.
177. Ngo, T.D.; Kashani, A.; Imbalzano, G.; Nguyen, K.T.Q.; Hui, D. Additive Manufacturing (3D Printing): A Review of Materials, Methods, Applications and Challenges. *Compos. Part B Eng.* **2018**, *143*, 172–196. [CrossRef]
178. Tambe, S.; Jain, D.; Agarwal, Y.; Amin, P. Hot-melt extrusion: Highlighting recent advances in pharmaceutical applications. *J. Drug Deliv. Sci. Technol.* **2021**, *63*, 102452. [CrossRef]
179. *ASTM F2792-12A*; Standard Terminology for Additive Manufacturing Technologies. ASTM International: West Conshohocken, PA, USA, 2012.
180. Flynn, J.M.; Shokrani, A.; Newman, S.T.; Dhokia, V. Hybrid additive and subtractive machine tools – Research and industrial developments. *Int. J. Mach. Tools Manuf.* **2016**, *101*, 79–101. [CrossRef]
181. Shahrubudin, N.; Lee, T.; Ramlan, R. An Overview on 3D Printing Technology: Technological, Materials, and Applications. *Procedia Manuf.* **2019**, *35*, 1286–1296. [CrossRef]
182. Egan, P.F. Integrated Design Approaches for 3D Printed Tissue Scaffolds: Review and Outlook. *Materials* **2019**, *12*, 2355. [CrossRef]
183. Arefin, A.; Khatri, N.; Kulkarni, N.; Egan, P. Polymer 3D Printing Review: Materials, Process, and Design Strategies for Medical Applications. *Polymers* **2021**, *13*, 1499. [CrossRef]
184. Walker, J.L.; Santoro, M. Processing and production of bioresorbable polymer scaffolds for tissue engineering. In *Bioresorbable Polymers for Biomedical Applications*; Woodhead Publishing: Cambridge, UK, 2017; pp. 181–203.
185. Huang, J.; Qin, Q.; Wang, J. A review of stereolithography: Processes and systems. *Processes* **2020**, *8*, 1138. [CrossRef]
186. Moroni, L.; Boland, T.; Burdick, J.A.; De Maria, C.; Derby, B.; Forgacs, G.; Groll, J.; Li, Q.; Malda, J.; Mironov, V.A.; et al. Biofabrication: A guide to technology and terminology. *Trends Biotechnol.* **2018**, *36*, 384–402. [CrossRef] [PubMed]
187. Selimis, A.; Mironov, V.; Farsari, M. Direct laser writing: Principles and materials for scaffold 3D printing. *Microelectron. Eng.* **2015**, *132*, 83–89. [CrossRef]
188. Zahid, U.M.; Aamir, M.N.; Zaman, S.; Rehman, M.; Javaid, Z.; Hussain, H.; Mohsin, N.U.A. Preparation, Optimization and Ex Vivo Release Studies of Risperidone Orodispersible Films. *Lat. Am. J. Pharm.* **2020**, *39*, 686–693.
189. Ali, M.S.; Vijendar, C.; Kumar, S.D.; Krishnaveni, J. Formulation and evaluation of fast dissolving oral films of diazepam. *J. Pharmacovigil.* **2016**, *4*, 210. [CrossRef]
190. Raju, P.N.; Kumar, M.S.; Reddy, C.M.; Ravishankar, K. Formulation and evaluation of fast dissolving films of loratidine by solvent casting method. *Pharma Innov.* **2013**, *2*, 31–35.
191. Zhang, C.; Liu, Y.; Li, W.; Gao, P.; Xiang, D.; Ren, X.; Liu, D. Mucoadhesive buccal film containing ornidazole and dexamethasone for oral ulcers: *in vitro* and *in vivo* studies. *Pharm. Dev. Technol.* **2017**, *24*, 118–126. [CrossRef] [PubMed]
192. Lancina, M.G.; Shankar, R.K.; Yang, H. Chitosan nanofibers for transbuccal insulin delivery. *J. Biomed. Mater. Res. Part A* **2017**, *105*, 1252–1259. [CrossRef]
193. Clitherow, K.H.; Murdoch, C.; Spain, S.G.; Handler, A.M.; Colley, H.E.; Stie, M.B.; Nielsen, H.M.; Janfelt, C.; Hatton, P.V.; Jacobsen, J. Mucoadhesive Electrospun Patch Delivery of Lidocaine to the Oral Mucosa and Investigation of Spatial Distribution in a Tissue Using MALDI-Mass Spectrometry Imaging. *Mol. Pharm.* **2019**, *16*, 3948–3956. [CrossRef]
194. Li, X.; Kanjwal, M.A.; Lin, L.; Chronakis, I.S. Electrospun polyvinyl-alcohol nanofibers as oral fast-dissolving delivery system of caffeine and riboflavin. *Colloids Surf. B Biointerfaces* **2013**, *103*, 182–188. [CrossRef]
195. Alkahtani, M.; Aodah, A.; Abu Asab, O.; Basit, A.; Orlu, M.; Tawfik, E. Fabrication and Characterization of Fast-Dissolving Films Containing Escitalopram/Quetiapine for the Treatment of Major Depressive Disorder. *Pharmaceutics* **2021**, *13*, 891. [CrossRef]
196. Ponrasu, T.; Chen, B.-H.; Chou, T.-H.; Wu, J.-J.; Cheng, Y.-S. Fast Dissolving Electrospun Nanofibers Fabricated from Jelly Fig Polysaccharide/Pullulan for Drug Delivery Applications. *Polymers* **2021**, *13*, 241. [CrossRef] [PubMed]
197. Almeida, A.; Possemiers, S.; Boone, M.; De Beer, T.; Quinten, T.; Van Hoorebeke, L.; Remon, J.; Vervaet, C. Ethylene vinyl acetate as matrix for oral sustained release dosage forms produced via hot-melt extrusion. *Eur. J. Pharm. Biopharm.* **2011**, *77*, 297–305. [CrossRef]
198. Pimparade, M.B.; Vo, A.; Maurya, A.S.; Bae, J.; Morott, J.T.; Feng, X.; Kim, D.W.; Kulkarni, V.I.; Tiwari, R.; Vanaja, K.; et al. Development and evaluation of an oral fast disintegrating anti-allergic film using hot-melt extrusion technology. *Eur. J. Pharm. Biopharm.* **2017**, *119*, 81–90. [CrossRef] [PubMed]
199. Bhagurkar, A.M.; Darji, M.; Lakhani, P.; Thipsay, P.; Bandari, S.; Repka, M.A. Effects of formulation composition on the characteristics of mucoadhesive films prepared by hot-melt extrusion technology. *J. Pharm. Pharmacol.* **2018**, *71*, 293–305. [CrossRef]
200. Palem, C.R.; Battu, S.K.; Maddineni, S.; Gannu, R.; Repka, M.A.; Yamsani, M.R. Oral transmucosal delivery of domperidone from immediate release films produced via hot-melt extrusion technology. *Pharm. Dev. Technol.* **2012**, *18*, 186–195. [CrossRef]
201. Palem, C.R.; Dudhipala, N.; Battu, S.K.; Goda, S.; Repka, M.A.; Yamsani, M.R. Combined dosage form of pioglitazone and felodipine as mucoadhesive pellets via hot melt extrusion for improved buccal delivery with application of quality by design approach. *J. Drug Deliv. Sci. Technol.* **2015**, *30*, 209–219. [CrossRef]
202. Elbadawi, M.; Nikjoo, D.; Gustafsson, T.; Gaisford, S.; Basit, A.W. Pressure-assisted microsyringe 3D printing of oral films based on pullulan and hydroxypropyl methylcellulose. *Int. J. Pharm.* **2021**, *595*, 120197. [CrossRef] [PubMed]
203. Wu, M.; Zhang, Y.; Huang, H.; Li, J.; Liu, H.; Guo, Z.; Xue, L.; Liu, S.; Lei, Y. Assisted 3D printing of microneedle patches for minimally invasive glucose control in diabetes. *Mater. Sci. Eng. C* **2020**, *117*, 111299. [CrossRef]

204. Tagami, T.; Yoshimura, N.; Goto, E.; Noda, T.; Ozeki, T. Fabrication of Muco-Adhesive Oral Films by the 3D Printing of Hydroxypropyl Methylcellulose-Based Catechin-Loaded Formulations. *Biol. Pharm. Bull.* **2019**, *42*, 1898–1905. [CrossRef]
205. Tian, P.; Yang, F.; Xu, Y.; Lin, M.-M.; Yu, L.-P.; Lin, W.; Lin, Q.-F.; Lv, Z.-F.; Huang, S.-Y.; Chen, Y.-Z. Oral disintegrating patient-tailored tablets of warfarin sodium produced by 3D printing. *Drug Dev. Ind. Pharm.* **2018**, *44*, 1918–1923. [CrossRef]
206. Cho, H.W.; Baek, S.H.; Lee, B.J.; Jin, H.E. Orodispersible Polymer Films with the Poorly Water-Soluble Drug, Olanzapine: Hot-Melt Pneumatic Extrusion for Single-Process 3D Printing. *Pharmaceutics* **2020**, *22*, 692. [CrossRef] [PubMed]
207. Databridge Market. Available online: https://www.databridgemarketresearch.com/reports/global-oral-drug-delivery-market#:~{}:text=The%20oral%20drug%20delivery%20market,period%20of%202022%20to%202029 (accessed on 20 November 2022).
208. Future Marketinsights. Available online: https://www.futuremarketinsights.com/reports/oral-controlled-release-drug-delivery-technology-market (accessed on 20 November 2022).
209. Grandview Research. Available online: https://www.grandviewresearch.com/industry-analysis/buccal-drug-delivery-systems-market-report#:~{}:text=The%20global%20buccal%20drug%20delivery,and%20development%20in%20this%20field (accessed on 20 November 2022).
210. Transparency Market Research. Available online: https://www.transparencymarketresearch.com/oral-transmucosal-drugs-market.html (accessed on 20 November 2022).
211. Biospace. Available online: https://www.biospace.com/article/oral-transmucosal-drugs-market-buccal-mucosa-segment-held-major-share-of-the-market/ (accessed on 20 November 2022).

pharmaceutics

Review

Tailoring Lipid-Based Drug Delivery Nanosystems by Synchrotron Small Angle X-ray Scattering

Barbara Sartori * and Benedetta Marmiroli

Institute of Inorganic Chemistry, Graz University of Technology, Stremayrgasse 9/4, 8010 Graz, Austria
* Correspondence: barbara.sartori@tugraz.at

Abstract: Thanks to specific physico-chemical properties, drug delivery systems based on nanoparticles have proven to effectively transport delicate molecules for therapeutic purposes, protecting them from degradation, increasing their stability in the blood circulation and allowing to convey and release the transported substances in specific areas of the body. Nanoparticles obtained from biopolymers for applications in medicine and pharmaceutics have become particularly popular in recent years due to the enormous research effort in the field of vaccines to respond to the pandemic emergency. Among the various types of biopolymers used to produce nanoparticles for therapeutics, lipids have characteristics that make them biocompatible, with low toxicity and ease of clearance. They can be synthesized by designing their characteristics according to the foreseen administration path, or to the target of the transported drug. The analytical methods mostly used to evaluate the characteristics of lipid nanosytems for drug delivery involve studying their effects on cells, in vitro and in vivo. Although it is often considered a "niche technique" for research in the bio-related sciences, Small Angle X-ray Scattering (SAXS) is a versatile tool to study the structure of nanosystems based on lipids, both ex situ and in situ. Therefore, it allows to evaluate both the effect of the different synthesis parameters and of the exposure of lipid nanoparticles to physiological conditions, which is of fundamental importance to design efficient drug delivery systems. In this mini-review, we will report some recent examples of characterization and design of nanoparticles based on lipids, where SAXS has been a fundamental step both to guide the synthesis of nanomaterials with tailored characteristics, and to understand the interaction between nanomaterials and cells.

Keywords: lipid nanoparticles; drug delivery; SAXS; structural analysis; external stimuli response; in situ

1. Introduction

Interest in nanoparticles grew in the 1950–1960s, when the first studies on their potential applications for drug delivery started. The idea was to use the nanoparticles as sort of targeted bullets that could directly interact with the pathogens, and thus, improve the therapeutic effect of the administered drugs. On the other hand, nanomaterials could be used to incapsulate molecules or antigens and achieve a continuous, prolonged delivery to enhance the immune response, thus, making easier the administration of vaccines that usually required multiple injections [1]. Many molecules have been used since then to produce nanoparticles with therapeutic purposes: the first attempts included toxic substances such as acrylamide, and were then soon replaced by other compounds with better biocompatibility [2].

Since these pioneering works, in the following decades, further significative advances have made synthesizing biocompatible materials capable of delivering a number of therapeutic substances on site possible, allowing to target the release of drugs and to increase their bioavailability [3,4]. Nanoparticles for biomedicine are currently available in different formulations: depending on the chemical characteristics, they show special features which are convenient, for example, controlled drug release or selective interactions with the target. Ideally, nanoparticles with potential therapeutic use are non-toxic, non-immunogenic, stable during in vivo circulation and are able to deliver the encapsulated drug in the desired site [5]. Despite their huge potentiality, many nanomaterials still present problems in terms of toxicity, distribution and clearance in the body [6], e.g., in case of long-term administration, such as for treatment of chronic diseases [7].

The steps towards fully biocompatible and site-specific nanoparticles for medical applications are closely linked to the deep understanding of their interaction with biological targets, such as cells, biomembranes and proteins.

From this perspective, biopolymers are very attractive because of their affinity with cell membranes, low toxicity, ease of preparation and structural tunability that make them ideal materials for the design of controlled drug delivery systems [8–10]. Lipid nanoparticles, in particular, proved to be effective in enhancing the availability of even poorly soluble molecules due to their aptitude to promote the formation of solubilized phases during digestion [11,12]. In the 1970s, the potential of liposomes for therapeutic applications were started to be considered, but several years were necessary before this interest resulted in the first official approval for injectable formulation, in the early 1990s [13].

Later on, the synthetic technology improvements and new scientific evidences turned out in the design and production of lipid nanoparticles composed of solid matrix of nontoxic lipids or waxes with an averaged dimension of 100–300 nm, which were able to protect from degradation the incorporated drugs.

In the following years, the discovery of non-lamellar liquid crystalline nanoparticles —mainly hexosomes and cubosomes, which are suited for the loading of lipophilic molecules, and of lipid–drug conjugates, which improve the loading capacity of hydrophilic drugs—opened new exciting opportunities in the field of molecule delivery [14–16].

Cubosomes and hexosomes are, presently, widely studied for the production of stable, biocompatible, non-immunogenic drug delivery formulations. They are typically synthesized by coupling amphiphilic lipids such as Monooleyn (MO) or Phytantriol doped with a suitable stabilizing agent such as triblock copolymers—for example, Pluronic® F127—or with natural compounds, such as casein, ethyl cellulose, modified starch or food-derived emulsifiers, such as citrem [17].

The crystal structure of lipidic nanoparticles seem to have an impact on the effectiveness of drug release [18]: since the number of molecules released by cubosomes is higher than hexosomes due to the smaller pores and surface area of the latter, this structural characteristic can be used to tune the drug release in response to different external stimuli, such as temperature or pH, which induce a phase transition in the lipid mesostructure [19].

A common strategy to understand the therapeutic potential of nanoparticles is the study of their interaction with living cells [20] to improve absorption and drug release once internalized. The cargo particles should be tailored to the specific therapeutic goal and on the administration route. A detailed characterization of the mesostructure of non-lamellar liquid crystal nanoparticles, as well as an in situ study of their response to environmental conditions in vitro and in vivo, is thus, essential.

In situ time-resolved chemical and structural characterization is of paramount importance for the design and synthesis of lipid-based nanomaterials with potential therapeutic applications. In this respect, many analytical techniques are available, which allow for determining the crystalline structure, the chemical composition and the stability against aggregation in solution, to name a few [21,22].

Commonly used methods for the morphological characterization of lipid nanoparticles include Dynamic Light Scattering (DLS), which allows for investigating the size distribution; Zeta potential, which gives information on the surface charge and the stability of nanoparticles in a solution; or electron microscopy techniques such as SEM and TEM, which are useful to gain insights on the overall shape and the structural features of the nanoparticles in a localized area. These techniques complement nicely the structural information that could be obtained with Small Angle X-ray Scattering (SAXS).

In scattering techniques, the whole illuminated volume of the sample is investigated, thus, average values of the structural parameters can be obtained, in contrast to microscopy, where a single object or a small part of the sample can be investigated. Moreover, SAXS allows for in situ and time-resolved studies to be carried out in a much simpler and faster way than TEM and SEM, allowing for changing the sample environment in terms of temperature, pressure, chemical composition, for example. DLS gives details on the average size and size distribution of particles in a sample, but does not inform on the shape of the particles.

SAXS is an analytical technique that takes advantage of the interaction of X-rays with the electrons of the analyte. Once treated with the appropriate mathematical model, the scattered radiation pattern collected at the detector gives information on the supramolecular envelope of the studied material [23], thus, on the size, shape and monodispersity of the nanoparticles in the present case. It also allows for investigating changes in the crystalline structure in response to external stimuli. SAXS was developed in the late 1930s, and since then, it has been widely used for the characterization of the structure and of the dynamics of nanosized materials [24]. In the last few decades, it has been profitably applied to structural biology because it offers the unprecedented possibility to elucidate the structure of macromolecules in a solution as well as to study the structural properties of biomembranes in an environment that mimics physiological conditions [25]. Thus, it is a powerful tool to investigate in situ both lipid nanoparticles structures and their behavior when exposed to biologically relevant environment. Small and Wide-Angle X-ray Scattering (SWAXS) data can be acquired simultaneously to investigate the intramolecular distances and interactions to evaluate the eventual crystalline structure of the loaded molecules [26], or to elucidate the perturbative effect of drugs on lipid model biomembranes [27].

Nowadays, excellent bench apparatuses for SAXS and WAXS analysis are available. On the other hand, the time resolution that is necessary to obtain a good signal-to-noise ratio with such instruments may be of minutes or even hours, depending on the electron density contrast of the analyte. Therefore, for time-resolved studies, e.g., to monitor the phase transition in response to a change in pH, synchrotron SAXS is the technique of choice because it requires just a few seconds of exposure, thus, increasing the efficiency of the measurements while limiting radiation damage on the exposed sample.

SAXS analysis is often coupled with SANS (Small Angle Neutron Scattering) for the structural and composition study of nanoparticles [28] and biomembranes [29]. Analyzing the sample with both techniques allows for refining the structural model of lipid nanoparticles [30]. While X-rays interact with electrons, neutrons interact with the nuclei of atoms: neutron radiation is non-destructive, thus, avoiding any radiation damage issue. The contrast variation method in SANS allows for determining the distribution of the components in complex nanoparticles, such as different lipids, excipient, stabilizers and even loaded molecules, e.g., mRNA in the recently developed COVID-19 vaccines [31–33].

In the present review, we will focus on synchrotron Small Angle X-ray Scattering (SAXS) experiments aimed to determine the best structure and morphology of lipid-based nanoparticles for therapeutics. We will report some case studies in which SAXS was of pivotal importance to optimize the efficiency of lipid nanoparticles as cargo systems, as well as to design new formulations with improved stability or tailored on the administration route.

Table 1 summarizes the systems described in this review grouped by their foreseen applications, and the objective of the SAXS investigations performed therein.

Table 1. Summary of the experiments described in the text, grouped by the foreseen application.

Application	System	In Situ/Ex Situ SAXS	Investigation on	Ref.	Outcome
pH-dependent drug release	Cubosomes (MO + Oleic acid + natural derived oil)	Ex situ formulation design	Stability and pH responsivity	[34]	Optimized formulation
	cubosomes(MO + fatty acetates)	In situ pH responsivity	pH and temperature responsivity	[35]	Improved formulation for pH-dependent phase change
	Cubosomes (MO + F127 + aminolipids)	In situ Investigate structural changes at different pH	pH responsivity	[36]	Improved formulation–pH-dependent phase change

Table 1. *Cont.*

Application	System	In Situ/Ex Situ SAXS	Investigation on	Ref.	Outcome
pH-dependent drug release	Cubosomes (MO + Oleic acid + natural derived oil)	Ex situ formulation design	Stability and pH responsivity	[34]	Optimized formulation
	cubosomes(MO + fatty acetates)	In situ pH responsivity	pH and temperature responsivity	[35]	Improved formulation for pH-dependent phase change
	Cubosomes (MO + F127 + aminolipids)	In situ Investigate structural changes at different pH	pH responsivity	[36]	Improved formulation–pH-dependent phase change
Oral drug delivery	cubosomes (Mo + F127 + polysaccharides)	Ex situ Formulation design	pH responsivity	[37]	Improved formulation–pH-dependent phase change
	Liposomes (chitosan-N-acetylcysteine)	Ex situ Effect of coating on NPs stability	Improvement on the interaction with intestine cells	[38]	Lipid bilayer integrity confirmed after polymer coating
	Self-emulsifying drug delivery systems	In situ structural changes during digestion	Poorly soluble drug cargo stability during lipolysis	[39]	Selection of excipient to improve SEDDS stability
Enable prolonged circulation in vivo	Liposomes/protein corona	Ex situ "Tailor-made" protein coating of NPs	Stability during circulation	[40]	Plasma protein coating induce structural changes in liposomes
Interaction betw.lipid-based NPs and live cells	cubosomes (Phytantriol–F127)	In situ Phase transition in contact with HUVE cells	Interaction between structured NPs and cells	[41]	Understanding the dynamics of vascular endothelial cells–lipid NPs interactions
Thermally triggered drug/nutrients delivery	cubosomes (MO + superparamagnetic iron oxide)	In situ Investigate phase transition	Temperature responsivity and magnetic trigger	[42]	Formulation of responsive magnetocubosomes
Active prostatic cancer cell targeting	PEG-ylated liposomes	Ex situ Effect of PEG chain length on protein corona	Stability, reduction in immune response	[43]	Formulation improvement of PEGylated cationic lipid NPs
Delivery of Doxorubicine for cancer treatment	Liposomes/biomolecular corona	Ex situ Stability after plasma incubation	Drug delivery efficiency	[44]	Commercial formulation stability assessment
mRNA transfection	Lipoplexes (DOTAP/RNA/Protamine)	Ex situ Formulation design	Physico-chemical particles characterization	[45]	Formulation improvement for better transfection efficiency
	Lipoplexes/Polysarcosin	Ex situ Particles design	Improvement of stability	[46]	Optimized formulation
DNA transfection	Lipoplexes (Cationic lipids/Cholesterol/DNA)	Ex situ Formulation design	Physico-chemical characterization	[47]	Optimized formulation
	Lipoplexes (Cationic lipid/DNA/PEG)	Ex situ Formulation design	Microfluidics for NPs synthesis	[48]	Optimized pDNA loaded NPs production
siRNA delivery	lipid shell/PLGA core/RNA	Ex situ Stability in physiological conditions	Design of aerosol drug delivery systems	[49]	Understanding the dynamics of lung cells–lipid NPs interactions

2. Phase Transition of Lipid Nanoparticles for Controlled Drug Delivery

Liposomes, cubosomes and hexosomes allow for incorporating lipophilic and hydrophilic drugs: their formulations can be stabilized with additives that increase the embedded drug stability and offer the possibility to trigger the release of the substances, making them ideal vectors for target drug delivery systems [50,51].

Adjusting the nanoparticle composition allows for creating drug delivery tools with specific responsive characteristics that can be investigated in situ with SAXS. With this information, lipid nanoparticles for therapeutic use can be designed according to the administration route, or to the target site.

It was recently demonstrated in an experiment in vivo that cubosomes obtained by coupling selachil alcohol with Tween® 80 as a stabilizer were more effective in delivering the poorly water soluble drug Phenytoin to the brain, with respect to hexosomes derived from the same lipid, but using Pluronic® F127 as stabilizing agent [52]. This better performance was attributed to the increased negative curvature of the lipid bilayer induced by Phenytoin in the presence of Tween80, which is more internalized in the lipid bilayer with respect to Pluronic F127. Usually, the lipid composition must be adjusted in order to change the lipid curvature to induce a phase transition. In the reported study, simple formulations composed of one single lipid were produced, and the phase-change was obtained by tuning the stabilizer.

Lipidic nanoparticles are particularly attractive for oral drug administration [53]. They protect the loaded drug from degradation in the gastrointestinal tract increasing the circulation time, and can be coupled with natural compounds, such as natural oils or fatty acids, to enhance pH responsiveness. Lipidic drug carriers must resist the very acidic stomach environment, and dissolve to release the loaded formulation once in contact with the neutral blood pH.

Understanding the dissolution mechanism is a critical aspect of delivery systems design: in this respect, SAXS offers the possibility to study in situ the lipolysis during digestion. Simultaneous Small and Wide-Angle X-ray Scattering (SWAXS) allows for thoroughly evaluating the extent of drug dissolution, as the changes in crystalline structure of the internalized drugs can also be observed [54]. The formation of novel structures during lipolysis—thus, during the digestion of the orally administered formulation—can affect the solubility of the nanoparticles, reducing the drug absorption. As recently reported, drug carriers based on natural surfactants have great potential for the production of Self Emulsifying Drug Delivery Systems (SEDDS). It was recently demonstrated in a study on coupled in situ SAXS and ex situ Cryo-TEM, that during in vitro lipolysis, SEDDS undergo a structural evolution that is correlated with the amount of the surfactant in the formulation [39].

Elucidating the fate of nanostructured lipid nanoparticles when they come into contact with target cells is of paramount importance. Recently, the phase-change of non-lamellar lipid nanoparticles when they interact with cells was studied in situ with SAXS. A 3D model simulating the endothelium of human umbilical vein (HUVE) was developed, and the evolution of cubosomes to hexosomes during circulation was investigated [41]. Umbilical cells were seeded on the inner surface of a glass capillary, and the cubosomes' dispersion was subsequently flown through it (Figure 1). The flow-through cell was exposed to X-rays, and the phase transition of cubosomes into hexosomes during contact with the cells was observed. These experiments demonstrated the potentiality of a structural study in real time.

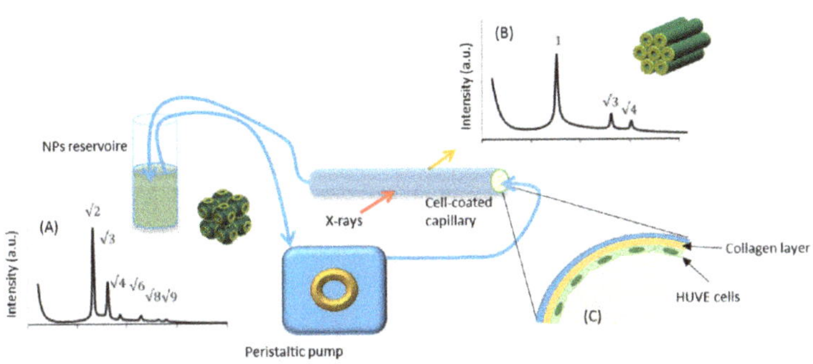

Figure 1. Scheme of the experimental setup. Phase transition from (**A**) cubosomes to (**B**) hexosomes, (the respective Bragg peaks indexing are indicated), when in contact with a glass tube coated with human umbilical vein endothelium cells (HUVE), was observed in situ with SAXS. (**C**) Detail of the observation cell coated with human umbilical vein endotheliümcells. (Adapted from [41]).

2.1. External Stimuli Responsive Lipid Nanoparticles

To design drug delivery nanoparticles that are responsive to external stimuli, understanding how to tune the phase transition is a critical aspect. It is usually induced by a change in temperature or in the pH of the medium, and it can be exploited to determine the drug-release extent.

By loading lipid nanoparticles composed of MO and oleic acid (OA) with the antitumor drug Doxorubicin and with a natural oil obtained from *Brucea javanica* (BJO)—a traditional medical herb with antiproliferative properties on cancer cells—Li and collaborators obtained pH-sensitive lipid nanoparticles capable of converting from hexagonal to cubic crystalline phase to emulsified microemulsions, in response to the lower pH in proximity and inside tumor cells. SAXS analysis demonstrated the pH responsiveness and the effect of Brucea Javanica oil on the nanoparticles' stability (Figure 2). At pH 7.4, hexosomes with an internal HII-phase structure were formed: at a lower pH (6.8), the crystalline structure changed to cubosomes, with an inner structure where two phases (Pn3m and Im3m) coexist [34].

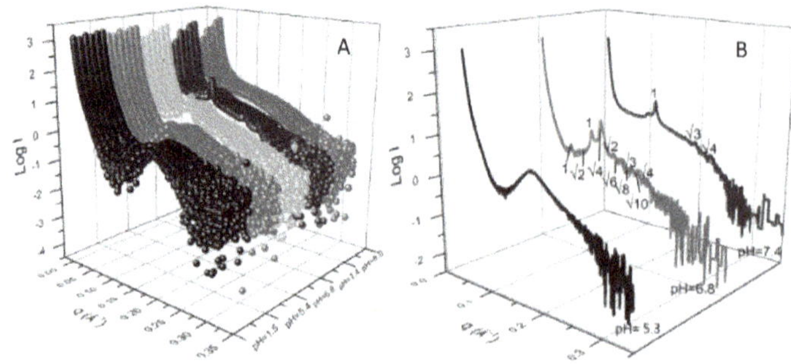

Figure 2. (**A**) SAXS patterns obtained from nanoparticles of different composition (front to back: MO:OA, MO:0.9OA:0.1BJO, MO:0.8OA:0.2BJO, MO:0.7OA:0.3BJO, MO:0.6OA:0.4BJO and MO:0.5OA:0.5BJO) collected at different pH (pH = 1.5, 5.3, 6.8, 7.4, and 8.0, front to back); (**B**) SAXS pattern of lipid nanoparticles loaded with oleic acid and Brucea javanica oil (MO:0.9OA:0.1BJO) collected at pH 7.4 (healthy cells), 6.8 (tumor cells proximity) and 5.4 (tumor cells endosomes). The Bragg peaks indexing of inverted hexagonal (pH 7.4) and inverted cubic (pH 6.8) liquid crystalline phases is indicated (Reproduced with permission from [34]).

By modifying their composition, it is possible to tailor the response to the external conditions and to create a "library" of pH- and temperature-responsive nanoparticles, whose crystalline structure can be tuned for fast (cubic) or slow (hexagonal) drug release. Recently, lyotropic liquid crystals with cubic Fd3M symmetry obtained from MO were doped with various types and amounts of fatty acids (FA) and fatty acetates (FAc) with different chain lengths (Figure 3). By adjusting the nanoparticles composition, it was possible to produce a collection of hybrid MO-FA–FAc micellar nanoparticles which exhibited a Fd3m structure. As demonstrated by in situ SAXS, these cubosomes underwent cubic to lamellar and cubic to hexagonal phase transitions in response to temperature and pH changes, respectively, depending on the chain length, the amount and the respective molar ratio of FA and FAc incorporated in the formulation [35].

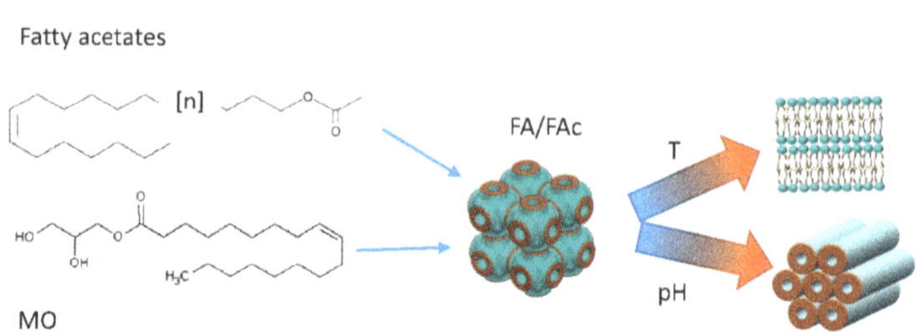

Figure 3. Cartoon representing the experiment reported in [35]. The phase transition from cubosomes to hexosomes of MO nanoparticles produced with different amounts of Fac and FA in different molar ratios can be controlled with temperature and pH variations.

In another experiment, a set of synthetic ionizable aminolipids was designed and used to dope MO nanoparticles stabilized with the surfactant Pluronic® F-127. SAXS allowed for investigating the phase transition of MO in response to different pH of the several formulations obtained by adjusting the nanoparticles' composition [36].

2.2. Hybrid Lipid-Based Nanoparticles

Lipids can be associated with other biocompatible molecules, such as polysaccharides, in order to induce the formation of structured mesophases with high stability and specificity for applications where bioadhesion is desired. Coupling MO with Pluronic® F127 as a stabilizer, chitosan-N-arginine and alginate lead to the formation of pH responsive cubosome nanoparticles. SAXS experiments allowed for determining the proper amount of stabilizer needed to obtain nanoparticles characterized by cubic symmetry, to gain insight in the phase transition in response to pH change as well as on the distribution of chitosan-N-arginine inside the lipid structure [37].

Lipids can be also coupled with inorganic materials, with a great potential for localized drug release. Caselli and collaborators studied the behavior of hybrid paramagnetic nanoparticles when exposed to a magnetic field: they exploited the capability of lipidic-ordered liquid crystals to change the phase from cubic to hexagonal, to trigger the localized drug release. The phase-change was accompanied by a massive water ejection that could promote the release of hydrophilic drugs embedded in the lipid nanoparticles. By coupling superparamagnetic iron oxide nanoparticles with 1-monoolein (GMO) nanoparticles, they produced hybrid cubosomes with Pn3m cubic symmetry, which changed to hexagonal in response to the heat generated by the magnetic nanoparticles exposed to a low-frequency alternating magnetic field [42]. With SAXS, it was possible to characterize in situ the lipid phase transition either during the exposure to temperature only, or to an alternating magnetic field. In the absence of a magnetic field, the phase transition from Pn3m to HII was observed at temperatures compatible with physiological conditions (below 40 °C) [55]. Both pure GMO cubosomes and hybrid nanoparticles were exposed to the alternating magnetic field: SAXS data confirmed the thermotropic effect of the superparamagnetic iron oxide on the crystalline structure of the hybrid cubosomes once exposed to the magnetic field, while no phase-change was observed in the pure cubosomes. SAXS data analysis also revealed a pearl-necklace-like arrangement of the hybrid magnetic cubosomes after 210 s of exposure to the magnetic field (Figure 4).

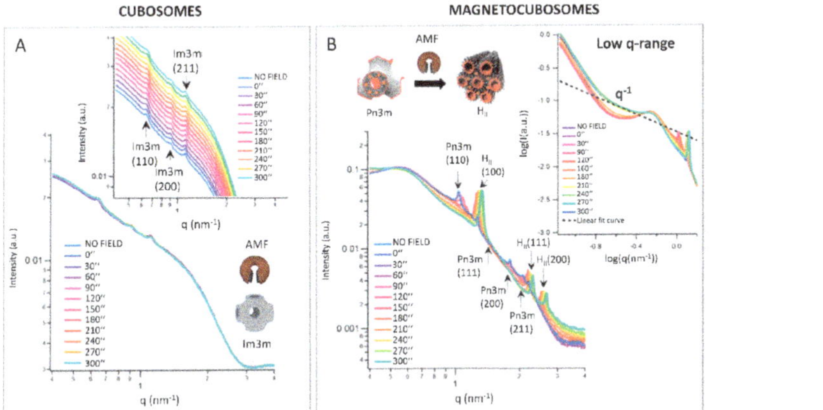

Figure 4. Response of (**A**) GMO cubosomes and (**B**) superparamagnetic cubosomes (showing a phase transition from Pn3m (cubic) to HII (hexagonal) crystalline phase) exposed to an alternated magnetic field. The Miller Indexes for the respective crystalline phases are indicated. Each SAXS profile corresponds to a different application time of the magnetic field. Inset in (**A**) highlights the scattering peaks of cubosomes. Inset in (**B**) represents a magnification of the low-q region of the scattering profile. The pearl-necklace-like arrangement of magnetocubosomes after 210 s of exposure to the magnetic field is evidenced by the appearance of a q^{-1} scalar law. (Reproduced with permission from [42]).

3. Functionalization with Polyethylene Glycol (PEG)

Functionalization of lipid nanoparticles with PEG has several advantages in the design of drug delivery systems [56]: PEG is commonly used to coat nanoparticles aiming to prevent aggregation and avoid phagocytosis, thus, increasing their residence time in the circulating blood; it also interferes in the formation of molecular corona on the nanoparticles [43]. PEG can be incorporated in the formulation as a pure component, or covalently linked to one of the lipids present in the formulation (PEGylated lipids). PEG creates a hydrophilic shell on the surface of lipid nanoparticles, which inhibits the desorption of drugs, particularly hydrophobic molecules that are encapsulated in the nanoparticle core. PEG molecules grafted on the surface also allow for accommodating a larger amount of hydrophilic drugs into core–shell nanoparticles, because they do not occupy space inside the nanoparticle [57].

Liposomes are promising tools for the delivery of antitumoral drugs. In order to reach the therapeutic target, they must resist the extracellular matrix (ECM). Aiming to mimic such an environment, Bandara and coworkers studied the effect of exposure to hydrogels composed of gelatin, alginate, PEG and the commercial ECM Matrigel® on DOPC (1,2-dioleoyl-sn-glycero-3-phosphocholine) liposomes. The structural features of the liposomes before and after exposure to the hydrogel were studied with SAXS. Their results showed that once exposed to the extracellular matrix, DOPC liposomes turn to multilamellar vesicles, and form aggregates of micrometer size. Adding a small amount of PEGylated DOPE (1,2-Dioleoyl-sn-glycero-3-phosphoethanolamine) to the liposome formulation, the formation of a PEG corona reduced the extent of liposomes aggregation, which has important implications for stability and immune responses [58].

In a systematic high-throughput study conducted by Sarode et al., it was demonstrated that the PEG-to-lipid molar ratio and length of the PEG carbon tail have significant effects on the particle size and on the loaded molecule delivery. SAXS data demonstrated that when a shorter-chained, negatively charged PEGylated lipid is used, the crystalline structure of the resulting nanoparticle is more ordered, with respect to longer-chained PEGylated lipids with a neutral charge, resulting in less-effective drug delivery efficiency [59].

4. Coated Nanoparticles

4.1. Biomolecular Corona

In vivo, lipid vectors for drug delivery adsorb proteins, lipids and other plasmatic biomolecules on their surface in a non-specific manner once injected: depending on the nanoparticles' properties, such as their charge, hydrophilicity, dimensions and morphology, the several components present in plasma interact and build a coating. The resulting biomolecular corona might interfere with the nanoparticle–cell interaction, induce an immune response and alter the drug-release efficiency [60].

Liposomes commonly used as vectors for therapeutic molecule delivery can be unilamellar vesicles, where the hydrophilic drug molecules are loaded in the central aqueous core of the nanoparticle, or multilamellar vesicles, which can host lipophilic drugs within the lamellar structure. The majority of liposomes commercially available for drug delivery purposes are Small Unilamellar Vesicles (SUV) [61]—although they are stable during circulation, the interaction of proteins with the SUV surface provokes the formation of multilamellar structures and of liposome clusters: the change in the crystalline structure implies the rupture of the vesicles, which may result in the release of the loaded drug.

The formation of larger aggregates may impede the cellular uptake of the particles. The latter effect, however, becomes less evident as the concentration of proteins attached to the nanoparticles surface increases, possibly due to the repulsive interaction of charged proteins, as demonstrated by Caracciolo and coworkers [44]. In a recent experiment, they used a commercial formulation of liposomal Doxorubicin, a drug used for cancer treatment, to study the effect of biomolecular corona on nanoparticle stability and drug release, incubating the liposome–drug formulation in the concentration of human plasma that mimics the physiological conditions. TEM, DLS and Zeta potential experiments demonstrated the formation of biomolecular corona already at a low plasma concentration, which led to an increased particle size and to a change in the particle surface charge. SAXS characterization of the drug-loaded nanoparticles before and after incubation in human plasma demonstrated the absence of the multilamellar structure, showing that the presence of the biomolecular corona did not lead to the formation of a multilamellar structure; moreover, a fluorescence-lifetime imaging microscopy analysis on the drug-loaded liposomes after incubation revealed that the amount of Doxorubicin in the incubated particles was significantly lower with respect to the pristine ones, suggesting that the exposure to human plasma might degrade the liposomes, altering their persistence in the blood circulation.

On the other hand, protein coating on liposome carriers (namely, protein corona), is effective in protecting the nanoparticles from degradation in blood, reducing the interaction with leukocytes and prolonging the circulation time in vivo. Giulimondi and colleagues recently demonstrated the protective effect against uptake by blood macrophages of pre-adsorbed human plasma (HP) protein corona on liposomes obtained from cationic, neutral and anionic lipids. SAXS experiments allowed for evaluating the different effects on the mesostructure of increasing concentrations of plasma proteins adsorbed on the surface of liposomes—these observations demonstrated that the nanostructure of liposomes pre-coated by HP protein corona depends on the surface charge of the uncoated liposomes. Following these results, the authors showed that the protein corona of pre-coated cationic liposomes is not affected by the interaction with human plasma, suggesting that the nanoparticles' ability to evade sequestration by leukocytes during circulation remains stable. The authors demonstrated that at a low concentration, human plasma protein corona prevents the internalization of cationic liposomes into macrophages, but does not have the same effect on neutrally charged or anionic liposomes. Their findings suggest the possibility to use drug-loaded cationic liposomes coated with low-concentration protein corona to prolong blood circulation and facilitate drug delivery to the target sites [40].

4.2. Others

Lipid nanoparticles can be coated with substances that enhance their adhesion properties to cells. Moreover, they may be used as vectors to facilitate the transport of smaller nanoparticles into the cells. Recently, chitosan-N-acetylcysteine-coated unilamellar liposomes were reported to improve the bioavailability of Seleniun nanoparticles in the small intestine. The availability of –SH groups on the coated liposomes' surface is critical for their adhesion to the intestine cells; in fact, -SH groups are possibly interacting with mucine glycoproteins via a disulfide bridge formation. As demonstrated by in vivo observations, uncoated particles did not adhere to the cells surface, confirming the efficiency of the chitosan coating. In the reported study, SAXS evidenced that the chitosan coating did not affect the particles integrity, which is necessary to guarantee the interaction with the cells surface and subsequent diffusion of Selenium nanoparticles into the intestinal mucosa [38].

5. Encapsulation Strategies for Gene Delivery

The therapeutic use of nucleic acids is a tricky process because they are very sensitive and can easily degrade during circulation in the blood. Encapsulation of nucleic acids into nanoparticles increases their circulation time.

Lipid-based systems designed for gene delivery such as lipoplexes, which are composed of RNA/DNA and cationic lipids, are promising tools for gene delivery into cells [47].

Understanding the structure of the loaded compound with respect to the cargo particle is particularly important when the molecules to be delivered are nucleic acids, such as in the case of the new therapies based on mRNA [45,62,63].

The traditional synthetic method to produce lipid–nucleic acid complexes is the self-assembly of lipid unilamellar vesicles obtained by the hydration of lipid films, and the subsequent passive diffusion of nucleic acid into them. This leads to an uncontrolled interaction between nucleic acids and lipids, and to poor loading. The microfluidics technique was recently used to improve the encapsulation efficiency of nucleic acids [48] Lipolexes were produced by injecting a lipid–PEG mixture and DNA plasmid into a commercial microfluidic device. The physio-chemical characterization of the obtained nanoparticles with SAXS showed the expected lamellar stacking of lipid layers as well as a different arrangement of the encapsulated DNA, with respect to that obtained by passive diffusion. In fact, in microfluidic-produced lipoplexes, DNA is not densely packed inside the lipoplexes. Loosely packed DNA may interact more easily with the cellular membrane and the release into the cell might be facilitated (Figure 5).

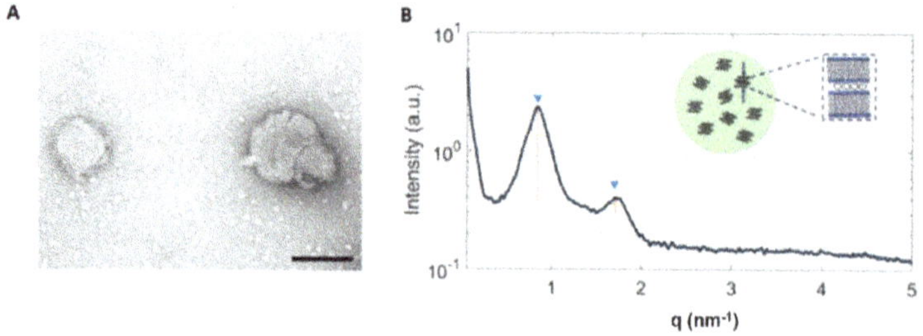

Figure 5. (**A**) TEM image of PEGylated nanoparticles (scale bar = 100 nm). (**B**) Scattering pattern of PEG-ylated nanoparticles. Inset: cartoon describing the nanostructure with multiple randomly oriented scattering domains. The broad peaks (▼) show the lamellar periodicity along the normal to the lipid bilayer, which includes the contribution of the membrane thickness and of the thickness of the water/DNA layer. (Reproduced with permission from Ref. [48]).

The administration route of nanoparticles for therapeutics is a challenge, especially when the therapy is directed against chronic diseases that require long administration times. In these cases, it is of pivotal importance to promote the delivery of the loaded drug to the site of disease, minimizing the interaction with other body districts and enhancing the internalization in the target cells. A recent study by Conte and collaborators [49] demonstrated the possibility to deliver siRNA to the lung via an aerosol route, overcoming the mucus barrier that may trap the inhaled nanoparticles and neutralize the therapeutic siRNA absorption. To reach this goal, revealing the interaction between the cells and the siRNA carrier particle composition was essential. They synthesized hybrid nanoparticles (hNPs) were composed of a core of poly(lactic-co-glycolic) acid (PLGA) and a lipid shell to encapsulate a pool of siRNA directed against the nuclear factor involved in severe acute and chronic inflammation, as in the case of patients affected by cystic fibrosis. They used SAXS to confirm the core–shell structure of the synthesized particles as well as their finite size and monodispersity. SAXS data analysis also allowed for determining the thickness of the lipid shell layer, which proved to be essential in promoting the internalization of nanoparticles into the cells. SAXS structural analysis of hNPs in contact with cellular mucus showed the presence of slightly larger nanoparticles, which were attributed by the authors to a rearrangement of the mucus network around the nanoparticles. Nevertheless, this does not affect the efficiency in terms of siRNA release in the cells, as demonstrated by in vitro experiments on Calu3, a model cell line for cystic fibrosis.

Polysarcosine is a synthetic compound based on repetitive units of the amino acid sarcosine, which can be used to functionalize the surface of lipidic nanoparticles. It has stealth-like properties similar to those of PEG, but polysarcosine-coated nanoparticles show a lower clearance in the blood with respect to PEG-ylated ones, and do not induce an immune response or hypersensitivity in patients [64,65]. Moreover, these nanoparticles show higher transfection efficiency if compared to PEG-ylated nanoparticles, which is desirable for the design of nanoparticles for mRNA delivery.

Recently, mRNA-loaded nanoparticles coated with polysarcosine were synthesized and tested as gene delivery tools in transfection experiments on hepatocarcinoma cell lines, as well as in in vivo experiments. SAXS analysis revealed information on the surface structure as well as on the internal organization of the nanoparticles and loaded RNA, which are important parameters to design nanoparticles tailored on the specific therapeutic application [46].

6. Outlook and Future Perspectives

The great push for research on the synthesis and optimization of lipid-based vectors for drug delivery that has taken place in recent years, also following the COVID-19 pandemic, has further shed a bright light on the potential of these nanomaterials for therapeutic or preventive applications. The environmental responses as well as the particle–cell interaction characteristics of the formulations proved the importance of properties such as dimension, structure and dosage of functional materials designed for therapeutic applications. In this perspective, synchrotron SAXS, with its in situ and time-resolved capabilities, has already proven to be an important technique to gain structural information, which is fundamental for drug delivery design and development.

Further studies are necessary to develop drug-loaded lipid nanoparticles and hybrid nanocomposites that will be able to maintain their integrity during circulation. In this respect, both in situ and ex situ SAXS are particularly suitable techniques.

Future challenges to be addressed in the near future include:

- The design of compounds able to improve the targeted release of drugs exploiting the properties of the molecular corona. The aim is to promote its interaction with specific cell receptors;
- The tailoring of lipid nanoparticles capable of reaching the disease sites and produce therapeutic activity selectively at the target organ;
- The achievement of an effective triggered drug release via the exposure to light, pH variation and temperature.

All these ambitious goals rely directly on the physio-chemical characteristics of the NPs—understanding the interaction with model membranes in situ, or by means of models reproducing the target organs or tissues, will help to achieve better functionality and performance.

Author Contributions: Conceptualization, B.M. and B.S.; writing—original draft preparation, B.S. and B.M.; writing—review and editing, B.S. and B.M. All authors have read and agreed to the published version of the manuscript.

Funding: This research received no external funding.

Institutional Review Board Statement: Not applicable.

Informed Consent Statement: Not applicable.

Data Availability Statement: Not applicable.

Conflicts of Interest: The authors declare no conflict of interest.

References

1. Kreuter, J. Nanoparticles—A Historical Perspective. *Int. J. Pharm.* **2007**, *331*, 1–10. [CrossRef] [PubMed]
2. Zielinska, A.; Filipa Carreiró, A.M.; Oliveira, A.N.; Pires, B.; Nagasamy Venkatesh, D.; Durazzo, A.; Lucarini, M.; Eder, P.; Silva, A.M.; Santini, A.; et al. Polymeric Nanoparticles: Production, Characterization, Toxicology and Ecotoxicology. *Molecules* **2020**, *25*, 3731. [CrossRef] [PubMed]
3. Waheed, A.; Aqil, M. Lyotropic Liquid Crystalline Nanoparticles: Scaffolds for Delivery of Myriad Therapeutics and Diagnostics. *J. Mol. Liq.* **2021**, *338*, 116919. [CrossRef]
4. Park, H.; Otte, A.; Park, K. Evolution of Drug Delivery Systems: From 1950 to 2020 and Beyond. *J. Control Release* **2022**, *342*, 53–65. [CrossRef] [PubMed]
5. Palanikumar, L.; Al-Hosani, S.; Kalmouni, M.; Nguyen, V.P.; Ali, L.; Pasricha, R.; Barrera, F.N.; Magzoub, M. PH-Responsive High Stability Polymeric Nanoparticles for Targeted Delivery of Anticancer Therapeutics. *Commun. Biol.* **2020**, *3*, 1–17. [CrossRef] [PubMed]

6. Parikh, P.K.; Parikh, N.H.; Patel, J.K.; Pathak, Y.V. Clearance Pathways and Tumor Targeting of Imaging Nanoparticles for Diagnostics. In *Pharmacokinetics and Pharmacodynamics of Nanoparticulate Drug Delivery Systems. In Pharmacodynamics of Nanoparticulate Drug Delivery Systems*; Springer: Cham, Switzerland, 2022; pp. 315–331.
7. Fonseca-Gomes, J.; Loureiro, J.A.; Tanqueiro, S.R.; Mouro, F.M.; Ruivo, P.; Carvalho, T.; Sebastião, A.M.; Diógenes, M.J.; Pereira, M.C. In Vivo Bio-Distribution and Toxicity Evaluation of Polymeric and Lipid-Based Nanoparticles: A Potential Approach for Chronic Diseases Treatment. *Int. J. Nanomed.* **2020**, *15*, 8609–8621. [CrossRef]
8. Gheorghita, R.; Anchidin-Norocel, L.; Filip, R.; Dimian, M.; Covasa, M. Applications of Biopolymers for Drugs and Probiotics Delivery. *Polymers* **2021**, *13*, 2729. [CrossRef]
9. Schneible, J.D.; Daniele, M.A.; Menegatti, S. Natural and Synthetic Biopolymers in Drug Delivery and Tissue Engineering. *Biopolym. Biomed. Biotechnol. Appl.* **2021**, *1*, 265–356. [CrossRef]
10. Byun, M.J.; Lim, J.; Na, S.; Dae, K.; Park, H.; Hyung, T.; Wooram, K.; Park, C.G.; Park, W. Advances in Nanoparticles for Effective Delivery of RNA Therapeutics. *BioChip J.* **2022**, *16*, 128–145. [CrossRef]
11. Nguyen, V.H.; Thuy, V.N.; Van, T.V.; Dao, A.H.; Lee, B.J. Nanostructured Lipid Carriers and Their Potential Applications for Versatile Drug Delivery via Oral Administration. *OpenNano* **2022**, *8*, 100064. [CrossRef]
12. Salentinig, S. Supramolecular Structures in Lipid Digestion and Implications for Functional Food Delivery. *Curr. Opin. Colloid Interface Sci.* **2019**, *39*, 190–201. [CrossRef]
13. Allen, T.M.; Cullis, P.R. Liposomal Drug Delivery Systems: From Concept to Clinical Applications. *Adv. Drug Deliv. Rev.* **2013**, *65*, 36–48. [CrossRef] [PubMed]
14. Attama, A.A.; Mumuni, A.-M.; Builders, P.F. Lipid Nanoparticulate Drug Delivery Systems: A Revolution in Dosage Form Design and Development. In *Recent Advances in Novel Drug Carrier Systems*; Ali Demir, S., Ed.; IntechOpen: London, UK, 2012. [CrossRef]
15. Muller, R.H.; Shegokar, R.; Keck, C.M. 20 Years of Lipid Nanoparticles (SLN & NLC): Present State of Development & Industrial Applications. *Curr. Drug Discov. Technol.* **2011**, *8*, 207–227. [CrossRef] [PubMed]
16. Chountoulesi, M.; Pispas, S.; Tseti, I.K.; Demetzos, C. Lyotropic Liquid Crystalline Nanostructures as Drug Delivery Systems and Vaccine Platforms. *Pharmaceuticals* **2022**, *15*, 429. [CrossRef]
17. Azmi, I.D.M.; Moghimi, S.M.; Yaghmur, A. Cubosomes and Hexosomes as Versatile Platforms for Drug Delivery. *Ther. Deliv.* **2015**, *6*, 1347–1364. [CrossRef]
18. Umar, H.; Wahab, H.A.; Gazzali, A.M.; Tahir, H.; Ahmad, W. Cubosomes: Design, Development, and Tumor-Targeted Drug Delivery Applications. *Polymers* **2022**, *14*, 3118. [CrossRef]
19. Zhai, J.; Fong, C.; Tran, N.; Drummond, C.J. Non-Lamellar Lyotropic Liquid Crystalline Lipid Nanoparticles for the Next Generation of Nanomedicine. *ACS Nano* **2019**, *13*, 6178–6206. [CrossRef]
20. Tian, X.; Chong, Y.; Ge, C. Understanding the Nano–Bio Interactions and the Corresponding Biological Responses. *Front. Chem.* **2020**, *8*, 446. [CrossRef]
21. Joudeh, N.; Linke, D. Nanoparticle Classification, Physicochemical Properties, Characterization, and Applications: A Comprehensive Review for Biologists. *J. Nanobiotechnol.* **2022**, *20*, 262. [CrossRef]
22. Salem, S.S.; Fouda, A. Green Synthesis of Metallic Nanoparticles and Their Prospective Biotechnological Applications: An Overview. *Biol. Trace Elem. Res.* **2021**, *199*, 344–370. [CrossRef]
23. Glatter, O.; Kratky, O. *Small Angle X-ray Scattering*; Academic Press: Cambridge, MA, USA, 1982.
24. Narayanan, T.; Konovalov, O. Synchrotron Scattering Methods for Nanomaterials and Soft Matter Research. *Materials* **2020**, *13*, 752. [CrossRef] [PubMed]
25. Franke, D.; Svergun, D.I. Synchrotron Small-Angle X-Ray Scattering on Biological Macromolecules in Solution. In *Synchrotron Light Sources and Free-Electron Lasers: Accelerator Physics, Instrumentation and Science Applications*; Jaeschke, E.J., Khan, S., Schneider, J.R., Hastings, J.B., Eds.; Springer International Publishing: Cham, Switzerland, 2020; pp. 1645–1672. [CrossRef]
26. Thakral, S.; Kim, K. Small-Angle Scattering for Characterization of Pharmaceutical Materials. *TrAC Trends Anal. Chem.* **2021**, *134*, 116144. [CrossRef]
27. Konstantinidi, A.; Naziris, N.; Chountoulesi, M.; Kiriakidi, S.; Sartori, B.; Kolokouris, D.; Amentisch, H.; Mali, G.; Ntountaniotis, D.; Demetzos, C.; et al. Comparative Perturbation Effects Exerted by the Influenza A M2 WT Protein Inhibitors Amantadine and the Spiro [Pyrrolidine-2,2′-Adamantane] Variant AK13 to Membrane Bilayers Studied Using Biophysical Experiments and Molecular Dynamics Simulations. *J. Phys. Chem. B* **2018**, *122*, 9877–9895. [CrossRef] [PubMed]
28. Di Cola, E.; Grillo, I.; Ristori, S. Small Angle X-Ray and Neutron Scattering: Powerful Tools for Studying the Structure of Drug-Loaded Liposomes. *Pharmaceutics* **2016**, *8*, 10. [CrossRef]
29. Semeraro, E.F.; Marx, L.; Frewein, M.P.K.; Pabst, G. Increasing Complexity in Small-Angle X-Ray and Neutron Scattering Experiments: From Biological Membrane Mimics to Live Cells. *Soft Matter* **2021**, *17*, 222–232. [CrossRef]
30. Kučerka, N.; Nagle, J.F.; Feller, S.E.; Balgavý, P. Models to Analyze Small-Angle Neutron Scattering from Unilamellar Lipid Vesicles. *Phys. Rev. E* **2004**, *69*, 9. [CrossRef]
31. Sebastiani, F.; Yanez Arteta, M.; Lerche, M.; Porcar, L.; Lang, C.; Bragg, R.A.; Elmore, C.S.; Krishnamurthy, V.R.; Russell, R.A.; Darwish, T.; et al. Apolipoprotein E Binding Drives Structural and Compositional Rearrangement of MRNA-Containing Lipid Nanoparticles. *ACS Nano* **2021**, *15*, 6709–6722. [CrossRef]

32. Arteta, M.Y.; Kjellman, T.; Bartesaghi, S.; Wallin, S.; Wu, X.; Kvist, A.J.; Dabkowska, A.; Székely, N.; Radulescu, A.; Bergenholtz, J.; et al. Successful Reprogramming of Cellular Protein Production through MRNA Delivered by Functionalized Lipid Nanoparticles. *Proc. Natl. Acad. Sci. USA* **2018**, *115*, E3351–E3360. [CrossRef]
33. Schoenmaker, L.; Witzigmann, D.; Kulkarni, J.A.; Verbeke, R.; Kersten, G.; Jiskoot, W.; Crommelin, D.J.A. MRNA-Lipid Nanoparticle COVID-19 Vaccines: Structure and Stability. *Int. J. Pharm.* **2021**, *601*, 120586. [CrossRef]
34. Li, Y.; Angelova, A.; Hu, F.; Garamus, V.M.; Peng, C.; Li, N.; Liu, J.; Liu, D.; Zou, A. PH Responsiveness of Hexosomes and Cubosomes for Combined Delivery of Brucea Javanica Oil and Doxorubicin. *Langmuir* **2019**, *35*, 14532–14542. [CrossRef]
35. Zhai, J.; Yap, S.L.; Drummond, C.J.; Tran, N. Controlling the PH Dependent Transition between Monoolein Fd3m Micellar Cubosomes and Hexosomes Using Fatty Acetate and Fatty Acid Additive Mixtures. *J. Colloid Interface Sci.* **2022**, *607*, 848–856. [CrossRef] [PubMed]
36. Rajesh, S.; Zhai, J.; Drummond, C.J.; Tran, N. Synthetic Ionizable Aminolipids Induce a PH Dependent Inverse Hexagonal to Bicontinuous Cubic Lyotropic Liquid Crystalline Phase Transition in Monoolein Nanoparticles. *J. Colloid Interface Sci.* **2021**, *589*, 85–95. [CrossRef] [PubMed]
37. Mathews, P.D.; Mertins, O.; Angelov, B.; Angelova, A. Cubosomal Lipid Nanoassemblies with PH-Sensitive Shells Created by Biopolymer Complexes: A Synchrotron SAXS Study. *J. Colloid Interface Sci.* **2022**, *607*, 440–450. [CrossRef] [PubMed]
38. Selmani, A.; Seibert, E.; Tetyczka, C.; Kuehnelt, D.; Vidakovic, I.; Kornmueller, K.; Absenger-Novak, M.; Radatović, B.; Vrček, I.V.; Leitinger, G.; et al. Thiolated Chitosan Conjugated Liposomes for Oral Delivery of Selenium Nanoparticles. *Pharmaceutics* **2022**, *14*, 803. [CrossRef] [PubMed]
39. Tran, T.; Siqueira, S.D.V.S.; Amenitsch, H.; Rades, T.; Müllertz, A. Monoacyl Phosphatidylcholine Inhibits the Formation of Lipid Multilamellar Structures during in Vitro Lipolysis of Self-Emulsifying Drug Delivery Systems. *Eur. J. Pharm. Sci.* **2017**, *108*, 62–70. [CrossRef] [PubMed]
40. Giulimondi, F.; Digiacomo, L.; Pozzi, D.; Palchetti, S.; Vulpis, E.; Capriotti, A.L.; Chiozzi Zenezini, R.; Laganà, A.; Amenitsch, H.; Masuelli, L.; et al. Interplay of Protein Corona and Immune Cells Controls Blood Residency of Liposomes. *Nat. Commun.* **2019**, *10*, 3686. [CrossRef]
41. Lam, Y.Y.; Hawley, A.; Tan, A.; Boyd, B.J. Coupling in Vitro Cell Culture with Synchrotron SAXS to Understand the Bio-Interaction of Lipid-Based Liquid Crystalline Nanoparticles with Vascular Endothelial Cells. *Drug Deliv. Transl. Res.* **2020**, *10*, 610–620. [CrossRef] [PubMed]
42. Caselli, L.; Mendozza, M.; Muzzi, B.; Toti, A.; Montis, C.; Mello, T.; Di Cesare Mannelli, L.; Ghelardini, C.; Sangregorio, C.; Berti, D. Lipid Cubic Mesophases Combined with Superparamagnetic Iron Oxide Nanoparticles: A Hybrid Multifunctional Platform with Tunable Magnetic Properties for Nanomedical Applications. *Int. J. Mol. Sci.* **2021**, *22*, 9268. [CrossRef]
43. Pozzi, D.; Colapicchioni, V.; Caracciolo, G.; Piovesana, S.; Capriotti, A.L.; Palchetti, S.; De Grossi, S.; Riccioli, A.; Amenitsch, H.; Laganà, A. Effect of PEG Lenght on the Intaraction with Biological Fluids.Pdf. *Nanoscale* **2014**, *6*, 2782–2792. [CrossRef]
44. Caracciolo, G.; Palchetti, S.; Digiacomo, L.; Chiozzi Zenezini, R.; Capriotti, A.L.; Amenitsch, H.; Tentori, P.M.; Palmieri, V.; Papi, M.; Cardarelli, F.; et al. Human Biomolecular Corona of Liposomal Doxorubicin: The Overlooked Factor in Anticancer Drug Delivery. *ACS Appl. Mater. Interfaces* **2018**, *10*, 22951–22962. [CrossRef]
45. Siewert, C.D.; Haas, H.; Cornet, V.; Nogueira, S.S.; Nawroth, T.; Uebbing, L.; Ziller, A.; Al-gousous, J.; Radulescu, A.; Schroer, M.A.; et al. Hybrid Biopolymer and Lipid Nanoparticles with Improved Transfection Efficacy for MRNA. *Cells* **2020**, *9*, 2034. [CrossRef] [PubMed]
46. Nogueira, S.S.; Schlegel, A.; Maxeiner, K.; Weber, B.; Barz, M.; Schroer, M.A.; Blanchet, C.E.; Svergun, D.I.; Ramishetti, S.; Peer, D.; et al. Polysarcosine-Functionalized Lipid Nanoparticles for Therapeutic MRNA Delivery. *ACS Appl. Nano Mater.* **2020**, *3*, 10634–10645. [CrossRef]
47. Caracciolo, G.; Amenitsch, H. Cationic Liposome/DNA Complexes: From Structure to Interactions with Cellular Membranes. *Eur. Biophys. J.* **2012**, *41*, 815–829. [CrossRef] [PubMed]
48. Quagliarini, E.; Renzi, S.; Digiacomo, L.; Giulimondi, F.; Sartori, B.; Amenitsch, H.; Tassinari, V.; Masuelli, L.; Bei, R.; Cui, L.; et al. Microfluidic Formulation of Dna-Loaded Multicomponent Lipid Nanoparticles for Gene Delivery. *Pharmaceutics* **2021**, *13*, 1292. [CrossRef]
49. Conte, G.; Costabile, G.; Baldassi, D.; Rondelli, V.; Bassi, R.; Colombo, D.; Linardos, G.; Fiscarelli, E.V.; Sorrentino, R.; Miro, A.; et al. Hybrid Lipid/Polymer Nanoparticles to Tackle the Cystic Fibrosis Mucus Barrier in SiRNA Delivery to the Lungs: Does PEGylation Make the Difference? *ACS Appl. Mater. Interfaces* **2022**, *14*, 7565–7578. [CrossRef]
50. Mehnert, W.; Mäder, K. Solid Lipid Nanoparticles: Production, Characterization and Applications. *Adv. Drug Deliv. Rev.* **2001**, *47*, 165–196. [CrossRef]
51. Khosa, A.; Reddi, S.; Saha, R.N. Nanostructured Lipid Carriers for Site-Specific Drug Delivery. *Biomed. Pharmacother.* **2018**, *103*, 598–613. [CrossRef]
52. Mohammad, Y.; Prentice, R.N.; Boyd, B.J.; Rizwan, S.B. Comparison of Cubosomes and Hexosomes for the Delivery of Phenytoin to the Brain. *J. Colloid Interface Sci.* **2022**, *605*, 146–154. [CrossRef]
53. Date, T.; Paul, K.; Singh, N.; Jain, S. Drug–Lipid Conjugates for Enhanced Oral Drug Delivery. *AAPS PharmSciTech* **2019**, *20*, 41. [CrossRef]

54. Khan, N.F.; Salim, M.; Binte Abu Bakar, S.Y.; Ristroph, K.; Prud'homme, R.K.; Hawley, A.; Boyd, B.J.; Clulow, A.J. Small-Volume in Vitro Lipid Digestion Measurements for Assessing Drug Dissolution in Lipid-Based Formulations Using SAXS. *Int. J. Pharm. X* **2022**, *4*, 100113. [CrossRef]
55. Mendozza, M.; Montis, C.; Caselli, L.; Wolf, M.; Baglioni, P.; Berti, D. On the Thermotropic and Magnetotropic Phasebehavior of Lipid Liquid Crystals Containingmagnetic Nanoparticles. *Nanoscale* **2018**, *10*, 3480–3488. [CrossRef] [PubMed]
56. Shi, L.; Zhang, J.; Zhao, M.; Tang, S.; Cheng, X.; Zhang, W.; Li, W.; Liu, X.; Peng, H.; Wang, Q. Effects of Polyethylene Glycol on the Surface of Nanoparticles for Targeted Drug Delivery. *Nanoscale R. Soc. Chem.* **2021**, *13*, 10748–10764. [CrossRef] [PubMed]
57. De Leo, V.; Ruscigno, S.; Trapani, A.; Di Gioia, S.; Milano, F.; Mandracchia, D.; Comparelli, R.; Castellani, S.; Agostiano, A.; Trapani, G.; et al. Preparation of Drug-Loaded Small Unilamellar Liposomes and Evaluation of Their Potential for the Treatment of Chronic Respiratory Diseases. *Int. J. Pharm.* **2018**, *545*, 378–388. [CrossRef] [PubMed]
58. Bandara, S.R.; Molley, T.G.; Kim, H.; Bharath, P.A.; Kilian, K.A.; Leal, C. The Structural Fate of Lipid Nanoparticles in the Extracellular Matrix †. *Mater. Horiz.* **2020**, *7*, 125. [CrossRef] [PubMed]
59. Sarode, A.; Fan, Y.; Byrnes, A.E.; Hammel, M.; Hura, G.L.; Fu, Y.; Kou, P.; Hu, C.; Hinz, F.I.; Roberts, J.; et al. Predictive High-Throughput Screening of PEGylated Lipids in Oligonucleotide-Loaded Lipid Nanoparticles for Neuronal Gene Silencing. *Nanoscale Adv.* **2022**, *4*, 2107–2123. [CrossRef] [PubMed]
60. Al-Ahmady, Z.S.; Hadjidemetriou, M.; Gubbins, J.; Kostarelos, K. Formation of Protein Corona in Vivo Affects Drug Release from Temperature-Sensitive Liposomes. *J. Control Release* **2018**, *27*, 157–167. [CrossRef]
61. Liu, P.; Chen, G.; Zhang, J. A Review of Liposomes as a Drug Delivery System: Current Status of Approved Products, Regulatory Environments, and Future Perspectives. *Molecules* **2022**, *27*, 1372. [CrossRef]
62. Eygeris, Y.; Patel, S.; Jozic, A.; Sahay, G.; Sahay, G. Deconvoluting Lipid Nanoparticle Structure for Messenger RNA Delivery. *Nano Lett.* **2020**, *20*, 4543–4549. [CrossRef]
63. Mitchell, M.J.; Billingsley, M.M.; Haley, R.M.; Wechsler, M.E.; Peppas, N.A.; Langer, R. Engineering Precision Nanoparticles for Drug Delivery. *Nature* **2021**, *20*, 101–124. [CrossRef]
64. Szebeni, J.; Muggia, F.; Gabizon, A.; Barenholz, Y. Activation of Complement by Therapeutic Liposomes and Other Lipid Excipient-Based Therapeutic Products: Prediction and Prevention. *Adv. Drug Deliv. Rev.* **2011**, *63*, 1020–1030. [CrossRef]
65. Szebeni, J. Complement Activation-Related Pseudoallergy: A Stress Reaction in Blood Triggered by Nanomedicines and Biologicals. *Mol. Immunol.* **2014**, *61*, 163–173. [CrossRef] [PubMed]

MDPI AG
Grosspeteranlage 5
4052 Basel
Switzerland
Tel.: +41 61 683 77 34

Pharmaceutics Editorial Office
E-mail: pharmaceutics@mdpi.com
www.mdpi.com/journal/pharmaceutics

www.ingramcontent.com/pod-product-compliance
Lightning Source LLC
Chambersburg PA
CBHW040742250926
51850CB00007BB/1387
* 9 7 8 3 7 2 5 8 5 1 3 9 3 *